The Responsive Brain

The Proceedings of the Third International Congress on Event-related Slow Potentials of the Brain

Bristol, England 13–18 August, 1973

Edited by

W. Cheyne McCallum
Burden Neurological Institute, Bristol, England

and

John R. Knott
University of Iowa (Emeritus), Boston University Medical Center, Boston, Mass., U.S.A.

John Wright and Sons Limited Bristol 1976

ISBN 0 7236 0443 6

Printed in Great Britain by Henry Ling Ltd., a subsidiary of John Wright & Sons Ltd., at the Dorset Press, Dorchester.

Preface

The Third International Congress on Event-related Slow Potentials of the Brain continued the essential plan developed in the first two: the verbal exchange of reactions and ideas by a group of scientists whose primary interests lie within this field of research. Over the period of time which has occurred since the 1967 meeting the ranks of those engaged in this type of investigation have multiplied and the areas of interest have expanded. The inevitable consequences have been formalization of meetings in order to handle the increased amount of information subject to discussion.

The Third Congress undertook a unique enterprise in scientific communication. In addition to the discussion of new material, an attempt was made to define areas of controversy and methodological difficulty and, *via* live experimental sessions, to focus the combined wisdom and experience of the participants upon their solution. This virtually uncharted area of interactive and co-operative science entailed considerable preparatory work by many individuals prior to the Congress and a major laboratory exercise during the meetings carried out through the medium of closed circuit television.

The title of this volume has been chosen to reflect the interest of the group in the brain as a dynamic entity, the organization and functions of which can be approached through a study of its electrical responsiveness to events in both the external and internal worlds.

The novel aspects of this Congress necessitated substantial funding. For this the Congress organizers are greatly indebted to the W. Clement and Jessie V. Stone Foundation, and to Mrs Donna Stone Bradshaw, who kindly opened the Congress on behalf of the Foundation.

A further essential feature in the success of this enterprise was the generous support and encouragement afforded by the Council of the Burden Neurological Institute and the extensive work undertaken by the Institute's Scientific Director, Dr Ray Cooper, and staff, at all stages of planning, organization and execution of the Congress. It was most fitting that the meeting should be held under the Institute's aegis and that such an international assemblage should be gathered in the laboratory in which the significance of slow potentials in human brain was initially discovered and emphasized by Dr W. Grey Walter and his collaborators.

The Editors of this Volume are also indebted to the Department of Psychiatry, University of Iowa, and to the Grass Instrument Company, for financial aid given to J.R.K. in the final stages of editorial effort in preparing the manuscript. Mrs Freda Darby, who laboured on the previous Proceedings, has again taken the efforts of the Editors and put them into an orderly assembly, for which we extend to her our heartfelt thanks.

January and November 1975 W.C.McC. & J.R.K.

Contents

Potential Changes associated with Motor Action, Reflex Responses and Readiness

Clinical and Developmental Aspects of Slow Potential Changes

Psychiatric Studies

Neurological Studies

Developmental Studies

INTRACEREBRAL RECORDING OF SLOW POTENTIAL CHANGES IN MAN AND ANIMALS

Animal Studies

Human Studies

Participants

Abraham, Lt-Col P. *Southampton, U.K.*

Becker, Dip-Ing W. *Ulm, German Federal Republic*

Blowers, Mr G. *Hong Kong*

Cohen, Dr J. *Chicago, U.S.A.*

Cooper, Dr R. *Bristol, U.K.*

Crow, Dr H. J. *Bristol, U.K.*

Deecke, Dr L. *Ulm, German Federal Republic*

Donald, Dr M. *Kingston, Canada*

Donchin, Professor E. *Urbana, U.S.A.*

Dubrovsky, Dr B. *Montreal, Canada*

Fenelon, Mr B. *Newcastle, Australia*

Fruhstorfer, Dr H. *Marburg, German Federal Republic*

Ganglberger, Dr J. *Vienna, Austria*

Grözinger, Dr B. *Ulm, German Federal Republic*

Hazemann, Dr P. *Paris, France*

Hillyard, Dr S. *La Jolla, California, U.S.A.*

Järvilehto, Dr T. *Helsinki, Finland*

Kamp, Mr T. *Utrecht, Netherlands*

Karrer, Dr R. *Chicago, U.S.A.*

Knott, Professor J. R. *Iowa City, U.S.A.*

Kohn, Dr H. *Piscataway, U.S.A.*

Leifer, Dr L. *Cambridge, U.S.A.*

Lelord, Professor G. *Tours, France*

Little, Mrs B. *Keele, U.K.*

Loveless, Dr N. *Dundee, Scotland*

Low, Dr M. *Vancouver, Canada*

Marsh, Dr G. *Durham, U.S.A.*

McAdam, Dr D. *Rochester, U.S.A.*

McCallum, Dr W. C. *Bristol, U.K.*

Näätänen, Dr R. *Helsinki, Finland*

Naitoh, Dr P. *San Diego, U.S.A.*

Otto, Dr D. *Chapel Hill, U.S.A.*

Papakostopoulos, Dr D. *Bristol, U.K.*

Peters, Mr J. *Iowa City, U.S.A.*

Rebert, Dr C. *Stanford, U.S.A.*

Ritter, Dr W. *New York, U.S.A.*

Storm van Leeuwen, Professor W. *Utrecht, Netherlands*

Tecce, Dr J. *Boston, U.S.A.*

Timsit-Berthier, Dr M. *Liège, Belgium*

Tueting, Dr P. *New York, U.S.A.*

Walter, Dr W. Grey, *Bristol, U.K.*

Weinberg, Dr H. *Vancouver, Canada*

Wilkinson, Dr R. *Cambridge, U.K.*

Winter, Mr A. L. *Bristol, U.K.*

Zappoli, Professor R. *Florence, Italy*

Abbreviations

Editorial note: The following standard abbreviations have been used in this volume. Other abbreviations are defined in the text when first used. A more extensive list of abbreviations used in event related potential research can be found in the *Proceedings of the Second Congress* (McCallum and Knott, 1973).

BP	*Bereitschaftspotential* (synonymous with Readiness Potential, RP).
CNV	Contingent Negative Variation.
DC	Direct current or, when used in relation to amplifiers, 'directly coupled'.
EDR	Electro-dermal response.
EKG	Electrocardiogram.
EMG	Electromyogram.
EOG	Electro-oculogram.
EP	Evoked potential.
ERP	Event related potential.
GSR	Galvanic skin response.
Hz	Hertz or cycles per second.
i.s.i.	Inter stimulus interval.
i.t.i.	Inter trial interval.
MP	Motor potential.
msec	Millisecond.
P3	A term used synonymously with P300. The latter is generally to be preferred.
P300	Positive wave with peak latency approximately 300 msec after stimulus.
R	Response.
RP	Readiness potential.
RT	Reaction time.
S	Stimulus.
S_1, S_2, etc.	Stimulus one, stimulus two, etc.
S	Subject.
sec	Second.
TC	Time constant.

President's Introduction

JOHN R. KNOTT

University of Iowa (Professor Emeritus), Iowa City, U.S.A.
Presently, Professor (Neurology), Boston University Medical Center, Boston, U.S.A.

The present assembly to view the responsive brain through its slow potential processes is the third in some 6 years, and marks only the ninth anniversary of the original publication of the existence of these phenomena in man, which was brought about through the serendipity of a key group of workers at the Burden Neurological Institute, led by W. Grey Walter, and including Ray Cooper, Vivian Aldridge, Cheyne McCallum and Arthur Winter. While to some it may appear that holding this Congress in Bristol is akin to carrying coals to Newcastle, none of us would regard it as such, since the phrase 'slow potentials'—or more properly, 'event-related slow potentials'—covers a far greater semantic field than does the term 'coals'. Further, it has been our experience in the past that when we carry our data to some common meeting point we all emerge with more than that which we brought, and we find that we each have had generated within us new programs of research, based upon the admixture of the 'coals' we all have carried and the various patterns of flames into which they have been fanned by our colleagues.

When one compares the Second (Vancouver) Congress with this one (and both with the original informal meeting in 1967 in Liège), there are three types of 'progress' apparent. First, slow potential techniques are being applied to a wider range of problems. Second, there is increasing technical elegance in experimentation. To a noticeable degree, experimental design is emerging from the primitive fixed foreperiod reaction time experiment into more complex behavioral situations. Third, and perhaps most important, is the fact that we, as a collection of scientists with some common interests in sub-electroencephalographic frequencies, have stopped being 'a CNV group' and are passing into a stage of being 'an event-related slow potential group'. As such the polarity and timing of the potentials seem less crucial, and the events have come to have a more critical implication, the goals being the obtaining of information about the responsiveness of the human brain. It is a sign of health that these are being considered collectively as indices of the complex processes of cerebration and not as competing items in the scientific market-place, one of greater value than another. We are not learning more and more about less and less, but we are learning more about *more.*

The unique feature of 'live' demonstrations of certain technical aspects of slow potential recording, utilizing closed-circuit television, will enable Dr Grey Walter to realize a wish he has expressed since 1967 (first in Liège, then in Vancouver), that we design and carry out a 'critical experiment' in order to satisfy ourselves regarding certain points involved in the generation of event-related cerebral slow potentials. To what extent 'experiment-by-committee' is viable will be seen—but the very execution of this attempt may itself be an important experiment.

We are, at the opening of this Congress, coming face to face with new sets of data, new ways by which these data may have been generated, better—or at least different—methods of analysing them and broader views of what may be regarded as the data themselves. The data are not, however, the end-point. The final goal is an understanding of the functional structuring of the human brain characteristics and its moment to moment activities. We should anticipate that if we keep our own brains responsive we shall be changed, ourselves, by the experiences we shall share with one another.

METHODOLOGY OF SLOW POTENTIAL CHANGES

Chairman's Opening Remarks on Present State of Methodology

R. COOPER

Burden Neurological Institute, Bristol, U.K.

Methodology in slow potential work can be separated into 4 areas concerned with: (1) characteristics of the signal; (2) characteristics of the electrodes; (3) characteristics of the amplifiers and display system; (4) signal analysis.

1. CHARACTERISTICS OF THE SIGNAL

These are well known—amplitude less than 100 μV, and bandwidth DC to 100 Hz. What is not yet known although it must be of considerable importance is the output impedance of the cortical generators and the distribution of current flow that enables us to record the potentials on the scalp.

2. CHARACTERISTICS OF THE ELECTRODES

The problems that still exist in scalp recordings are primarily due to slow drift of potential which is troublesome during DC recordings. It seems likely that the electrodes themselves do not change very much but that the potentials at the skin change as the conducting jelly soaks in. In addition, fluid may collect at the site of any abrasion that has been made to improve electrode contact and change the electrical potentials. Drift rates of more than 200 μV per minute are common even with well prepared and applied electrodes.

For intracerebral recordings, particularly in man, the polarization of the electrodes causes a major problem. Acceptable metals such as gold or stainless steel act like a capacitor and present a very high impedance to DC and low frequency activity (many megohms). When used with DC amplifiers the low frequency activity is attenuated by a factor determined by the values of the electrode impedance and the amplifier input impedance (*see* Cooper *et al*., 1974 or Cooper, 1971 for discussion). The low frequency activity is reduced 20 fold by an input impedance of 1 megohm if the electrode impedance is 19 megohm. Although the impedance might be reduced by changing the surface layer of the electrode by etching or chemical reaction (in the same way as silver chloride changes a silver electrode) we have used EEG amplifiers with input impedance of 100 megohm. Using such amplifiers the CNV and other slow phenomena can be recorded from cortical and subcortical electrodes without too much loss.

3. CHARACTERISTICS OF AMPLIFIERS AND DISPLAY SYSTEM

DC amplifiers are of such high stability that few problems remain. When using RC coupled amplifiers the time constant reduces the amplitude of the CNV and other slow phenomena (*Fig*. 1). The amount of attenuation depends

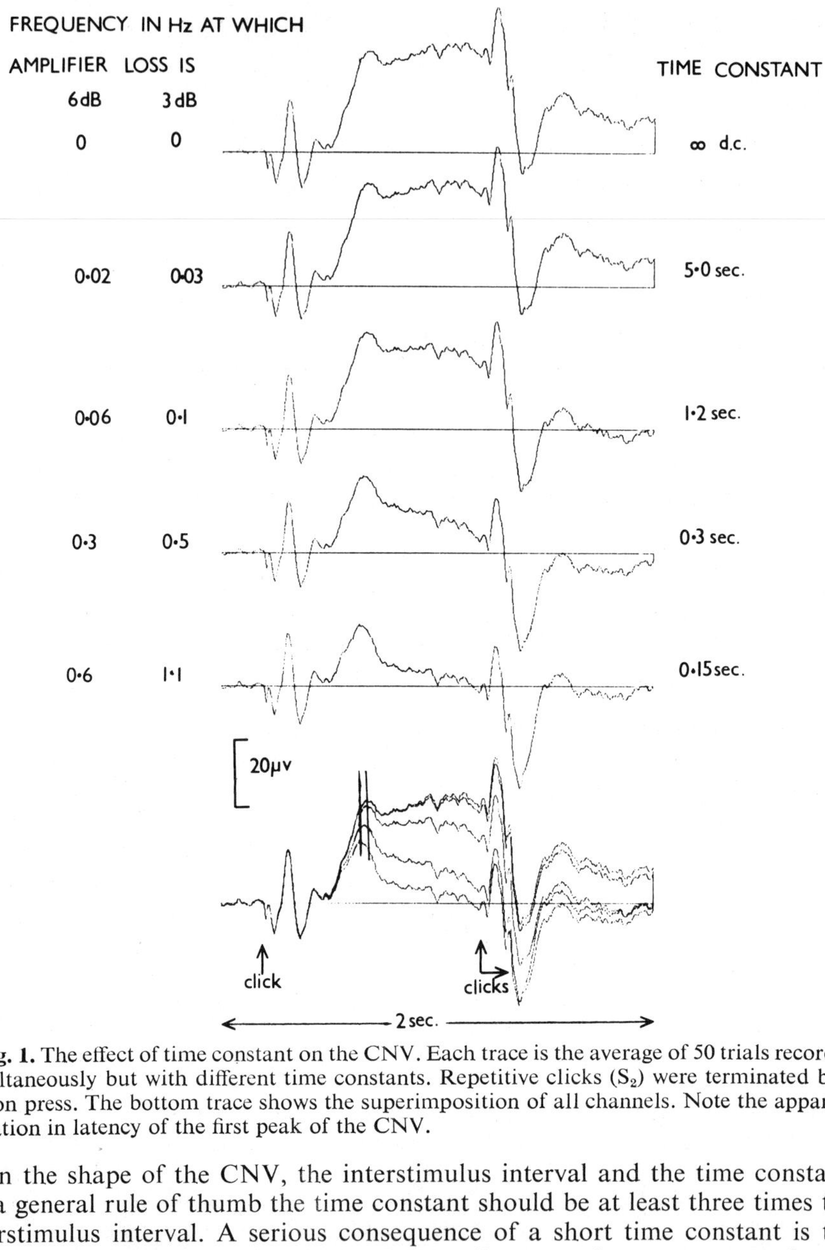

Fig. 1. The effect of time constant on the CNV. Each trace is the average of 50 trials recorded simultaneously but with different time constants. Repetitive clicks (S_2) were terminated by a button press. The bottom trace shows the superimposition of all channels. Note the apparent variation in latency of the first peak of the CNV.

upon the shape of the CNV, the interstimulus interval and the time constant. As a general rule of thumb the time constant should be at least three times the interstimulus interval. A serious consequence of a short time constant is the deflection below the baseline as the CNV collapses and measurement of such phenomena as P300 may be in serious error. Performance of amplifiers is often quoted as bandwidth rather than time constant although for CNV work in particular the time constant gives a more graphic description. The bandwidth/time constant relationship is illustrated in *Fig.* 2.

The high frequency cut off depends upon the experiment—whether, for example, cerebral evoked responses are being measured; 100 Hz is reasonable and allows most signals to be recorded.

At the present time practically all averaged slow potentials are written out as a voltage/time graph. For spatial distribution some contour display may be necessary, although such a display would imply that the CNV is a unitary phenomenon.

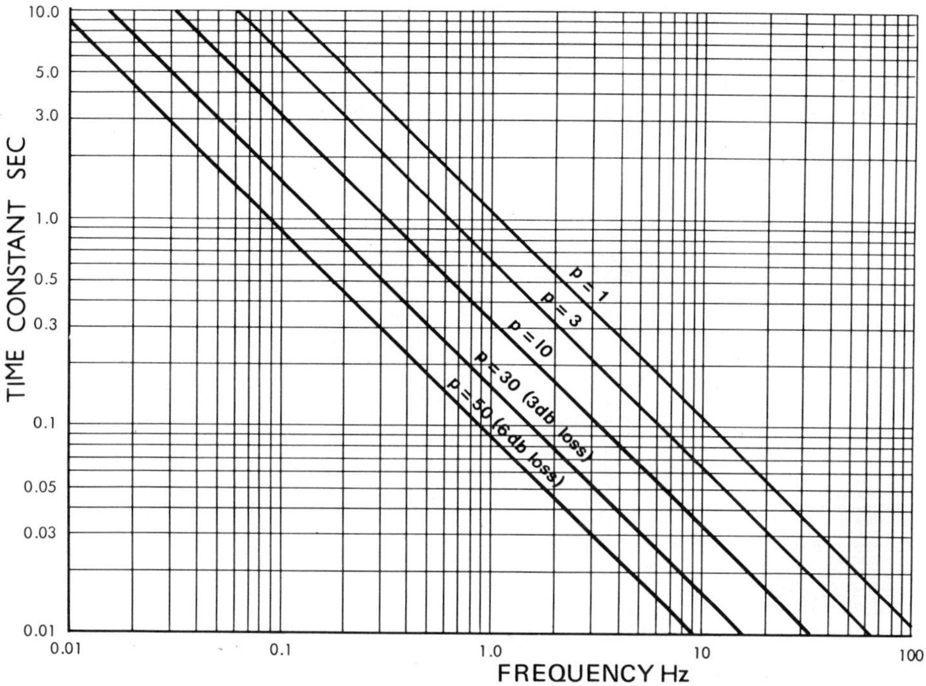

Fig. 2. The relationship between frequency and time constant for various percentage losses of sensitivity (p).

4. SIGNAL ANALYSIS

At present most event related potentials are extracted from the background activity by averaging techniques but the variability or significance of changes of averaged evoked responses are rarely calculated. One reason is that the hardware and software necessary for these calculations is considerable and the calculations make large demands on computer memory. This limitation can be circumvented by storing individual trials on magnetic tape and making the necessary calculations off-line, but this is time consuming. The problem does not end when the trial to trial variability has been calculated since this usually shows that the variability itself varies throughout the epoch of time sampled and it may be that we should study the variability of the variability—'little fleas have smaller fleas . . .'.

In cases where there is variability of both latency and amplitude averaging can be misleading. Pfurtscheller and Cooper (1975) describe a method of improving the accuracy of measurement of evoked responses using correlative techniques to estimate latencies and amplitudes. Such methods might lead to the use of a smaller number of trials before a given confidence limit is reached.

The need for reducing the number of trials before a 'clear' response is obtained is a fundamental problem of all psychophysiological experimentation. This problem has been approached in a different way by Weinberg and Cooper (1972) who, using correlative techniques, looked for particular patterns of activity in the ongoing EEG. The problem in this approach is to specify the pattern of activity that is being sought. Nevertheless in certain circumstances such methods might be preferable to averaging. The problems involved in the selection of methods for extracting signal from noise can be summed up by saying that in averaging confidence is gained at the expense of repetition whereas in correlative techniques confidence is gained by knowing what one is looking for.

Co-chairman's Opening Remarks on Measurement in AEP Studies*

EMANUEL DONCHIN

Department of Psychology, University of Illinois

I shall take as my topic the two-fold problem created by the need to combine the data obtained from different *Ss* used in an experiment: How do you combine the data so that *you* know what the results were? How do you communicate your conviction to others? Both are vexing problems in average evoked potential (AEP) research because the measurements made from our *Ss* are based on very complex waveforms. We must define some specific, consensual methods for reducing them to uniquely defined values which can then be handled by normally available conclusion-deriving techniques.

Our problem is severe because, even though there is some inter-subject consistency, the relationships between components of AEPs vary from one *S* to the other. For example, most EPs elicited by visual stimuli, of moderate to bright intensities, will have a negative peak with a latency between 75 and 120 msec, a positive peak between 90 and 120 msec and, frequently, another negative peak at about 160 msec. The experimental variables, over *Ss*, may affect similar components in similar ways. Nevertheless, any one *S*'s evoked responses will be sufficiently idiosyncratic to make it difficult to develop specific measures of the characteristics of the components and it is difficult to decide how to combine inter-subject data.

Several approaches to this problem have been employed. Often, one attempts an intuitive conclusion based on visual inspection of each *S*'s data. One can superimpose different AEPs in a variety of different combinations and quite often a clearly perceptible pattern appears in the data. However, it is difficult to communicate such conclusions convincingly. Alternatively, some unique characteristic of the AEP, such as the amplitude or latency of one component, is measured. Conclusions are then based solely upon an analysis of variance, without knowing what the original data in fact looked like.

Another approach is to average the data from all *Ss*, computing a mean AEP for each of the experimental conditions, thus drastically reducing the number of complex waveforms to be digested. However, computing across subjects may lead to inadequate measures of variability.

A measure of variability is preserved in discriminant analysis, which is a pattern recognition approach, similar in its underlying logic to the Recognition Index Technique described by Weinberg *et al.* (p. 34). All pattern recognition techniques involve the development of a decision rule, so that a score larger than a criterion value leads to one decision, and a score below the criterion value leads to the alternate decision. Pattern recognition techniques differ

* (*Editors' note:* A full description of the method of analysis described by Professor Donchin is to be found in Donchin, E. and Herning, R. I. (1975). A simulation study of the efficacy of stepwise discriminant analysis in the detection and relation of event related potentials. *Electroenceph. Clin. Neurophysiol.* **38,** 51–58.)

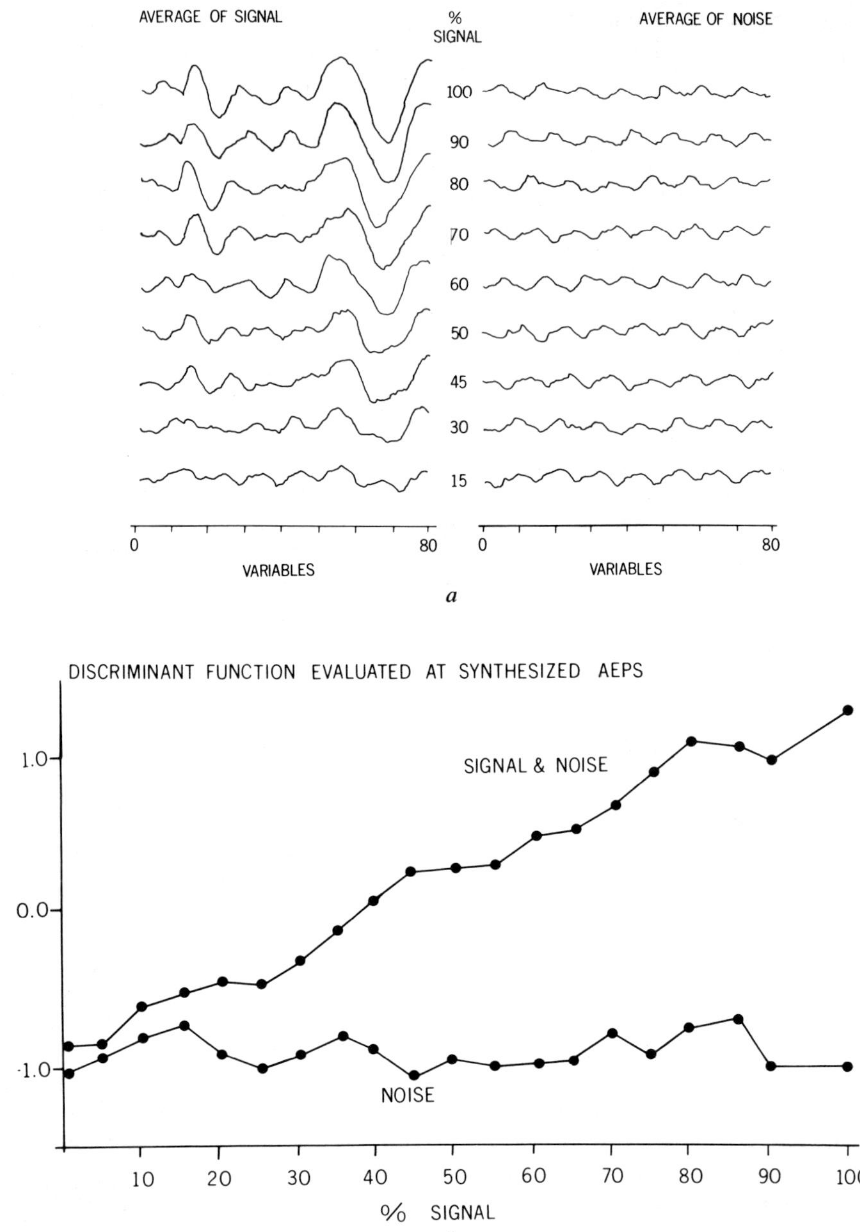

Fig. 1. *a,* Simulated evoked potential 'intensity' experiment. Each record in the left panel is a synthetic evoked response obtained by adding 5 damped sinusoids. The characteristics of the sinusoids are identical for all records in the left panel, however, the synthetic evoked response was successively multiplied by a fraction, to reduce its overall amplitude. On the right are averages of 100 realizations of a noise process with which the synthetic evoked response was mixed. A discriminant function was developed on the basis of the data obtained for the full amplitude signal and the corresponding noise realizations. *b,* The discriminant scores for the synthetic AEP shown in *Fig.* 1*a*. These are plotted against the relative amplitude of the signal.

in the logic underlying the selection of the criterion values and in computation of scores. The advantage of discriminant analysis (DA) as a pattern recognition technique is that it has a fairly well defined manner for determining the decision rules and the criterion values.

To explain the technique I shall describe a simulation study of evoked potential data which we have just completed.

Synthetic EPs were generated by summing five damped sinusoids, thus producing a waveform with five independent components, each of which could be independently manipulated. These were embedded in 'noise' generated by a Gaussian random process, which varied in frequency, power, etc. We were thus generating different synthetic 'experiments' in which known 'evoked response' differences were embedded in defined background noise. We then applied the DA technique to the data and determined if it correctly assayed the differences we had introduced.

An example of the simulation of an 'intensity' experiment, generating the two waveforms, is shown in the top row of *Fig.* 1*a*. On the left are the results of averaging one hundred synthetic records each containing an 'evoked response' mixed with noise. On the right are shown the average of one hundred similar noise processes in which an 'evoked response' was not embedded. We repeated this process several times, each time reducing the amplitude of the 'evoked response', and the signal waveform gradually disappeared from the 'average evoked response'.

To use the DA for analysing these data one starts with *known differences*, which are used as a baseline in terms of which the rest of the data are described. Thus data are not measured simply in terms of specific physical units; each average evoked response is described in terms of two preselected EPs representing conditions which define a dimension. For example, in an intensity experiment we know that the stimuli of maximal intensity should yield the largest possible differences between the EP and an average of ongoing EEG. Data obtained with other stimuli can be described in terms of a template which is derived from the two curves representing the maximal intensity EP and the ongoing noise.

The program used selects 6 of 80 time points used in developing the discriminant function. Six coefficients provided by the program are applied at these points. The result of these 6 multiplications is a decision score. *Fig.* 1*b* presents the results of applying this discriminant function to the simulated data shown in *Fig.* 1*a*. Each of the lines begins on the right with the score computed for the template condition. The score for noise alone is quite different from the noise score for signal noise. The rest of the scores were computed for data in which the signal was reduced by the percentages indicated on the abscissa, thus providing a measure of the intensity of the signal, and we have found a way of measuring the characteristics of an evoked response which reflects the aspect of interest in the data. This technique enables the scores for each of the individual trials recorded in an experiment to be computed and, by indicating the distribution of the scores, provides the information required about variability.

Fig. 2*a* shows that this technique works with real, as well as simulated, data. This is a series of evoked responses elicited by flashes with durations ranging from 12·8 to 0·100 msec and shows that, as the brightness of the flashes is reduced, the AEP amplitude is diminished. The EEG recorded just prior to the presentation of the stimulus has been used to represent 'noise' records. We

a

b

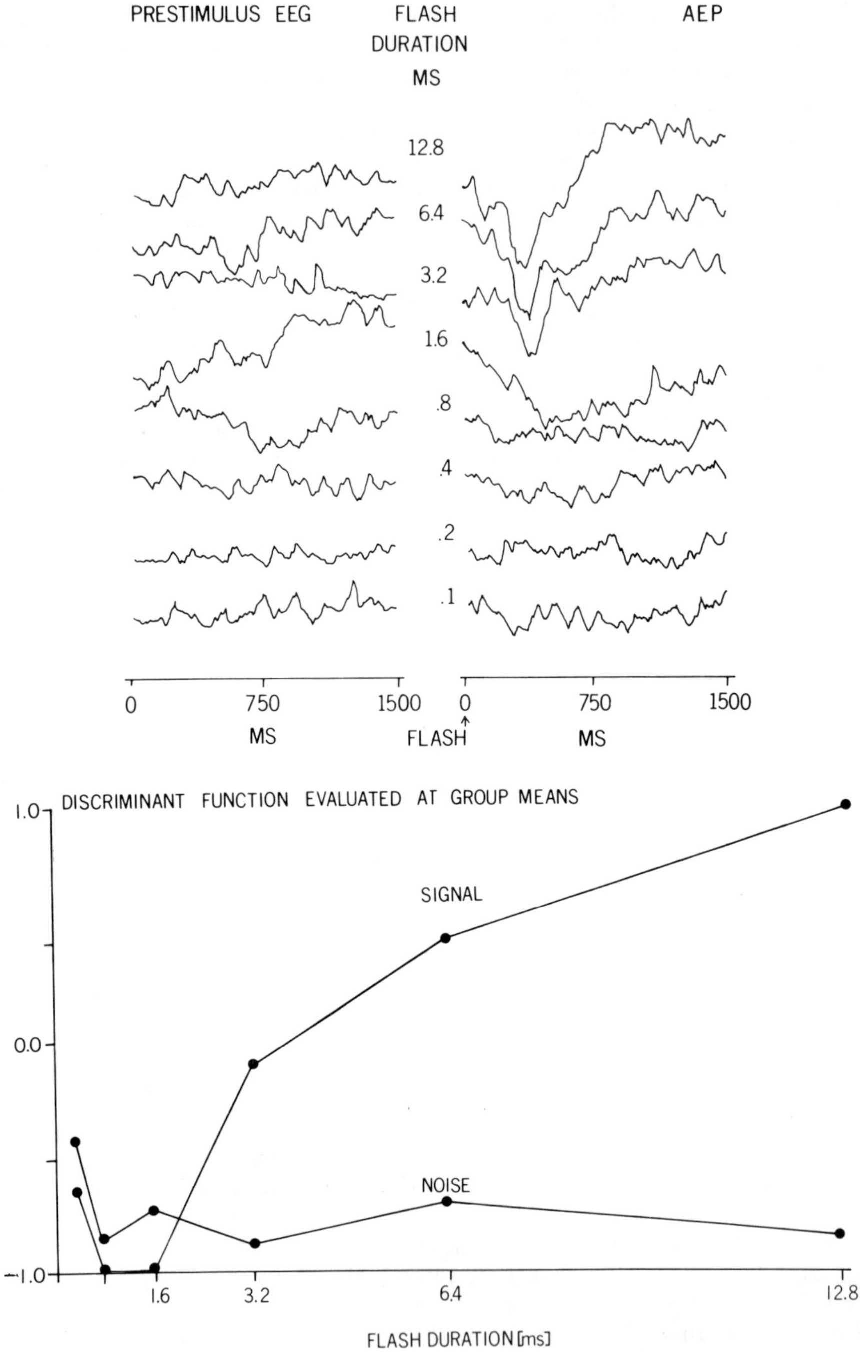

Fig. 2. *a*, Visual evoked responses elicited by flashes with duration ranging between 12·8 and 0·100 msec. The left hand column shows ongoing EEG activity just preceding stimulus onset. A discriminant function was developed on the basis of the data elicited by the 12·8 msec stimulus. *b*, Discriminant scores for the data shown in *Fig.* 2*a*, plotted against flash duration.

developed a discriminant function using the EP elicited by the 12·8 msec flash and the corresponding pre-stimulus EEG records, and applied the discriminant functions so obtained to the rest of the data. The results are shown in *Fig. 2b*.

In the stepwise discriminant analysis technique we have a very useful and powerful technique for measuring the characteristics of evoked responses, for summarizing the data obtained in an experiment in a meaningful way and for relating the between-subject and within-subject variability to our conclusions. The use of the technique is supported by a large body of statistical theory and the behaviour of the statistics thus obtained under many different conditions can be derived.

I have been advocating for some time the use of multivariate statistical techniques in the analysis of evoked response data, since the EP is a multivariate observation. Oddly, many evoked response investigators persist in handling the data as if they were recording an assortment of univariate observations. Various strategems are developed for combining basic univariate statistical tests in ways which look intuitively satisfactory. However, multivariate analysis combines univariate tests into one test. It seems to me that rather than develop a variety of palliative measures derived from the textbook statistics we were brought up on, it would behove us to make use of more advanced, more appropriate statistical techniques.

To some extent the reluctance to use multivariate techniques derives from the instrumentation requirements that may be imposed by such techniques, since it is necessary that the data, trial by trial, be in a computer-compatible form. Then the data can be put into statistical packages available at most University computer centres. Such transformation of data is expensive. However, the discriminant analysis described above used the BMD packages developed at UCLA, and this is readily available. It is hoped that more extensive use will be made of this available technique.

Topography of Evoked Potential Amplitude Fluctuations

MERLIN W. DONALD

Queen's University, Kingston, Ontario, Canada

It is a distinctive feature of some transient sensory EPs that they fluctuate widely in amplitude in a waking subject even though stimulus intensity is constant. A variety of EP phenomena share this characteristic: the P200 and P300 waves; the post-motor positivity; the positive overswing following S2 which often terminates the CNV; and EPs to 'missing' stimuli where the constant stimulus value is zero. It is possible that other EP phenomena fluctuate significantly but have been ignored because of the current tendency to study late positive waves to the exclusion of other aspects of the EP.

The distributions of late EP waves have been studied in the conventional manner of mapping potential gradients on the scalp from multiple-electrode averages. However, since these waves are notable principally as fluctuating processes, it was decided to map the distribution of their *fluctuations* as well, by subtracting successive topographical displays of the EP obtained at different amplitude levels. The subtractions are effectively a map of EP change. Pilot work with this technique suggested that EP amplitude fluctuations might have a different distribution from the EP itself at any given moment in time.

For maximum clarity in computing subtractions it was necessary to obtain as wide a range of EP amplitude fluctuations as possible without changing the effective stimulus. Unfortunately, strictly controlled experiments seldom produce EP amplitude changes of more than 10 or 20 per cent that can be attributed to such operationally defined variables as stimulus probability, instructions, or information load. The spontaneous EP fluctuations routinely observable within any experimental condition are often much larger, and for this reason it was decided to study spontaneous variation in EP amplitude first, where the most extreme changes could be found and located. Following this, experimentally-produced EP fluctuations were compared with the distributions of spontaneous ones.

METHOD

Somatosensory evoked responses were recorded from 8 normal adult male *Ss*, visual evoked responses from 2 male and 2 female. Individual subjects were studied in detail, some returning for as many as 15 recording sessions. Throughout all of the sessions *Ss* performed a sensory discrimination which required them to respond to a target stimulus embedded at random in a series of standard stimuli; no motor response was made to the standard stimuli, and evoked responses were averaged only to these stimuli. Interstimulus interval was 2 sec.

The EEG was recorded with silver disc electrodes and amplifier gains were set at 10^4 with upper and lower half-amplitude frequency settings of 100 and 0·5 Hz, respectively. Electrode impedances were kept below 3 kohm, and placements followed the 10–20 system except for a detailed contralateral

Rolandic array where electrodes were added at midpoints between P_4 and C_4, C_4 and F_4, and F_4 and Fp_2. Eye movements were monitored by referring the supra-orbital ridge to the external canthus of the right eye. Peripheral median nerve responses were recorded from the left elbow. Monopolar reference was always linked ears, with electrodes placed on the upper pinnae.

The standard somatic stimulus was a shock to the median nerve of the left wrist, with current fixed just below thumb twitch threshold. The somatic target was a weaker shock, adjusted to allow 80–90 per cent identification accuracy. The visual stimulus was a rectangular checkerboard pattern rear-illuminated by a diffused Grass PS2 strobe; the target was a double flash whose delay was adjusted to allow 80–90 per cent identification accuracy. The brightness of the white squares as seen by *S* was 4·4 ft L and of the black, 0·019 ft L. The individual squares subtended a visual angle of 30′. The stimulus was presented in either the upper or the lower half of the visual field with a continuously visible fixation cross in the centre of the visual field. A chin and forehead rest positioned *S*'s head relative to the stimulus, and an artificial pupil restricted the diameter of the stimulus beam to less than 2 mm. Ambient light level was not changed during recording.

Shock evoked responses were averaged by a PDP–12 computer and stored on digital tape 8 channels at a time. Visual evoked responses were averaged 4 channels at a time with a Fabri-Tek averager; the digital printout was then read onto tape off-line.

RESULTS

a. Spontaneous Amplitude Fluctuations in the Somatosensory EP. Each *S* produced standard somatosensory evoked response wave-forms with relatively little amplitude variability during initial recording sessions. These recording conditions did not produce any waves resembling P300 in the EP of any *S*. However, as *Ss* became highly over-practiced, large fluctuations began to appear in the 200 msec positive wave (P200) located near the vertex. For the *S* represented on the graph in *Fig.* 1*a*, these P200 fluctuations initially took the form of a gradual diminution of amplitude during each daily session; but on succeeding sessions the amplitude range of the P200 became larger, the fluctuations more rapid, and sudden increases as well as decreases were observed. Every *S* eventually showed the kind of amplitude lability demonstrated in *Fig.* 1*a*, several showing more extreme fluctuations. Note that identification accuracy was unchanged throughout these sessions. Peripheral nerve responses were also stable, and therefore the P200 changes were of central origin.

An apparent paradox in these P200 fluctuations was that even at moments when P200 was virtually extinguished from the averaged response, a bipolar electrode configuration spanning the Rolandic regions contralateral to the stimulated side revealed the presence of residual, differently distributed, late activity. These 'Rolandic' late waves remained remarkably stable in wave form and amplitude throughout the waxings and wanings of P200. In *Fig.* 1*a* (lower) a series of six successive evoked responses illustrates a 30-min period of rapid amplitude change in the P200 (left column) during which simultaneous bipolar Rolandic averages (right column) remained unchanged. In the bipolar averages, neither the early components (the sharp negative-positive waves at 20–60 msec latency) nor the late waves (positive and negative peaks at 90 and 160 msec) bore any proportionality to the vertex wave changes. These data contradicted the notion that the bipolar late waves simply reflected an occipital-frontal latency lag in the spread of the vertex wave, and suggested there was another late process focused over contralateral Rolandic regions, which is independent of the P200 while overlapping it in time.

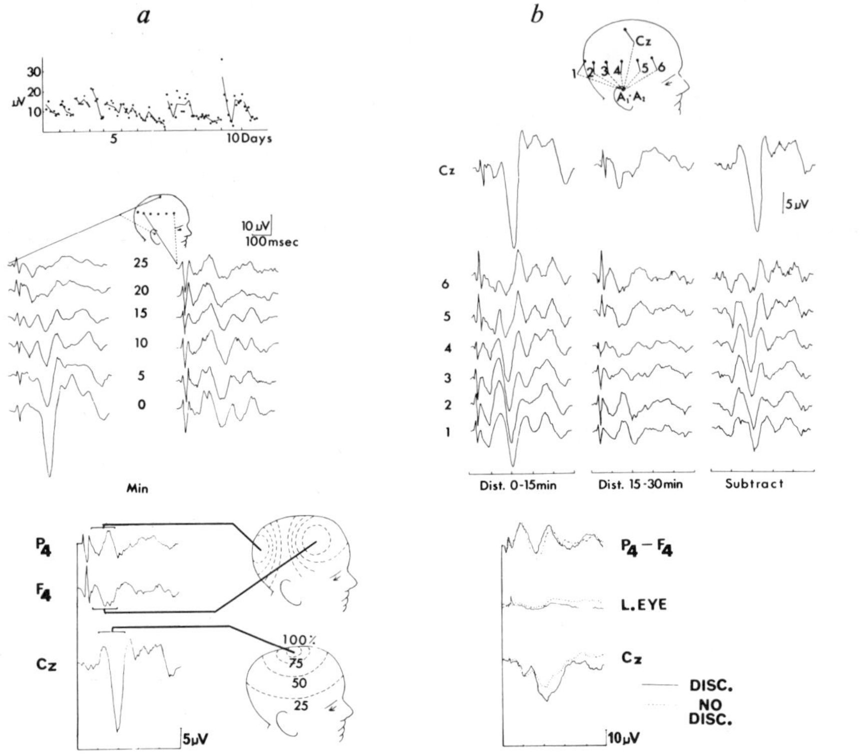

Fig. 1. *a. Upper* graph shows base-to-peak amplitude of P200, each dot representing 64 summations, recorded from one subject on 11 successive days. Breaks in curves, and marks on abscissa show where daily sessions ended. *Middle* graph: six successive evoked responses (64 summations per trace) recorded from one subject during a 30-min time period. P200 (left column) gradually extinguished while simultaneous bipolar trans-Rolandic (P_4–F_4) electrodes revealed late waves (right column) which remained stable throughout vertex wave fluctuations. *b.* Monopolar distribution of the responses shown in *Fig.* 1*a*. Left column: averages (192 summations) of first 3 responses, 0–15 min, when vertex wave was present; middle column: averages for 15–30 min (192 summations per trace) when vertex wave extinguished; note apparent polarity reversal of late waves, and similarity of monopolar waves to bipolar wave forms shown in *Fig.* 1*a*. Right column, subtraction of middle column from left column, showing distribution of vertex wave fluctuations. Averaging epochs = 500 msec, negative down for all traces.

Lower left. Schematic representation of wave form and distribution of two parallel late processes in the somatic evoked response. The Rolandic wave is biphasic, spanning the interval from 70 to 200 msec after stimulation, and reverses in polarity on scalp regions over the Rolandic fissure. The P200 process is also biphasic, spanning the interval from 100 to 250 msec, and has a central distribution. *Lower right.* Comparison of discrimination response trials (disc) with no-response trials (no disc). Sampling epochs = 500 msec, negative down for all traces.

b. Topography of Fluctuations in Somatosensory EP. To separate the distributions of these two simultaneous late processes sequential topographical displays were computed from monopolar electrode arrays during rapid fluctuations in P200. At moments when P200 was extinguished, the distribution of the Rolandic late waves could be seen clearly. Then by subtracting, at each electrode, evoked responses with no visible P200 from those with fully developed P200, it was possible to compute what was added to the former responses when P200 grew larger in amplitude, this being the true distribution of the fluctuating P200 process. *Fig.* 1*b* illustrates these points, showing the topography of contralateral parietal and frontal scalp regions for the data in the lower part of *Fig.* 1*a*. The left column of *Fig.* 1*b* shows the initially large vertex wave; its distribution over the right side of the scalp is typical of the standard somatic EP, with late negative-positive waves appearing at all electrodes in varying degrees. The middle column shows what happens to this distribution when the vertex wave extinguishes; the late waves at parietal electrodes 1 and 2 now resemble the bipolar response in *Fig.* 1*a*, while the same wave form appears reversed in polarity at frontal electrodes 5 and 6, the central electrodes nearest the Rolandic fissure showing almost no response. These Rolandic late waves are distributed in roughly the same way as the somatic early components, suggesting a source in or near to the primary cortical hand area. The right column shows a subtraction of the second from the first column, yielding the distribution of the fluctuating P200 process. This is now seen as a biphasic negative-positive wave, fairly symmetrical over pre- and post-Rolandic scalp regions.

The distributions of the P200 and Rolandic late processes are illustrated with a schematic estimate calculated from 15 electrode locations (*Fig.* 1, lower left). The P200 process spans the interval from 100 to 250 msec after stimulation, while the Rolandic late waves span from 70 to 200 msec. As far as can be gathered from these electrode placements the P200 fluctuations are maximal at the vertex, but this remains to be confirmed with a more detailed lateral electrode array.

c. Comparison of Spontaneous and Induced EP Fluctuations. Amplitude changes in P200 were experimentally induced in five *Ss* by contrasting the standard discrimination task with a condition in which *S* was instructed to stop responding to the weak stimuli in the series. The effect of this instruction was to reduce P200 amplitude 10–20 per cent below the level established in the 'respond' condition in all the *Ss*. The Rolandic EP amplitude measured with bipolar P_4–F_4 leads was not significantly changed by the instruction (*Fig.* 1, bottom right). The induced P200 fluctuations therefore appear to be similar in distribution though smaller in magnitude than the range of spontaneous P200 fluctuations observed over an 11-day period.

d. Fluctuations in the Visual EP. A similar analysis of the visual EP was attempted. The posterior scalp distributions shown by the four *Ss* were in agreement with the data of Jeffreys (1971): the lower visual field stimulus produced a positive peak at O_Z of about 80 msec latency, negative at 110 msec and positive at 180 msec. The upper visual field stimulus produced peaks of similar latency but opposite polarity, peaking in amplitude at P_Z.

The central and frontal electrodes showed a different wave form from those reported by Jeffreys, and large fluctuations in amplitude, presumably reflecting differences in procedure introduced in this study—a longer i.s.i., randomization of the stimulus series, and requiring *S* to perform a discrimination.

An analysis of the anterior-posterior distribution of spontaneous amplitude fluctuations to the lower field stimulus in one *S* is shown in *Fig.* 2. The left and centre columns are superimpositions of the 4 traces with the most positive (left) and the most negative (centre) peak amplitudes in the triphasic late wave peaking at C_Z. The right column shows a superimposition of 4 corresponding subtractions, showing the latency and distribution of the major EP fluctuations in this *S*. The wave form shown in the subtraction consistently showed a positive peak at about 200 msec peaking in amplitude near C_Z. Virtually no spontaneous fluctuation in amplitude was observed in the occipital components, as shown by the O_Z subtractions. A detailed description of these results is in preparation. In principle they confirm the analysis of somatosensory EP change, i.e. the presence of a stable response near or in primary cortex, and a parallel and simultaneous fluctuating process in central scalp regions.

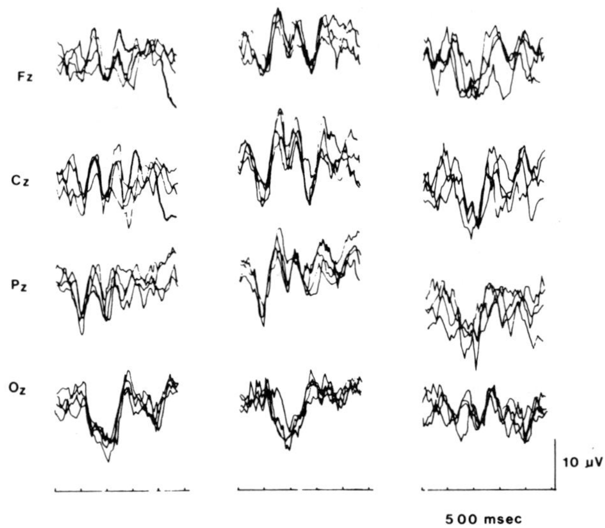

Fig. 2. Spontaneous fluctuations in amplitude of visual EPs during a vigilance task. Right column shows subtractions of first two columns, showing the anterior posterior distribution of EP amplitude fluctuations during task performance.

SUMMARY

The distribution of spontaneous fluctuations in EP late wave amplitude was determined by subtraction of successive topographical displays of the EP obtained during periods of rapid EP amplitude change. Most EP fluctuations during this task occurred in central scalp regions 100–200 msec after the stimulus; EP amplitude changes produced by changing *S's* instructions were smaller in magnitude but similar in distribution. Late waves 70–180 msec in latency located in or near primary sensory regions did not fluctuate significantly at any time during the experiment.

Supported by U.S. Public Health Service NIMH Grant M-05286, NSF Grants GB–3919 and GB–5782 and by the Veterans Administration Continuing support provided by NRC (Canada) grant A–8612 and by Queen's University Advisory Research Council Grant 274–073–0. I thank Dr W. R. Goff for his support and advice and T. C. Fisher, J. Jasiorkowski and David Ross for technical assistance.

Further Thoughts on Measurement of 'the' CNV

JON F. PETERS, JOHN R. KNOTT and CHARLES E. HAMILTON*

University of Iowa, Iowa City, Iowa 52242, U.S.A.

Concern is often expressed over the multiplicity of methods employed for quantifying the CNV (or P300) (McCallum and Knott, 1973). The question of what constitutes a satisfactory 'baseline' (Rousseau, 1969; Donchin, 1973) from which to measure is the key issue. Generally, some pre-S_1 interval is sampled and either the baseline is established 'visually' or set to zero by a computer.

Once the baseline has been determined, the CNV has been quantified as: peak amplitude (*anywhere* in the S_1–S_2 interval) from baseline (Irwin *et al.*, 1966); mean amplitude in the 150 msec period just prior to S_2 (McCallum and Walter, 1968); area under the curve which may or may not include the evoked response to S_1 and S_2 (Low and McSherry, 1968); and mean amplitude in each of 10 segments in the interval beginning 450 msec after S_1 to S_2 onset (Van Veen *et al.*, 1973). In addition, McAdam and Rubin (1971) measured the CNV as the difference between a positive peak called 'P302' and the average level occurring during the 300 msec ending 100 msec before S_2. A similar measure has been used by Kopell *et al.* (1972) and Costell *et al.* (1972).

Convincing arguments have been marshalled both for and against each method. Since fluctuations in the ongoing EEG which ride on top of the averaged CNV make accurate measurement of peak amplitude difficult, most authors visually fit a smoothed curve onto the CNV to measure peak amplitude. Area measures, on the other hand, average out irregularities due to the ongoing EEG. The question then becomes, Over what interval should this area be obtained, i.e. from S_1 onset to S_2 onset, or some interval after S_1 or S_2, or beyond? Inclusion or exclusion of the evoked responses to each stimulus may profoundly influence such measures. Measurement of the mean amplitude in some fixed pre-S_2 epoch incompletely describes the CNV, as shown by the data of Van Veen *et al.*, the CNV having different temporal courses in various subject populations. Amplitude does have merit, however, as a measure of single trial data (McCallum and Papakostopoulos, 1973). Hamilton *et al.* (1973) have presented data based on both baseline to peak negativity, and peak positivity to peak negativity amplitude measures of the CNV. The two measures showed disparate results in the same group of subjects.

The present investigation compared, on the same set of data, all of the methods of measurement presented above. Previous reports have correlated peak amplitude scores with that of area, but no attempt has been made to relate *all* commonly used measures of the CNV.

METHOD

Thirteen males (mean age 21·5) served as naïve paid volunteer *Ss* ($5·00).

* Visiting Professor. Permanent Address: Department of Psychology, Coe College, Cedar Rapids, Iowa 52402.

Silver–silver chloride electrodes were attached to the vertex and both mastoid processes (linked for reference), to record the CNV. Similar electrodes placed above and below the eye monitored eye movements. All electrode sites were vigorously abraded with Redux® to control for skin potentials (Picton and Hillyard, 1972).

CNVs and eye movements were recorded with Grass 7P–1 chopper stabilized pre-amplifiers in a Grass Model 7 polygraph. A PI 6100 FM tape recorder was used to store the data for later off-line analysis on a PDP–12. Only those trials which the on-line paper record showed to be free of eye movements and other artefacts were averaged along with a calibration pulse (Emde, 1964). Typically, 10–12 trials constituted each average.

A discriminative reaction time paradigm was employed with S_1 a 1500 and S_Δ an 800 Hz tone each at 50 dB SRL. Both S_1 and S_Δ were followed 1500 msec later by S_2, a flash from a Grass PS–2 photic stimulator (intensity '1'). The *Ss* were instructed to respond as fast as possible to S_2 only when it followed S_1 (Resp), not when it followed S_Δ (NResp). Twenty response and 20 non-response trials were given in random sequence for a total of 40 trials. The i.t.i. varied randomly from 15 to 30 sec.

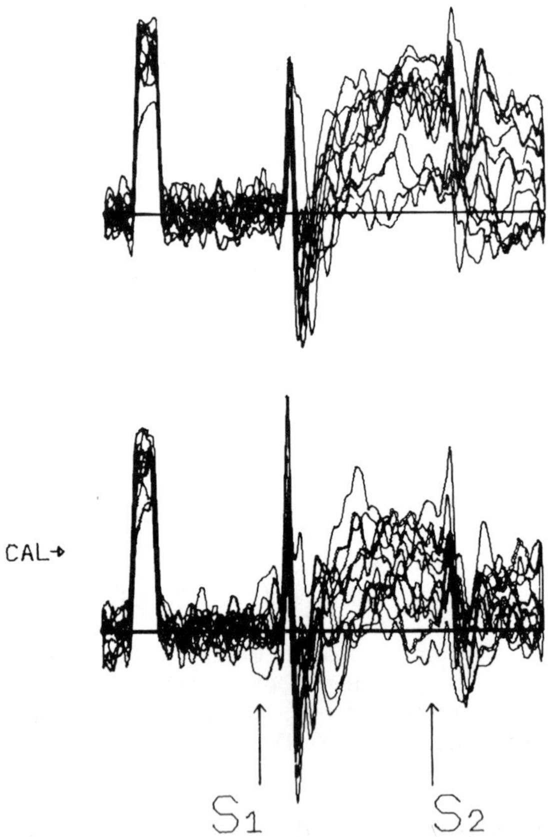

Fig. 1. Overlay of individual subjects averaged CNVs for the response (*top* trace) and non-response (*lower* trace) conditions. Calibration is 25 μV, negative up. Average is over 256 points in 4096 msec.

During data collection the S was seated in a comfortable armchair inside a darkened sound-damped chamber. *Ss* were told to fixate their gaze on a cross drawn on the centre of the PS–2 lamp. The warning stimuli (S_1 and S_Δ) were delivered from a speaker directly behind the lamp.

All measurements of the CNV were made with a cursor program for the PDP–12. Baseline was set at the mean of the activity in the interval between the calibration pulse and S_1 onset. The following measurements were then made: (1) peak amplitude (anywhere in the S_1–S_2 interval) from baseline as determined by a visual smoothing of the interval (PA); (2) peak positivity (to S_1) to peak negativity (same negative point as in (1) (PPA); (3) area from S_1 onset to S_2 onset (A_1); (4) area beginning 450 msec after S_1 to S_2 onset (see Rebert and Knott, 1970) (A_2); (5) mean amplitude from baseline in the 200 msec epoch preceding S_2 (MA); and (6) mean amplitude in each of 19 segments beginning 450 msec after S_1 to the end of the average (SA).

RESULTS

Fig. 1 is an overlay of the averaged CNVs recorded from all *Ss* in each condition.

Table 1A represents the PA, PPA and MA amplitude data averaged across all *Ss*. Significant response v. non-response differences (related t tests, 1 tail, df=12) were found for the PA (t=3·270, P<0·005) and MA (t=4·645, P<0·005) measures but not for the PPA (t=1·454) measure. *Table* 1B contains the mean data for the area measures A1 and A2 expressed in microvolts. Examination of the area data showed significant response v. non-response differences for both the A_1 (t=3·180, P<0·005) and A_2 (t=3·570, P<0·005) measures.

Table 1. A. Mean amplitudes in microvolts for the Peak Amplitude (PA), Peak-to-Peak Amplitude (PPA) and Mean Amplitude (MA) methods of amplitude measurement. Standard deviations (SD) are also presented. *See* text for details of measurement

	PA		*PPA*		*MA*	
	Resp	*NResp*	*Resp*	*NResp*	*Resp*	*NResp*
Mean	16·32	11·26	30·21	25·86	14·51	7·61
SD	6·33	5·56	8·87	13·16	6·94	6·87

B. Mean area in microvolts for the two area measures employed. Standard deviations (SD) are also given. *See* text for details of measurement.

	A_1		A_2	
	Resp	*NResp*	*Resp*	*NResp*
Mean	85·04	76·85	59·00	51·10
SD	10·44	11·14	9·65	9·52

The segment analysis (SA) data are shown in *Fig.* 2. The distribution of significantly different points (F tests P<0·05) between the two curves reveals a markedly different temporal course of CNV for the two conditions.

Pearson-product moment correlations obtained between the three amplitude and two area measures are listed in *Table* 2.

Significant correlations were found between all measures involving the baseline reference. However, none is seen between the amplitude measure based on peak positivity to S_1 (PPA) and any of the other measures.

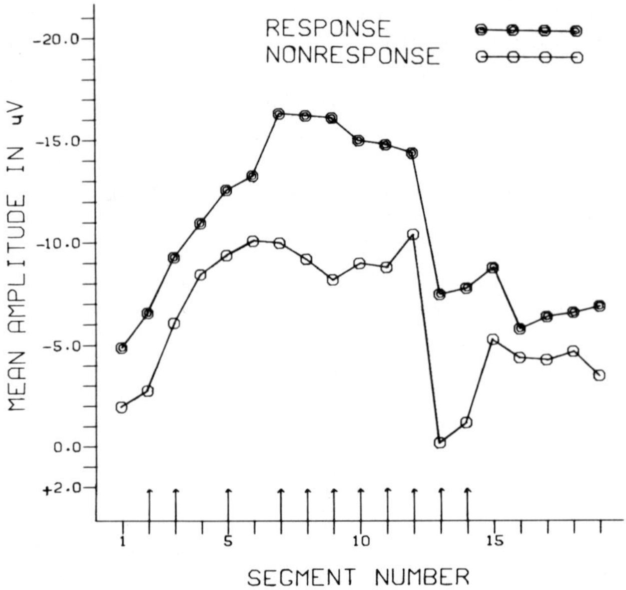

Fig. 2. Group mean CNV amplitude in each of 19 segments commencing 450 msec after S_1 and going to the end of the average. Arrows on the segment axis indicate significant differences (F tests $P<0{\cdot}05$).

Table 2. Pearson-product moment correlations for the 3 amplitude and 2 area measurements

			RESPONSE CONDITION		
	PA	*PPA*	*MA*	A_1	A_2
PA		0·59	0·91*	0·91*	0·93*
PPA			0·44	0·51	0·60
MA				0·90*	0·93*
A_1					0·98*
			NON-RESPONSE CONDITION		
	PA	*PPA*	*MA*	A_1	A_2
PA		0·59	0·85*	0·92*	0·94*
PPA			0·40	0·36	0·46
MA				0·84*	0·86*
A_1					0·98*

* $P<0{\cdot}01$ (2 tail)

DISCUSSION

Of the five methods of CNV amplitude measurement presented, all but the one based on the peak positivity to S_1 (PPA) reflected the expected response v. non-response difference (Irwin *et al.*, 1966). Absence of this difference in the PPA measure is quite possibly due to intra-subject variability in the size of peak positivity, as evidenced by the large standard deviation in this measure and in the area measure which incorporated the EP to S_1 (A_1). Measures of this nature are confounded by the inability to determine whether the CNV or the evoked response varied with the paradigm. Such an objection is not uncommon in evoked response research (Friedman *et al.*, 1973). However, in cases where the CNV never recrosses the baseline in the S_1–S_2 interval, this method is apparently the only alternative! (*See* discussion in the P300-CNV demonstration, p. 219.)

The high correlation of all measures derived from the baseline reference agrees with previously reported data (McCallum and Papakostopoulos, 1973).

Examination of the CNV as a voltage change which may peak at different times may be of considerable importance as a tool for investigating psychiatric problems, particularly for prolonged CNVs.

SUMMARY

Six measures of the CNV were compared. All measures based on a pre-S_1 baseline showed comparable results. The measure based on the peak positivity was unrelated to the others. This latter technique has an added problem of difficulty in determining whether the CNV or EP varied.

ACKNOWLEDGEMENT

The authors are indebted to Maurice I. Mendel, Ph.D., Department of Otolaryngology and Maxillofacial Surgery for use of the PDP–12.

Implications of Cross-modality Stimulus Permutations for the CNV

G. H. BLOWERS, G. C. ONGLEY and J. C. SHAW

Graylingwell Hospital, Chichester, Sussex, U.K.

The current state of CNV research is such, that increasingly, investigators are speculating on the possible underlying psychological manifestations of the phenomenon. This paper is one of a series (Blowers *et al.*, 1973) in which certain psychological labels have been questioned and some experimental situations have been re-investigated. While the result of this study is in agreement with previous reports in the literature its modified experimental approach and somewhat inconclusive results will hopefully promote further interpretation.

In the experiment to be reported here, consideration was given to the sensory modalities in which both the conditional (S_1) and imperative (S_2) stimuli are normally presented. Most CNV studies employ clicks or tones and flashes of light given in the auditory and visual modalities respectively. Although it has been claimed that stimulus modality has no significant effect upon the waveform (Walter, 1967), no study involving more than two modalities has been reported in this context and investigations have been unsystematic. The present study deals with the presentation of conditional and imperative stimuli in permutations of three modalities using a constant i.s.i., somewhat longer than is normally employed in a CNV paradigm. The aims of the study were twofold:

1. To determine whether the waveform measures of amplitude and latency are modality dependent.
2. To verify the establishment of long latency potentials in a CNV paradigm (S_1–S_2–R) in an extended interval.

It was hypothesized that if modality of stimulus elicits no significant effect upon the CNV, there should be no significant difference in waveform across the different conditions.

METHOD

Subjects

Nine female *Ss*, all of whom had participated in previous CNV studies, were recruited as volunteers. None, however, had any knowledge of the real purpose of this experiment.

Stimuli

Stimuli used were tones (1000 cps frequency presented through a loudspeaker positioned 3 feet behind the *S*'s chair), flashes (from bulbs set behind translucent Perspex set in goggles and worn by the subject; the intensity of the light was sufficient for the subject to detect with his eyes closed) and a vibrating steel tip for tactile presentation of stimuli through the subjects' fingers. The steel tip was attached to a piezo-electric crystal driven by a low frequency oscillator.

Thus, three types of stimulus were presented: auditory (A), visual (V) and tactile (T). Each type of stimulus was presented in both conditional (S_1) and imperative (S_2) situations and this gave rise to nine permutations of stimulus pairs:

S_1—S_2
A—A
V—V
T—T
A—T
T—A
V—T
T—V
A—V
V—A

Each permutation of a stimulus pair represented one condition and each condition was presented to every subject. Each condition comprised eight stimulus pairs or trials.

Apparatus

Silver–silver chloride electrodes of a type designed for DC recording were attached to the vertex (Cz), each mastoid and the centre of the forehead. The vertex was referred to the centre of a 25 kohm potentiometer connected between the forehead electrode and the paired mastoids. This enabled some of the ocular potentials that might be contaminating the CNV recordings to be balanced out (McCallum and Walter, 1968). The EEG signals, stimulus markers and subject response marker were recorded on a Van Gogh polygraphic recorder. The signal from the vertex derivation was recorded using both DC and AC amplifiers, other derivations were recorded only with AC amplifiers. The AC amplifiers had a time constant of 6 sec with a high frequency 3dB point at 35 Hz. Only the signals from the vertex derivation recorded with AC amplifiers are used in the analysis described here because of problems of drift encountered on the DC channel. However, the DC recordings do support the results obtained on the AC channel.

Experimental Procedure

After the electrodes were affixed, *Ss* were seated in reclining easy chairs and faced an off-white blank wall in a room of constant illumination. Eyeblink potentials were balanced out and *Ss* were instructed in the format of the experiment, i.e. that they would receive different permutations of stimuli. *Ss* were then told that a trial runthrough of all the permutations in the sequence to be presented in the experiment proper, would take place, during which time they should pay attention to the stimuli without pressing the key. They were then presented with 3 trials of each of the 9 permutations. At the end of this runthrough (familiarization), *Ss* were informed that the sequence would be repeated, but that this time they were to press the key to the second of each stimulus pair, thus terminating it. *Ss* were informed of each permutation of stimuli or condition just prior to receiving it, both in the familiarization and the experimental phases. Thus, for example, with an auditory–visual (A–V) condition, the experimenter informed the subject 'You will now receive a series of short tones followed by long flashes; using each tone as a warning you will turn off each flash as soon as it appears'. The interstimulus interval was constant throughout the experiment at 5 sec while the intertrial interval was varied randomly by the experimenter (from 5 to 30 sec).

Analysis of Data

For each trial, an 11-sec. epoch of EEG signal beginning 0·4 sec before S_1 onset was digitised and stored on magnetic tape on a PDP–12 computer. The baseline was established for each epoch by computing the mean amplitude of the signal in the 0·4 sec preceding S_1. Trials were averaged in groups of eight. Signal averaging was controlled by the computer program and averages were stored digitally on magnetic tape for further analysis.

Three characteristics of each wave form were measured, two of amplitude and one of latency. The two amplitude measures were calculated by taking 8 consecutive sample points on the wave form (a total interval of about 170 msec), and calculating their mean amplitude from the baseline. In one case the 8 points started at the maximal amplitude within the i.s.i. and this measure is referred to as the 'peak amplitude'. In the other case they were the 8 sample points immediately prior to S_2 onset, the 'standard amplitude'. The third measure, 'latency of averaged peak amplitude', was the interval between S_1 onset and the first sample point of the 'peak amplitude'. The first of the 8 points was selected manually, using a cursor program on the averaged wave form presented on the computer display. The average of the 8 samples was then determined automatically. These measures are illustrated in the example of a typical averaged wave form shown in *Fig*. 1.

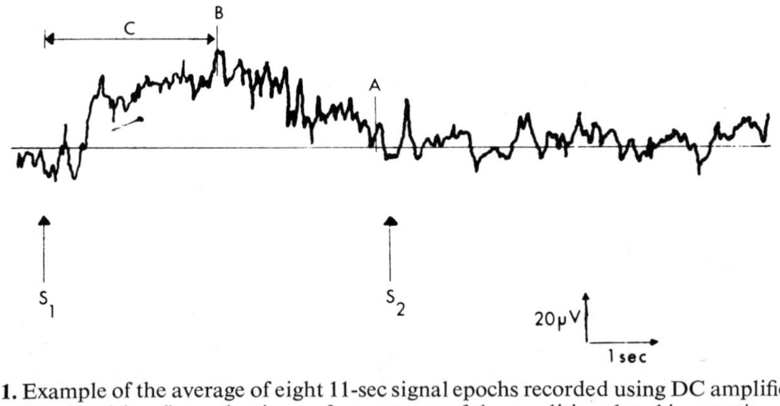

Fig. 1. Example of the average of eight 11-sec signal epochs recorded using DC amplification. S_1 (visual) and S_2 (visual) are the times of occurrence of the conditional and imperative stimuli A indicates the 'standard amplitude', B shows the point at which the 'peak amplitude' is measured and C shows the 'latency of averaged peak amplitude'.

Trials on which artefacts occurred were rejected from the averages. Reaction times to key pressing on individual trials were computed and the log of each calculated.

RESULTS

The results of the means and standard deviations of the waveform measures for the 9 *Ss* are shown in *Table* 1. The values for the standard amplitude show much fluctuation over conditions and the mean amplitudes are somewhat low. The peak amplitudes are also subject to variation although the mean amplitudes appear larger. An interesting point about the peak latencies is that they fall well short of the imperative stimulus.

Table 1. Means and Standard Deviations (S.D.) of CNV amplitudes (microvolts), latencies (seconds) and Log Reaction Times (n=9)

Signal Feature	*Statistic*	*Conditions*								
		A—A	*A—V*	*A—T*	*V—V*	*V—A*	*V—T*	*T—T*	*T—A*	*T—V*
'Standard' Amplitude	Mean	−6·89	−1·11	−10·0	+0·89	−0·67	−7·56	−3·11	−5·56	+2·22
	S.D.	9·96	17·09	15·33	11·36	9·33	17·49	5·84	7·47	10·60
Peak Amplitude	Mean	−17·11	−16·89	−17·11	−10·22	−12·44	−16·0	−10·66	−10·89	−11·11
	S.D.	6·94	13·23	17·50	7·58	9·10	14·46	9·17	8·01	9·85
Peak Latency	Mean	1·80	2·08	2·64	2·36	1·87	2·62	2·67	2·29	2·74
	S.D.	1·61	1·59	1·40	1·49	1·24	1·51	1·38	1·47	1·62
Log Reaction Time	Mean	2·470	2·541	2·539	2·568	2·408	2·527	2·60	2·494	2·488
	S.D.	0·058	0·066	0·069	0·032	0·278	0·149	0·114	0·085	0·251

The most appropriate statistical model for this experiment with 9 treatments and 9 conditions was considered to be a Latin Square analysis with F ratios computed for *Ss*, treatments and order of treatments. Applying the standard amplitude measures to the model gave F ratios for treatments (conditions) and order of presentation of conditions as being non-significant. The same result was also achieved for peak amplitude and latency measures—the only significant F ratio calculated in each case was for *Ss*.

The log RT scores showed a significant effect for *Ss* and also for treatments (F=2·306; *P*=0·05; d.f.=8 and 56). On application of a test for the deviation of a set mean from a population mean (Guilford, 1950), the treatments (sets) proving significant were the 'V–A' permutation which was lower than the population mean for treatments (*P*=0·01) and the 'T–T' permutation which was significantly higher (*P*=0·05) than the population mean. An interesting observation is that the three permutations containing the tactile stimulus in the imperative (S_2) position all had long reaction times although only one proved significant.

Product moment correlations of peak amplitude with latency scores taken over all treatment conditions showed a significant negative correlation (r=−0·238; *P*=0·05; d.f.=79) while peak amplitude and latency of peak amplitude both failed to reach significance when each was correlated with RT. After partialling out the various third variables, the same correlations and non-correlations held. That is to say, peak amplitude was significantly correlated with latency, holding reaction time constant ($r=-0{\cdot}219$; $P=0{\cdot}05$; d.f.$=78$). Peak amplitude with RT, holding latency constant, and peak latency with reaction time holding peak amplitude constant, were non-significant. Within-treatment correlations proved non-significant.

DISCUSSION

The results suggest that the modality in which the stimuli are presented is a non-contributory factor in CNV genesis, agreeing with previous reports. The results also show that within a CNV paradigm, potentials can be generated for long periods and that the peak amplitude may occur well short of the imperative stimulus. This finding has important implications for reconsidering the relevance of the concept of temporal expectancy supporting earlier work (Blowers *et al.*, 1973) and perhaps an alternative explanation of these potentials should be sought. The RT data suggest that different levels of activity are occurring in different conditions. However, auditory RTs are usually lower than visual RTs (Woodworth and Schlosberg, 1954) and the lengthened RT to the double tactile permutation might be due to an interval needed to develop an adequate stimulus level at the sensory nerve ending. It seems likely therefore that the RT changes can be explained by peripheral physiological factors.

The similarities of the latencies of peak amplitudes over conditions and the lowered amplitudes prior to S_2 suggest initially that 'arousal' might be a dominant factor affecting these potentials whose amplitudes appear to be time locked to S_1. However, if this were the case, one would have predicted that the potentials in different conditions would have differed significantly from each other in amplitude or latency because *Ss* had all had exposure to the auditory stimuli on previous sessions with the consequent possibility of some degree of habituation. Visual and tactile stimuli were in this respect 'novel' and might have been expected to possess greater alerting properties (Sokolov, 1963). However, no significant differences were found.

We are left thinking in terms of an 'attention' model based on a selection mechanism operating at a central level. Selection here can be thought of as a filtering of sensory input (e.g. Treisman, 1964) or a biasing of stored sensory representations (e.g. Norman, 1969) and would operate across all sensory inputs irrespective of their modality. A negative finding, such as the appearance of no significant difference in amplitudes or latency in conditions where stimuli are presented in the same or different modalities, tends to support a model for these long latency potentials based on a selection process. It also allows experimenters using such a paradigm to avoid the difficulty posed by Näätänen (1967) of having to demonstrate the existence of reduced EPs to stimuli not attended to, when attempting to show that EPs are a measure of selective attention. As CNV potentials have no specific cortical sites one cannot very well demonstrate reduced CNV activity in modalities in which stimuli are not attended to. The interpretations therefore remain speculative in view of both the difficulty of constructing a paradigm to demonstrate selective attention and the support given to the null hypothesis by the experimental findings.

SUMMARY

The Contingent Negative Variation was measured in nine subjects using an i.s.i. of 5 sec. Each subject endured nine conditions. Each condition consisted of varying the modality in which the conditional (S_1) and imperative (S_2) stimuli were given. Signals were averaged in groups of eight. Measures of three characteristics of the averaged wave forms were collected together with reaction times. All data collected was treated by a Latin Square Statistical analysis. No significant differencs in wave form characteristics occurred across the nine conditions. Reaction times were significant in two conditions. The psychological implications of these findings are discussed in terms of a selective attention model.

Distribution of Response to Non-signal Stimuli

N. E. LOVELESS

Department of Psychology, University of Dundee, Dundee, Scotland

INTRODUCTION

Two ERPs have been associated with 'orienting responses'. Ritter *et al.* (1968) found that unpredictable changes in stimulation evoked a prominent P300 response; but Haider *et al.* (1968) found that similar changes produced a negative response of 250/350 msec latency. Cohen (1969) also reported that when a novel stimulus is first presented, a negative wave lasting 300/500 msec develops, but habituates unless it is given significance as in the CNV situation; in this case it merges into the potential preceding response, so that it is impossible to separate 'orienting' and 'expectancy' components within the traditional CNV paradigm. However, Loveless and Sanford (1973) found that these components were easily distinguished when the i.s.i. was extended to several seconds. The relationship between this negative 'orienting' response and P300 is obscure, especially since the latencies attributed to them vary within ill-defined and overlapping ranges.

It seems possible that part of the confusion depends on distributional factors. While vertex recordings may show both positive and negative features (e.g. Sutton *et al.*, 1967), these may have different origins. Vaughan and Ritter (1970) state that the field of P300 centres in the mid-parietal region. Cohen (1969) states that while the terminal amplitude of CNV is usually greatest at the vertex, its amplitude peaks earlier in the frontal leads. This suggests that the 'orienting' component of CNV may be of frontal origin. The present investigation examined the distribution of ERPs evoked by unpredictable changes in stimuli to which no response was required, in a situation intended to minimize habituation.

METHOD

Tones of 200, 400, 600 and 800 Hz were presented in random order, with the constraint that each occurred 5 times in a block of 20 trials. Three such blocks were presented without a break. The i.t.i. was of the order of 10 sec, but very irregular. *S* was told only that he would hear some tones. He was instructed to fixate a visual target, and asked to minimize blinking and eye-movements. The EEG was recorded from scalp electrodes at Fp_z, F_z, C_z and intermediate (Fp'_z, F'_z) placements, and from P_z; reference was to linked mastoids. Vertical EOG was also recorded. TC was 5 sec. A 2-sec epoch of signal covering each trial was fed into a PDP–12 computer, which stored the result on digital tape for subsequent averaging.

Eleven undergraduate *Ss* were tested, but the records of 5 were discarded because they persistently blinked or made gross eye-movements in response to stimuli. For the remaining 6 *Ss*, the EOG for each trial was scrutinized by two independent investigators, and only trials which both agreed to show minimal eye-movement were regarded as acceptable. Eye-movement was most frequent

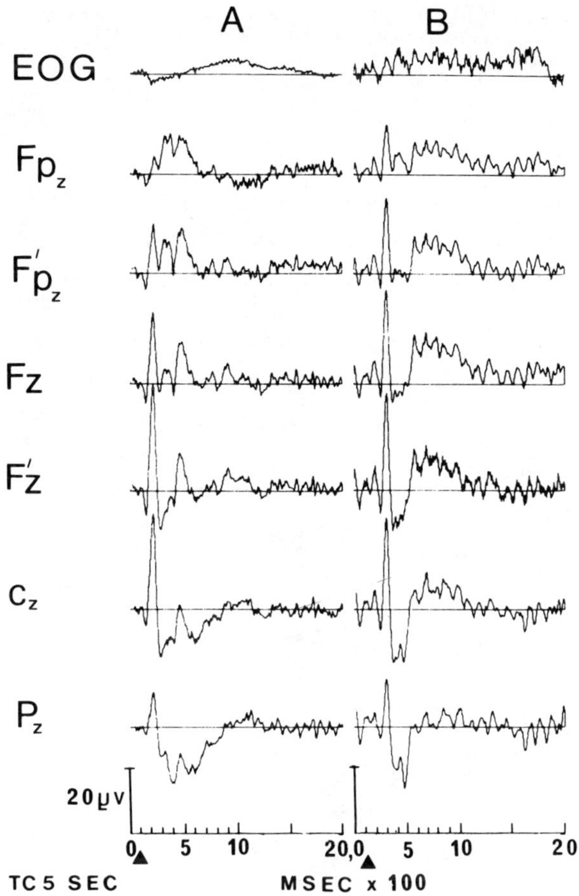

Fig. 1. Anterior-posterior distribution of averaged ERPs to unpredictable tones from 2 *Ss*.

in the first few trials, and in the last block of 20, when *Ss* were becoming restless. Data were therefore taken only between the 10th and 40th trials, over which period habituation effects appeared to be slight. There were no discernible differences in the ERPs to the four tones, so that responses could be pooled. It was thus possible to form averages of 10 acceptable trials from each *S*. Baselines were normalized to a mean value of the activity for 100 msec preceding the stimulus.

RESULTS

Averaged ERPs from two *Ss* are shown in *Fig.* 1. Distribution of the major peaks, averaged across all *Ss*, is shown in *Fig.* 2.

DISCUSSION

Such eye-movements as occurred on the selected trials were gradual and inconsistent in direction, so that their average appears relatively flat.

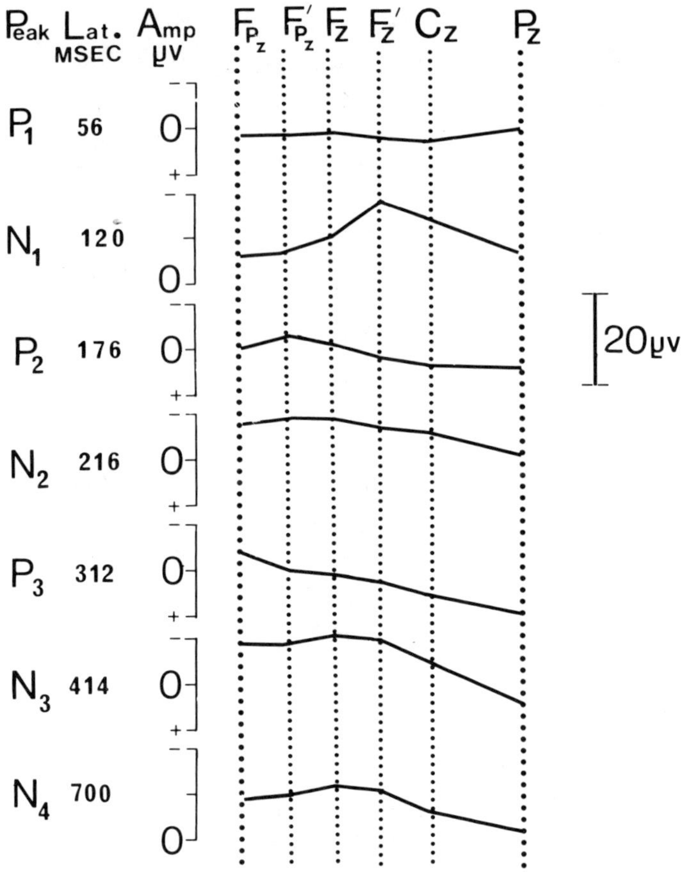

Fig. 2. Distribution of the major peaks of ERPs to unpredictable tones, averaged across 6 *Ss*.

Comparison of waveforms suggests nevertheless that the Fp lead shows some contamination, but this is barely discernible on the more posterior leads.

It seems clear that ocular effects cannot be responsible for the negative 'orienting response', which appears to be maximal near F_z. Its initial peak N3 (latency 410 msec) is at least as great at F_z as at Fp, and in most subjects appreciably greater (*Fig.* 1B). N4 (700 msec) has an identical distribution. The waveform between N3 and N4 varies between *Ss*. At one extreme it may be a plateau (*Fig.* 1B), at the other there may be a marked positive depression (*Fig.* 1A).

All the late components show a general anterior-posterior gradient, forming a prolonged positive wave which is most conspicuous at P_z, where the frontal complex is most attenuated (*Fig.* 1A). This is reminiscent of the suggestion by Karlin (1970) that all late positive components may be reflections of a common process. The relationship between P2 and P3 is unclear, but they may be difficult to distinguish (Roth, 1973), and are similarly affected by stimulus uncertainty (Friedman *et al.*, 1973). Vaughan and Ritter (1970) claim that their topography is different, but the difference appears slight, and it must be recognized that their apparent amplitude at any point will depend on the interaction of central positive and frontal negative processes. It is indeed open to question whether some, at least, of the positive 'peaks' represent anything but such intersections. For instance, the latency of P2 tends to increase systematically from the anterior to the posterior leads in a way which is consistent with its being the intersection of N2 with a central positive wave. Again, where there is a positive depression between N3 and N4, this tends to occur earlier in the posterior leads (*Fig.* 1A). On the other hand, P3 is commonly a well-marked peak of constant latency in the frontal leads; but more centrally, attenuation of the frontal complex and summation with the positive wave tend to blur this component. The possible ambiguity of 'P300' measures has been noted, among others, by Wilkinson and Lee (1972) who report a prolonged positive wave following N1, whereas in most of their subjects a discrete P300 could not be discerned at the vertex. It may be relevant that N2 is not always visible at the vertex either (Roth, 1973).

The most parsimonious interpretation of the present data, therefore, would be that all genuine peaks following N1 form part of a frontal-dominant complex which interacts with a central-dominant slow positive wave. A similar response to signal stimuli has been reported by Järvilehto and Fruhstorfer (1970). This interpretation also seems consistent with the report by Rebert and Knott (1970) that reinforcement increases the negativity of all peaks following N1, as recorded at the vertex.

SUMMARY

The 'orienting response' to unpredictable stimuli appears to include a frontally-dominant negative complex and a centrally-dominant slow positive wave.

ACKNOWLEDGEMENTS

This research was supported by a grant from the Science Research Council. The author is grateful for the assistance of Mr B. L. Potter.

Cortical Responses Evoked by Thermal Stimuli in Man

H. FRUHSTORFER, H. GUTH and U. PFAFF

Institute of Physiology, University of Marburg, Marburg/Lahn, German Federal Republic

The study of thermally evoked cortical potentials in man is a new approach to the investigation of the human anterolateral system. This approach may contribute to our understanding of the neural processes underlying thermal sensations.

In primates, two kinds of specific cutaneous thermoreceptors have been identified:

Cold receptors—show a dynamic increase in their steady state activity on cooling of the skin; their receptive terminals are situated at the level of the epidermal basement membrane and their afferent axons are partly myelinated, partly non-myelinated.

Warm receptors—respond to warming of the skin with an increase in their steady state activity, are probably situated in deeper layers and their axons are exclusively non-myelinated. In addition to these specific thermoreceptors there exist temperature sensitive mechanoreceptors, particularly the fast-conducting cold sensitive SA I receptors (for details on thermoreceptors, *see* Hensel, 1973). It is safe to assume that man has a similar set of temperature sensitive receptors.

Little is known of the significance of warm and cold receptors for temperature sensation. The hypothesis that cold sensations are elicited by stimulation of cold receptors whereas warm sensations are elicited by stimulation of warm receptors is not generally accepted. One possibility to test this hypothesis is the recording of cortical potentials evoked by thermal stimulation. Due to the greater depth of warm receptors and the slower afferent conduction along their axons, the latency of warm stimulus responses should be greater than that of cold stimulus responses, and the magnitude of the latency difference should be a function of the length of the afferent pathway.

In this study, cortical potentials were recorded from the vertex and from above the specific sensory area, following thermal stimulation of the lip and hand. The results are compared with those obtained from a parallel study on thermal reaction times (Fruhstorfer *et al.*, 1972).

METHOD

Fifty-seven paid students (37 males, 20 females) were *Ss* in these experiments, which took place during daytime in a room kept at a temperature 20–22°C. During an experimental session, *S* sat in a comfortable chair reading. A water-circulated thermode of 18 mm diameter was placed, with minimal pressure, either on his right upper lip or on the glabrous skin of his right thenar, close to the border of the hairy skin. The thermode kept the adjacent skin at a temperature of 30°C and presented either warm or cold pulses of 500 msec duration and approximately 10°C amplitude at a rate of 1 stimulus/30 sec (*Fig.* 1).

The EEG was recorded with chlorided silver disc electrodes of 10 mm diameter from C_Z–A_Z (TC=4·0 sec), and from a point halfway between the left external meatus and C_Z, with F_5 as a reference (TC=0·3 sec).

The EEG was fed on-line into a PDP–12 computer, which presented the stimuli and recorded the responses evoked; 500 samples from each channel were taken within 1 sec, starting 100 msec before the stimulus onset. The two EEG curves were displayed for 10 sec during which time the operator had to decide whether they contained artefacts (e.g. transients caused by blinks or movement, etc.) and should be discarded, or whether they should be summated. For each type of stimulus (i.e. warm or cold) and each stimulus location (i.e. lip or hand), 25 good responses were summated during one session.

Peak amplitudes of the summated responses were measured on the computer oscilloscope. Additionally, the voltage integrals of the prominent components were calculated, using the average prestimulus voltage as a baseline. The point in time which divided each integral into two equal halves was denoted as latency of the component.

RESULTS

1. *Lip stimulation.* With cold stimulation, distinct vertex responses were obtained from 52 of the *Ss*, whereas only 44 *Ss* produced vertex responses to warm stimuli (*Fig.* 1). Although the responses generally resembled those obtained under tactile or auditory stimulation, they were of smaller amplitude, longer latency and longer duration (*Table* 1). In all *Ss*, the latency of N1 was shorter with cold stimulation as compared with warm. During a session, the single responses showed a pronounced reduction in amplitude; this phenomenon, however, was not further quantified in this study.

   From the electrode above the specific sensory cortex, distinct responses were recorded from 24 *Ss* with cold stimulation and in 20 *Ss* with warm stimulation (*Fig.* 1). Although in wave form these responses resembled the

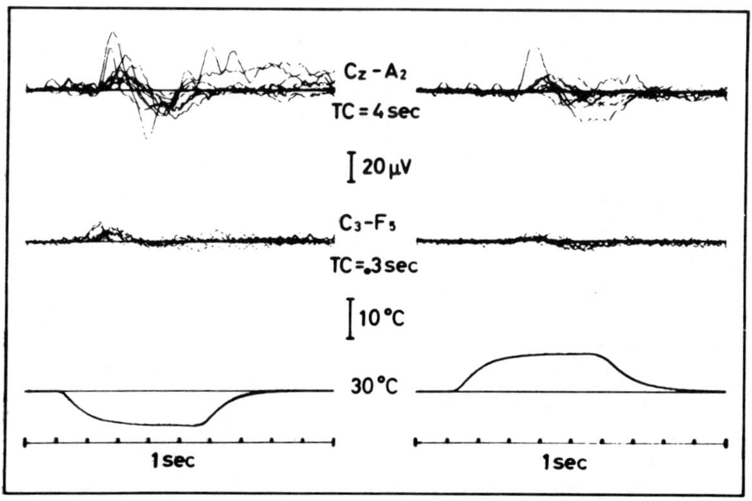

Fig. 1. Typical average evoked responses to pure cold (*left*) and warm (*right*) stimulation from 10 *Ss*. For the exact location of the C_3 electrode *see* Method section. The skin temperature change during stimulation is shown on the lowest trace.

vertex response, they were of smaller amplitude, and their latencies were significantly shorter ($P<0{\cdot}01$, sign test). The latency of the prominent negative component amounted to 180±34 msec with cold stimulation and to 308±37 msec with warm stimulation. Whether or not these responses correspond to those recorded from above the specific sensory cortex with thermal stimulation (Kenshalo, personal communication) or tactile stimulation (Franzen and Offenloch, 1969) cannot be concluded from these data.

Table 1. Latencies and peak amplitudes of the vertex response obtained with thermal stimulation (C=cold, W=warm) of lip (L) and hand (H). Means and standard deviation given

	N_1			P_2		
	n	*Latency* (*msec*)	*Amplitude* (μV)	*n*	*Latency* (*msec*)	*Amplitude* (μV)
LC	50	228± 48	7·0±7·4	52	377± 70	9·0±6·6
LW	44	343± 83	5·8±4·8	43	503± 87	7·2±4·8
HC	13	396± 71	4·8±2·6	13	551± 85	8·6±7·2
HW	8	602±146	4·0±2·4	8	771±191	7·2±8·6

2. *Hand stimulation.* In 15 *Ss* with large vertex responses to lip stimulation, the responses to hand stimulation were additionally recorded. Thirteen *Ss* produced distinct vertex responses with cold stimulation and 8 *Ss* with warm stimulation. These responses had smaller amplitudes and longer latencies than the responses elicited with lip stimulation. Distinct responses from above the specific sensory area were obtained from 5 *Ss*. These had similar characteristics to those obtained with lip stimulation.

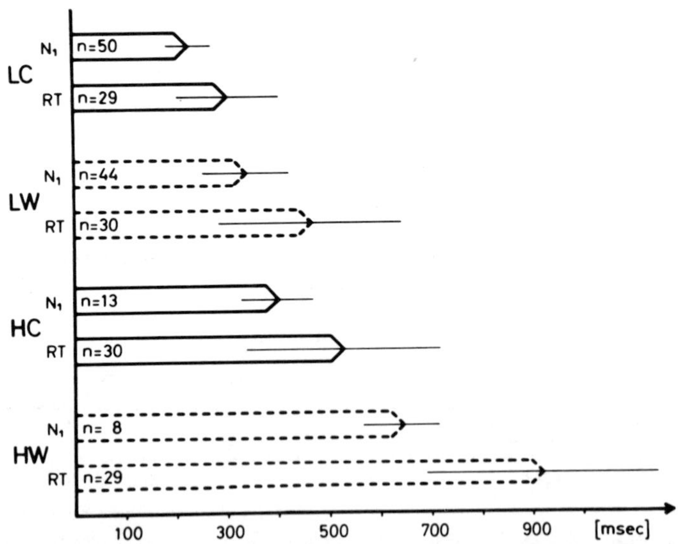

Fig. 2. Latency of vertex response (component N1) and duration of RT under thermal stimulation (C=cold, W=warm) of lip (L) and hand (H). Means and standard deviation given.

Transmission Velocity in the Afferent Pathways

The hand–lip latency difference of the vertex response (component N1) was 174±58 msec with cold stimulation and 309±57 msec with warm. Assuming an average difference of 75 cm in the length of the afferent pathways, the transmission velocity in the hypothetical cold pathway amounts to approximately 4·3 m/sec, that in the warm pathway to 2·4 m/sec.

DISCUSSION

This study has shown that typical vertex responses can be elicited with pure thermal stimuli. The responses evoked by warm stimuli have longer latencies than those evoked by cold stimuli, this difference being greater the longer the afferent pathway. Thus the results are in good agreement with the hypothesis that two different afferent pathways serve cutaneous cold and warm sensibility.

Data comparable to the present results have been obtained from parallel experiments in which reaction time to thermal stimulation of lip and hand was measured (*Fig.* 2). The conduction velocities in the hypothetical afferent pathways calculated from these studies, correspond well with the conduction velocity measured in cutaneous cold and warm receptor neurons of the primate (Hensel and Iggo, 1971).

The fast conducting cold sensitive mechanoreceptors do not appear to play a significant role either in eliciting vertex responses or in evoking sensations to thermal stimulation. From the latency difference of vertex responses elicited by tactile stimulation of the lips and the finger tips (Fruhstorfer, 1966), it can be concluded that the pathways carrying mechanoreceptor information conduct with a velocity of approximately 30 m/sec. Furthermore, chemical nerve block experiments (Zenz *et al.*, 1973) have shown that, in the majority of subjects, a dissociated loss of cold and warm sensibility occurs where warm sensibility is lost before cold; at the time when cold sensibility is blocked but the thicker myelinated fibres are still conducting, cold stimuli are neither perceived as cold nor as tactile. Probably the activity increase of cold sensitive mechanoreceptors elicited by cold stimulation is not enough to overcome the central threshold for sensation.

SUMMARY

In 57 *Ss*, cortical potentials evoked by pure thermal stimulation of the lip and hand were recorded from the vertex and from above the specific sensory area. In nearly all *Ss* vertex responses were obtained which resembled those elicited by tactile or auditory stimulation. They were, however, of smaller amplitude, longer latency and longer duration. The responses evoked by warm stimuli had longer latencies than those evoked by cold stimuli, this difference being greater the longer the afferent pathway. The results support the hypothesis that two different afferent pathways serve cutaneous cold and warm sensibility.

ACKNOWLEDGEMENT

Supported by the Deutsche Forschungsgemeinschaft (Fr 265/2).

Pattern Discrimination in the Spatial Distribution of the Contingent Negative Variation

H. WEINBERG

Simon Fraser University, Vancouver, Canada

D. PAPAKOSTOPOULOS

Burden Neurological Institute, Bristol, U.K.

Two new techniques were used to study the distribution of the CNV recorded from the scalp of humans. One technique was for the suppression of eye movements and the elimination of them from the average data (Papakostopoulos *et al.*, 1973). The other technique was for the recognition of differences in the pattern of CNVs recorded from different sites (Weinberg and Cooper, 1972a, b). The latter technique utilized an average vertex CNV as a template against which CNVs from the vertex, left and right frontal, central and parietal sites were matched.

The result showed that the frontal CNV was different in form from the vertex, central and parietal CNVs; that vertex, central and parietal CNVs were similar in form. An analysis of the amplitudes of CNVs showed that the frontal CNVs were significantly smaller than the CNVs from all other sites; that the vertex CNV was significantly larger than the parietal or central CNVs and that the central CNV was significantly larger than the parietal CNV. There were no differences in amplitude between hemispheres within Condition A (responding with the right hand) or Condition B (responding with the left hand) or between Conditions A and B. Subjects were, in different conditions, required to respond with the left hand and with the right. No significant differences in amplitude between hemispheres were found either within or between conditions.

Note: The above work has, since the congress, been reported more fully in Weinberg, H. and Papakostopoulos, D. 1975. *Electroenceph. Clin. Neurophysiol.* **39,** 21–28.

SLOW POTENTIALS AND BEHAVIOUR

Cognitive Aspects (including CNV-P300 Studies)

Opening Remarks and Review of Selected Cognitive Aspects: 1971–1973

DALE W. McADAM

Departments of Psychology and Neurology, University of Rochester, U.S.A.

The task of reviewing the recent literature on cognitive aspects of slow potentials and behaviour is a horrendous one. First, it is an ever-growing literature, and that which is written today will be incomplete, if not wrong, tomorrow. Second, the reviewer is faced with an incredibly broad spectrum of behaviours which have been coupled with a variety of sometimes interdependent, sometimes independent, potentials. Third, as a consequence of often grand experimental designs, subtleties occur which cannot be fully explored or appreciated in so brief a format as this. Last, we are dealing with macro-potentials and with macro-concepts; both admit to a variety of measurements, analyses and interpretations. To constrain either the potentials or the concepts too tightly with rigorous definitional bonds is to lose their generality. To allow them free rein is to deny the good advice of Flourens, 'The art of isolating simple facts is the whole art of experiment.'

Among the ERPs which have been investigated in cognitive situations, the later components of sensory EPs have run a very close race with the CNV in popularity. Chief among these has been P300, a positive wave having a peak latency of about 300 msec. Several studies have dealt with the problem of the relationship of P300 to CNV offset and have tended to the conclusion that the two phenomena may have different neurophysiological substrates, different distributions, different sensitivities to stimulus and response manipulations, and a good deal in common. This should be no surprise; both phenomena are often recorded conjointly, and could be expected to co-vary to the extent that they share common elements of what might be called their 'cognitive context'. At the same time, experiments have been designed which will tend to minimize those common elements, and when this is done, the two phenomena tend to dissociate. Suffice it to say that experiments which have taken this latter course (of which Donald and Goff, 1971, Papakostopoulos and Crow, p. 201, and McCallum *et al.*, p. 205, are models) have clearly demonstrated that there is no necessary link between them and that they can be made to operate independently. Tueting and Sutton (1973), Tueting and Sutton, (p. 71) and Donchin *et al.* (p. 76) have shown P300 and CNV to shift their interdependent relationship as a function of the experimental situation. Wilkinson (p. 88) cautions that the interactions between CNV and P300 can go undetected in so far as uncontrolled

CNVs may be present as a function of intertrial interval regularities or simply of *S* expectancy.

Results of a number of studies have led their authors to conclude that a chief correlate of P300 is orienting (Tueting *et al.*, 1971; Ritter *et al.*, 1972; Peters *et al.*, 1972; Sandler and Schwartz, 1973; Roth and Kopell, 1973; and Friedman *et al.*, 1973). This conclusion has usually flowed from the observation that P300 amplitude is inversely related to stimulus probability. Some of the above authors, along with others, have taken a more general view and have noted that stimulus probability alone is not the sole determiner of P300 variations, but that in addition the importance of the stimulus to *S*, i.e. its salience, is the chief factor in P300 manipulation (Tueting *et al.*, 1971; Sandler and Schwartz, 1973; Jeness, 1972a, b; Friedman *et al.*, 1973; Wilson *et al.*, 1973; Ford *et al.*, 1973; Leifer *et al.*, p. 65; Marsh *et al.*, p. 122). A third emphasis which is largely a special case of salience, is that laid upon decision making (Hillyard *et al.*, 1971; Wilkinson and Lee, 1972; and Rohrbaugh *et al.*, 1927) and a fourth stresses inhibitory brain processes (Papakostopoulos and Crow, p. 201). Each of these 'explanations' (or 'favoured correlates') is credible, and the choice is largely determined by the experimental situation. It is worth noting that they overlap considerably, that all are more general or more special versions of a selective attention hypothesis; indeed, selective attention has been appealed to by many of the above authors.

While they may be gratuitous, two cautions may be in order. First, just as not all CNVs are the same, i.e. there are variations in form and distribution as well as amplitude, not all P300s are the same, for they also vary considerably in form and distribution as well as amplitude; it is helpful if multiple brain sites are monitored, and it is essential that examples of the potentials be presented along with other critical elements such as time constants, measurement schemes and multichannel recordings. (*See* Hillyard *et al.*, p. 81) for a model of the power of these elements.) Second, while it is reasonably well established that the CNV and P300 may be independent, a parallel finding of independence of P300 and the RAP of the RRP has not been forthcoming. Indeed some P300 variations (e.g. Rohrbaugh *et al.*, 1972) are just what one would predict for the RAP; the response requirements of the situation should be spelled out fully, and RTs should be presented.

CNV concomitants of 'learning' continue to be elusive, although some interesting evidence is beginning to be seen. Pavlovian conditioning of the CNV has been demonstrated in two rather unexpected ways. Dolce and Sannita (1973) reported a remarkably low voltage 'CNV-like shift' recorded during deep coma*. The shift appeared only after many (120–150) repetitions of the paired stimuli. Similarly, Lelord *et al.* (p. 120) show CNVs with passive evocation of a response either by percussion or by electrical stimulation as a conditioned phenomenon in awake *Ss*. Using a verbal learning situation, Billinger *et al.* (p. 58) provide some minimal evidence for an inverted-U function relating CNVs to trials and point up some subtle performance effects in this situation. Rebert (p. 191) has presented some evidence from monkeys showing monotonic increasing CNV in a learning situation. Leifer *et al.* (p. 65) and Marsh *et al.* (p. 122) in studies of a number of related waves with learning (*see above*) both

*(Editors' note: This finding has since been established as resting on a methodological error.)

report unremarkable changes in the CNV over trials and as performance improved.

Approaches continue to be made to problems of stimulus and/or response characteristics as they may influence the CNV. Modality of stimulation, a characteristic which was dismissed as being of relatively little importance early on in CNV research, has appeared as a considerable variable, both as it influences the CNV and as it may determine reaction time (RT). Gaillard and Näätänen (p. 40) demonstrate stimulus modality influences on both RT and the CNV, the latter being affected both in amplitude and in shape by this manipulation. Blowers *et al.* (p. 20) also showed important effects of stimulus modality on RT, although their CNV measures did not vary with this treatment.

Feedback to the subject about his responses has seen some attention since the last congress (Peters *et al.*, 1970; Walter *et al.*, 1972; Rebert, 1972b; Weinberg, 1972; and Rebert and Sperry, 1973). The results of these studies tend toward the conclusion that it is not feedback, *per se*, but rather 'expectant attention' (Weinberg, 1972) which is the important factor. That this latter factor is important is surely no surprise, for its potency has been demonstrated repeatedly.

Indeed, engaging the *S* in the task remains the most solid way of predictably manipulating his CNV. Within limits, as the *S*'s involvement in subsequent stimuli and responses is increased up to a point, his CNV will likewise increase; the converse also tends to hold. The ways of manipulating that engagement have been legion, and many of them are represented here. Tecce and Cole (p. 59) used amphetamine, and found that when the agent acted as an 'upper', CNVs increased, while when it acted paradoxically as a 'downer', CNVs became smaller. Costell *et al.* (1972) used pictures of nude males, nude females, and a clothed 'neuter' figure as both S_1 and S_2 in a CNV situation, and found evidence that the CNV may be an indicator of sex object preference. That it is not necessary to resort to drugs or sex to manipulate engagement is also apparent. McCallum (p. 46) presents the results of an ingenious study which allowed vigilance to wax and wane in a carefully controlled fashion, and which shows the CNV to wax and wane along with it. Blowers *et al.* (1973) manipulated temporal expectancy by varying the i.s.i., and found effects on CNV amplitude and peak latency which paralleled predicted *S* engagement.

Where the relationship between the CNV and *S* engagement becomes discontinuous, that discontinuity is very striking. The most striking demonstrations of these discontinuities have involved manipulation of constitutional anxiety, situational anxiety, sex, or all three. All of these studies have built upon the groundbreaking observations of Knott and Irwin (1968) who showed that CNV amplitudes became smaller rather than larger as stress and constitutional anxiety were combined. They postulated a 'ceiling' hypothesis to account for this data, and subsequent studies have tended to bear them out. Low and Swift (1971) reported that low anxious *Ss* tended to have higher CNVs than high anxious *Ss*, and that the CNV was reduced with stress. When their *Ss* were run a second time, presumably having habituated to the stress, the situation and the experimenters, their CNVs increased. Knott and Peters (p. 51) used a Pavlovian 'experimental neurosis' paradigm to demonstrate that stress has differential effects on male and female *Ss*, the latter's CNVs behaving more like those of more anxious *Ss*. Since CNV values became more similar for all groups and conditions as stress was increased, one might also conclude from their work that stress is the great equalizer between the sexes. In a second study, Peters and Knott (p. 52) looked at post-S_2 large negative potentials, and showed that these are much larger during stress for the female *Ss*, with males showing significant increases in this negative wave only on non-response trials. Perhaps the most viable approach to the combinations of arousing and attention provoking factors which are illustrated by this work is that which was provided by Tecce (1972) who, in reviewing the considerable body of data dealing with these factors, concluded that the relationship holding between attention and the CNV was a monotonically increasing one while that holding between arousal and the CNV assumed an inverted-U shape.

Man is at his cognitive best when he is using language; however we have fallen short in cognizing many of the electrophysiological concomitants of those aspects of language behaviour which are usually characterized as 'higher processes'. Most of the experimentation which has been done to date has been concerned with answering very basic questions about whether there is an electrical concomitant to language. While a considerable literature concerning language reception has developed, we will not concern ourselves with it here except to say that lateralization of evoked potentials to linguistically relevant stimuli is a fairly robust phenomenon, and that intrahemispheric localization of language evoked potentials has been reported on occasion. Our emphasis will be on electrical phenomena which accompany language production or which accompany a 'linguistic set'.

The earliest attempts at finding an electrical correlate of language production were reported by Ertl and Schafer (1967) and by Schafer (1967) using AC recording in an RRP setting. They subsequently reported that they felt that their results were heavily contaminated by lip EMG potentials (Ertl and Schafer, 1969). Using DC recording, McAdam and Whitaker (1971) were able to demonstrate reliable prearticulation readiness potentials which were lateralized and which tended to be localized over Broca's area. Partial confirmation of these results were provided by Morrell and Huntington (1972) who studied a wide variety of sites and used a number of controls for 'artefactual' potentials. They found that the maximal shift occurred over temporal-parietal areas. Later analyses showed that when left-right differences are taken, they are largest over inferior frontal areas (Morrell, personal communication). Grözinger *et al.* (p. 103) have also investigated prearticulation potentials, and have raised the caution that these potentials may be influenced by respiratory activity which accompanies speaking. They further report that respiratory activity tends to be maximal over that hemisphere showing the most alpha activity. When this piece of information is juxtaposed with the growing number of findings that alpha activity tends to be *depressed* over the dominant hemisphere during linguistic tasks (*see*, for example, McKee *et al.*, 1973), there is some cause for concern that the lateralization of the potentials preceding speech may be to some degree confounded by the R-waves. Zimmermann and Knott (p. 164) provide some data about prearticulation potentials in stutterers and non-stutterers which, while not explicitly ruling out respiratory effects, tend to support the credibility of the brain origin of the potentials.

CNV techniques have also been applied to the analysis of localization of brain function for language production. Low *et al.* (p. 165) had the good fortune of being able to correlate their CNV data with amytal tests of hemispheric dominance in a number of their *Ss*, and report a remarkably good fit between the two. Zimmermann and Knott (p. 164) have studied prearticulatory CNVs in stutterers and non-stutterers, and their results provide support for theoretical positions holding that stuttering may be associated with occasions of waning hemispheric dominance. Marsh *et al.* (p. 122) looked at CNVs in a mixed spatial-verbal task. The extent to which each of these tasks contains elements of the other may have confounded their results to some degree. In any case the attempt highlights the important difficulty of purifying the modality—in a 'higher function' sense—in this research. Even with these problems, they were able to show both intra- and inter-hemispheric differences which were task related.

In summary, while this body of work is very exciting, only the most tentative of conclusions may be drawn from it at this time. Many very basic questions about the behavioural correlates of the potentials remain to be answered fully. At the same time, it is not inappropriate to begin to explore some of the questions dealing with 'higher functions' in language behaviour and to tackle problems proceeding from linguistic and psychological analyses of this quintessential human endeavour.

In conclusion, I would like to offer one feeling, one quotation, and one exhortation. First, the feeling: I ask your charitable understanding for the efforts of those of us who would venture into biopsychological reaches where the shades of mentalism lurk and occasionally capture us. We will try to be hard-nosed enough to give you a clear description of what we did and what we found. Our concepts and behaviours will be molar ones and our designs will be grand and complicated; we will often trip over some of the finer points. Bear with us. Second, the quotation, a favourite definition from *The Devil's Dictionary*, by Ambrose Bierce: 'mind (n): A mysterious form of matter secreted by the brain. Its chief activity consists in the endeavor to ascertain its own nature, the futility of the attempt being due to the fact that it has nothing but itself to know itself with.' Last, the exhortation: Theorize! Operationalize! Recognize! Cognize!

Modality Effects on the Contingent Negative Variation in a Simple Reaction-time Task

A. W. GAILLARD and R. NÄÄTÄNEN*

Institute for Perception TNO, Soesterberg, The Netherlands

A slow-potential change, called the CNV (Walter, 1964; Walter *et al.*, 1964), develops in the brain between two successive signals when the first signal (S_1) serves as a warning while the second (S_2) requires either an overt or covert response. A variety of stimuli in several modalities have been used as S_1 and S_2. Several authors have asserted that both the intensity and the modality of the stimuli are unrelated to the amplitude of the CNV. 'The intensity and modality of the conditional and imperative stimuli bear no relation to the E wave provided that the significance of their association is preserved; it is thus entirely non-specific in relation to the provenance of the stimuli but specifically related to the stimulus contingency and the attitude of the subject' (Walter, 1967, p. 125). Low *et al.* (1966) found no differences in the amplitude of the CNV between three levels of stimulus intensity, using either tones, clicks or flashes as S_2, while S_1 always was a flash of a constant intensity. However, Connor and Lang (1969) found that the amplitude of the CNV was larger with loud tones (80 dB) than with faint tones (50 dB), S_1 and S_2 being identical.

Furthermore, in the study of Low and his associates (1967) the amplitude of the CNV tended to increase as S_2 (a click) intensity approached threshold and it was shown by Rebert *et al.* (1967) that a larger CNV occurs when a faint tone is signalled by S_1 than when a moderate tone is signalled.

Whereas there are studies on the effect of the intensity of S_2 on the CNV, the effect of the modality of S_2 on the CNV has remained unexplored (except for the Low *et al.* study, 1966, mentioned above). And the effect of neither the intensity nor the modality of S_1 on the CNV has been investigated. The present study is an attempt to discover the possible effects of the modality of both S_1 and S_2 on the CNV in a simple RT task. The modality of the stimuli was systematically varied between the auditory and the visual. Three electrode positions were used which made it possible to investigate also the topographical differences of the CNV related to the modality of the RT task.

METHOD

Subjects and apparatus

The *Ss* were 4 male students from the University of Utrecht and 4 male employees of the Institute for Perception (aged 21–38 years); two other *Ss* were discarded from the analysis for having too much α-rhythm.

The *Ss* were tested in a sound-proof and electrically shielded cubicle (Amplifon); a nixie-tube display and a loudspeaker were placed in front of the *S* at a distance of about 1·5 metres and a response key was mounted on the sloped top of a table, which provided an armrest for the *S*. The auditory signals

* Present address: Institute of Psychology, University of Helsinki, Ritarik. 5, Helsinki 17, Finland.

consisted of a 1000 Hz tone of 70 dB (duration 40 msec) and the visual signals of a standard red nixie-tube with all elements activated, also with a duration of 40 msec. The luminance of this stimulus was 204 cd/m^2, while the surrounding background had a luminance of 174 cd/m^2.

The EEG was recorded from vertex (C_z), occipital (O_2) and temporal T'_3 silver–silver chloride disc electrodes with a common reference electrode over the right mastoid. The T'_3 position was 1 cm anterior to T_3. Vertical eye movements were recorded from above and below the right eye. An electrode on the left ear served as the ground. After AC amplification (TC 6 sec) the EEG and EOG signals were stored on magnetic tape (Philips Analog-7).

The timing of signals, including the triggers, and measurement of RTs (in msec) was done by means of the PSARP-equipment (Van Doorne and Sanders, 1968).

Procedure

After at least one practice session, *Ss* were given four 'modality'-conditions in a repeated latin-square design. That is, the session was divided into three series, while each series contained the following conditions:

1. Visual S_1 –Visual S_2 (VV);
2. Auditory S_1–Visual S_2 (AV);
3. Visual S_1 –Auditory S_2 (VA);
4. Auditory S_1–Auditory S_2 (AA).

Each of these four conditions consisted of a block of 50 stimulus pairs (S_1–S_2), wherein five catch trials were randomly interspersed. On catch trials S_2 was omitted and *S* had to withhold his reaction.

The i.s.i. was 100 msec and the i.t.i. 8 sec. The RT in hundredths of a second was read to *Ss* through an intercom after each response. To reduce eye movements, *Ss* were trained during the practice session to fixate on the nixie-tube (also with auditory stimuli), to suppress eye movements during the S_1–S_2 interval and to blink well after the response was given.

Analysis of Data

The EEG signals from the vertex, occipital and temporal channels were averaged by a CAT-400b computer over 16 consecutive trials. Trials with eye movements and catch trials were excluded from these averages. From these graphs the following measures were taken for each of the three EEG-leads used:

a. '*Early CNV*'. The mean of five equispaced points (40 msec apart) within the period about 600–760 msec after S_1, relative to a baseline established by sampling five equispaced points, 40 msec apart, 160–0 msec before S_1.

b. '*Late CNV*'. The mean of five equispaced points (40 msec apart) within the period 840–1000 msec after S_1, relative to the same baseline.

RESULTS

In *Fig*. 1 the early and late CNV amplitudes from the three EEG-derivations (C_z, O_2, T'_3), as well as the RT, are plotted out for the different conditions. As expected, the (late) CNV amplitudes were largest at the vertex (mean 20·3 μV) while those from the O_2 and T'_3 channels were rather small (means 5·1 μV and 4·3 μV, respectively) (*see Table* 1). *Table* 2 gives the main results of the analysis of variance performed on these variables.

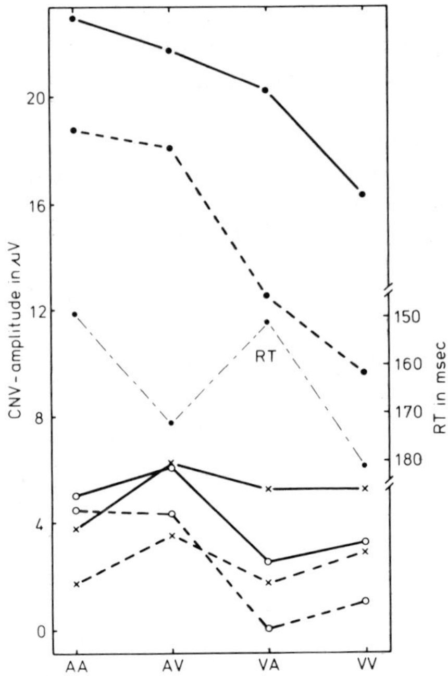

Fig. 1. CNV amplitude and RT for the four conditions (*see* text) across *Ss*. Solid line = late CNV; dotted line = early CNV. Filled circles = C_z; open circles = T'_3; crosses = O_2 (TC = 6 sec).

Table 1. The mean amplitudes of the early and late CNV (in μV negative) for the four conditions and the three derivations across *Ss* and series. The upper value refers to the first block of 16 trials, the lower value to the second block of 16 trials within series

		AA	*AV*	*VA*	*VV*
C_z	early	20·23	18·5	13·87	8·12
		17·35	17·76	11·20	11·22
	late	24·15	21·23	20·58	14·94
		22·0	22·34	19·91	17·72
O_2	early	2·71	3·68	3·02	3·42
		0·62	3·39	0·48	2·35
	late	4·93	5·79	5·92	5·67
		2·57	6·63	4·6	4·8
T'_3	early	5·6	4·81	1·12	1·18
		3·39	4·02	+1·23	0·85
	late	6·13	6·63	3·59	3·16
		3·89	5·37	1·47	3·4

RT

The RTs were greatly influenced by the modality of S_2 (*see Table* 2) which is consistent with the literature (*see*, e.g., Woodworth and Schlosberg, 1954; Sanders and Wertheim, 1973). Also the modality of S_1 had an influence on the RT: all the *Ss* showed slower visual RTs with a visual S_1 than with an auditory S_1 (sign-test; $P=0{\cdot}0039$). However this effect did not reach the 0·05 level of statistical significance in the analysis of variance (*Table* 2). A similar S_1 effect was not observed with respect to the auditory RTs, i.e. RTs to an auditory S_2 were independent of the modality (auditory or visual) of S_1.

Table 2. Summary of the analysis of variance

	median RT	*CNV amplitude*					
		C_Z-early	*C_Z-late*	*O_2-early*	*O_2-late*	*T'_3-early*	*T'_3-late*
S_1 aud. v. vis.	3·93	20·98‡	27·36‡	<1	<1	39·21†	3·54
S_2 aud. v. vis.	95·68‡	1·20	10·79†	5·15*	3·17	<1	<1
$S_1 \times S_2$: aud. v. vis.	1·59	<1	2·75	<1	3·31	<1	<1
Sequence	<1	<1	1·38	<1	<1	—	—
Series	1·01	<1	<1	<1	1·15	<1	<1

$^*P<0{\cdot}05$, † $P<0{\cdot}01$, ‡ $P<0{\cdot}001$.

The Amplitude of the Vertex CNV

The amplitude of the CNV derived from the vertex was greatly influenced by the modality of S_1: considerably larger negative shifts were obtained with an auditory S_1. This result involves both the early and the late CNV (*see Table* 1).

Fig. 2 shows that the shape of the vertex CNV was also dependent on the modality of S_1. When S_1 was auditory the negative shift was already large (18·6 μV), at the time when the early CNV was measured (680 msec), but did not grow appreciably thereafter; the mean difference between the amplitudes of the early and the late CNV was only 4·0 μV. When S_1 was visual, the vertex CNV showed a slow initial rise (the amplitude of the early vertex CNV being 11·0 μV) but increased by as much as 7·2 μV during the last part of the i.s.i., i.e. between the two measures taken (*Fig.* 2).

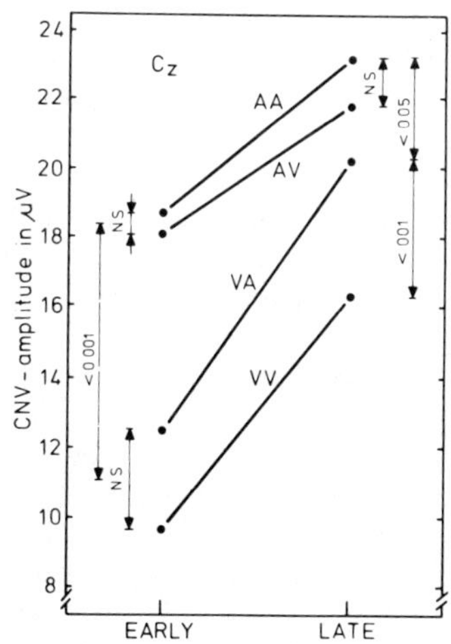

Fig. 2. The amplitude of the early and late vertex CNV for the four conditions (*see* text) across *Ss* (TC=6 sec).

The amplitude of the vertex CNV was also affected by the modality of S_2, higher values being obtained for the auditory S_2. This modality effect was significant only for the late vertex CNV (*Table* 2), and was due mainly to the larger growth of the amplitude between the early and late measuring periods in the VA condition as compared with the VV condition.

To sum up, an auditory modality S_1 (compared with a visual modality) seems particularly to enhance the amplitude of the early CNV, but exerts some influence also on the amplitude of the late CNV. Evidence was also obtained for an enhancing effect on CNV amplitude of an auditory modality S_2 but this evidence was much weaker.

Amplitudes of Temporal and Occipital CNVs

CNV amplitudes measured at the temporal and occipital areas were small; this observation is consistent with the results of topographical studies (e.g. Low *et al.*, 1966; Cohen, 1969).

As shown in *Fig.* 1, the temporal CNV was influenced by the modality of S_1 in a similar way to the vertex CNV, i.e. a large CNV was observed with an auditory S_1. This effect was significant only for the early but not for the late CNV (*Table* 2). The modality of S_2 had no effect on the temporal CNV.

By contrast the modality of S_2 had a significant effect on the occipital CNV but not that of S_1. A larger CNV amplitude was observed when S_2 was visual (*see Fig.* 1). This effect was, however, significant only for the early CNV (*Table* 2).

Thus, some indications of a topographical shift of the CNV related to the modality of the stimuli was obtained: (1) the significantly larger early occipital CNV when a *visual* S_2 was used (*see also* Vaughan and Ritter, 1973) and (2) the significantly larger early temporal CNV when an auditory S_1 was used. However the generally low mean amplitudes of the occipital and temporal CNVs made conclusions about topographical distribution somewhat uncertain.

DISCUSSION

The modality of the stimuli used had a clear influence on the amplitude of the CNV in the simple RT task. For the vertex and temporal derivations the modality of S_1 had a large effect on the CNV amplitude especially when the amplitude was measured in the early phase (600–760 msec after S_1). No effect of the S_1 modality on the CNV was observed in the occipital data.

The large initial rise obtained with an auditory S_1 made the CNVs resemble rectangular 'type A' CNVs in the classification of Tecce (1972). He suggested that the early CNV is related to 'arousal processes', supporting this notion with evidence that (1) a type A CNV is produced, i.e. a differential effect on the early CNV is observed, after the intake of amphetamine (Walter, 1964); (2) *Ss* required to clench their fist during the i.s.i. show a relatively greater increase in the early than in the late CNV. The present results might possibly be interpreted to mean that the enhancing effects on the early CNV of an auditory stimulus are produced by its arousing properties (*see* Bertelson and Tisseyre, 1969; Sanders and Wertheim, 1973; Trumbo and Gaillard, 1973). Coincidentally, some evidence was obtained for a facilitatory effect on the RT of an auditory S_1 notwithstanding the relatively long i.s.i. (1 sec). The fact that the enhancing effect was less prominent in the late CNV suggests that the arousing effect of the auditory S_1 decreased during the latter part of the i.s.i.

Thus, an auditory S_1 seems to be associated with a tendency towards the A-type CNV and a visual S_1 towards a B-type CNV (long negative ramp). We prefer to interpret the effect of the modality of S_1 obtained in terms of arousing properties of an auditory stimulus rather than of the orienting response (*see* Weerts and Lang, 1973): No significant differences were observed, when the amplitudes of the early and late CNV, obtained from the first block of 16 trials were compared with those from the second block of 16 trials within series, neither for an auditory S_1 nor for a visual S_1.

Whether the modality effects observed are 'real', i.e. specific to the modality or only due to stimulus intensities used has still to be investigated. A crucial experiment is difficult to design due to the fundamental problem of comparing stimulus intensities across modalities. Up to the present time even the influence of stimulus intensity on the CNV has not been systematically enough investigated.

Future investigations of stimulus modality effects will need to pay particular attention to the nature of the task, the i.s.i., electrode positions and the duration of experimental conditions and tasks.

SUMMARY

The effects of the modality (visual and auditory) of S_1 and S_2 on the CNV measured from three channels (C_Z, O_2, T_3') were systematically investigated. With an auditory S_1 the amplitudes of the vertex and temporal CNV were enhanced especially in the early phase.

Also the modality of S_2 had an influence on the CNV: higher amplitudes were obtained with an auditory S_2 for the vertex CNV and with a visual S_2 for the occipital CNV.

ACKNOWLEDGEMENT

This study was supported by the Netherlands Organization for the Advancement of Pure Research (ZWO).

Brain Slow Potential Changes and Motor Response in a Vigilance Situation

W. C. McCALLUM

Burden Neurological Institute, Bristol, U.K.

Slow potential research in man has centred around the contingent negative variation (CNV). The majority of experiments have employed variations on a foreperiod reaction time task in which paired warning and imperative stimuli are presented at well spaced, and usually irregular, intervals.

In this somewhat artificial situation small islands of S_1–S_2 activity are separated by periods of inactivity during which the subject's psychological and behavioural state is relatively poorly controlled. The CNV is considered to be non-habituating provided that *S* maintains his involvement, attention or motivation towards the experimental task. However, this contention has seldom been put to the test.

In the current experiment a continuous involvement task has been employed, designed to sustain the *S*'s participation throughout and to be able to measure, by an error score, any lapses of attention.

METHOD

Twenty-seven paid *Ss* between the ages of 17 and 45 were tested. Additionally two patients having intracerebral gold electrodes, implanted in connection with clinical procedures, were given an abbreviated form of the same test procedure.

During the main experimental task each *S* watched a single digit numeric display located 2 metres from his eyes. Numbers between 1 and 8 changed at the rate of one per second in a pseudo random sequence, the subject having to respond as rapidly as possible with a button press whenever he observed a sequence of three odd numbers. This task constituted a variation on a vigilance situation used experimentally by Gale *et al.* (1971).

Control sequences were added at the beginning and end of the experiment during which *S* watched the display but made no response. Additional sequences involved the presentation of an irrelevant 80 dB click midway between certain number changes in both high vigilance and low vigilance (control) conditions.

Finally for comparison with previous CNV experimentation short sequences of number changes were presented at irregular intervals between 5 and 15 sec. Numbers changed four times in each sequence, always ending on an even number. As in the main experimental condition, *S* was instructed to press the button if the changes were through a sequence of three odd numbers.

The following sequences could be sampled: (A) odd-odd-odd, (B) odd-odd-even, (C) odd-even-even or (D) even-odd-even. The number preceding and following each sequence was automatically set to be even.

Three-second samples of appropriate sequences were coded and stored by the computer for subsequent averaging.

The experimental paradigm consisted of:

Stage 1 (control)—No response. 27 selected trials stored.

Stage 2 —No response. Irrelevant click presented during 18 selected and stored trials.

Stage 3 —Short sequences of changing numbers at irregularly spaced intervals. Button press to occurrence of 3 consecutive odd numbers. 27 selected trials stored.

Stage 4 —Main experimental sequence. Continuous presentation. Button press to occurrence of 3 consecutive odd numbers. 108 selected trials stored.

Stage 5 —Similar to Stage 4 but with interposed irrelevant click during 18 selected and stored trials.

Stage 6 (control)—As for Stage 1.

Stage 7 —As for Stage 2.

EEG was recorded from 8 scalp silver–silver chloride electrodes at positions Fp_z; F_3; F_4; C_3; C_z; C_4; P_3 and P_4, referred to linked mastoids. The C_z channel was compensated for eye movements as described by McCallum and Walter (1968). Recording was by a modified 16-channel Schönander Mingograph recorder with a time constant of 5·0 sec and the experiment was under on-line control of a PDP–12 computer, which was also used for the off-line processing of data. The on-line part of the experiment lasted approximately 1 hour 20 min for each subject. Subjects fixated the number display throughout, brief relaxation for the eyes being allowed between some of the stages.

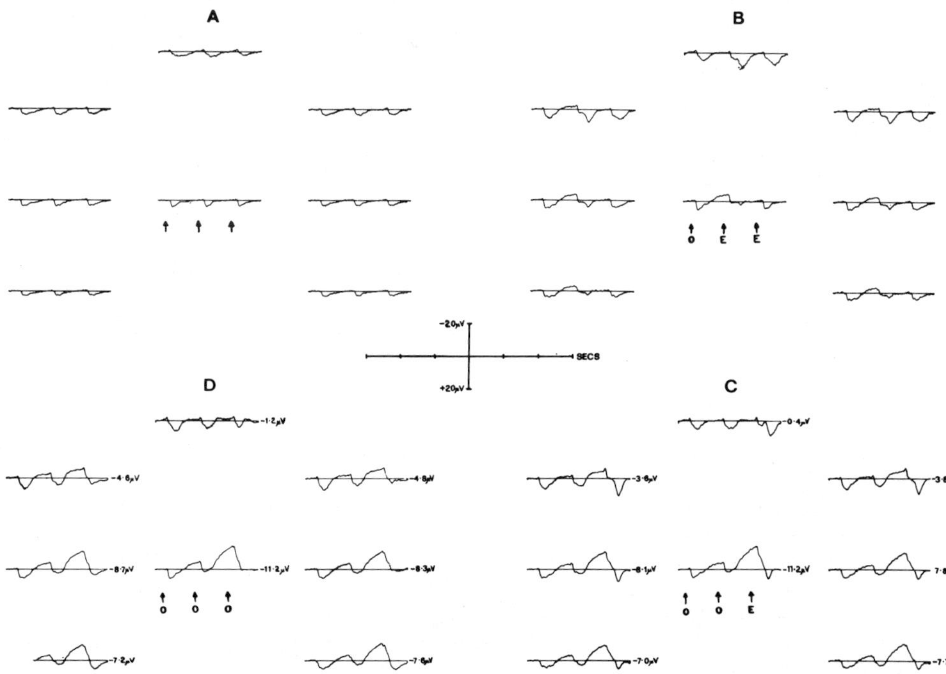

Fig. 1. The average of averages for 27 normal subjects (N=27×36) in A: Resting state (i.e. no response made to number changes), and in B, C and D: Conditions in which *S* had to respond to sequences of 3 odd numbers. Arrows indicate number changes. 'O' indicates that the number was odd; 'E' that it was even. Each set of 8 traces shows the distribution over the head. Microvolt values indicated in C and D refer to amplitudes of the second slow negative wave.

The 2 patients from whom intracerebral recordings were made were given a modified experimental paradigm consisting of Stages 1, 4 and 6 slightly abbreviated. Recording was from 12 cortical, 150-micron, gold electrodes. These extended in an anterior posterior line from the frontal pole to the postcentral gyrus of the right hemisphere in 1 patient and of the left hemisphere in the other. Electrodes were 4-mm long and spaced at 4-mm intervals.

RESULTS

In the control condition (Stages 1 and 6), where no button pressing response was required, a small positive evoked potential with a peak latency of about 224 msec and a mean amplitude of 4 μV was observed to each number change. This component did not appear to habituate. Between these responses the baseline remained at or very close to zero (*see Fig.* 1A).

In the main experimental condition (Stage 4) a small, but consistent negative slow potential shift, averaging −3 μV at the vertex, followed the occurrence of a single odd number (*Fig.* 1B). A larger negative slow potential shift, having a mean vertex amplitude of −11 μV, followed the occurrence of two sequential odd numbers (*Fig.* 1C and D). Amplitude measures were based on average values over a period of 200 msec immediately preceding a number change. Baselines were normalized to be the mean level of activity over 700 msec prior to the first number change in the sequence sampled. Following the response condition, i.e. odd-odd-odd, the negative shift terminated and tended to return directly to baseline. In the odd-odd-even (no response) condition a late positive potential appeared following termination of the shift. The mean latency of this component was 440 msec after the appearance of the even number (*see Fig.* 2A). Mean RT for all *Ss* in the response condition showed a similar latency (470 msec), suggesting that the late positive component might reflect response inhibition. A product moment correlation between mean late positive com-

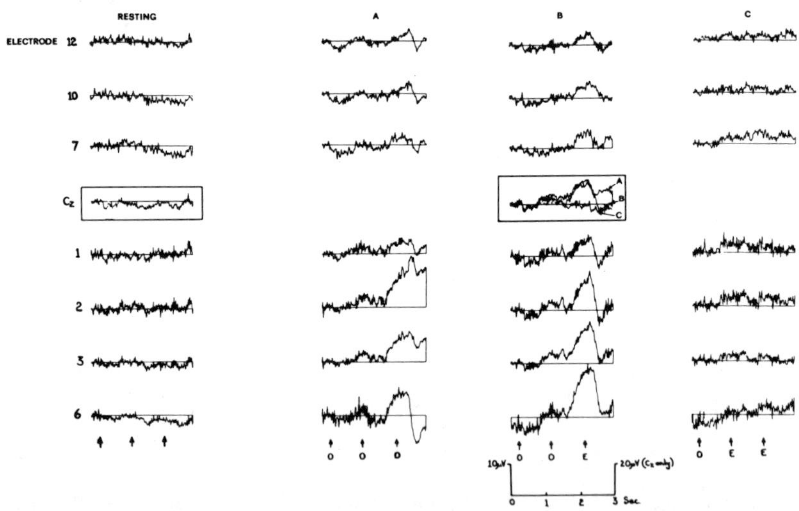

Fig. 2. Average recording (N=20) from intracerebral gold electrodes extending subdurally over the right hemisphere from a prefrontal location (electrode 12) to the postcentral gyrus (electrode 6). Simultaneous recordings from Scalp Vertex are shown in boxes. Traces under A, B and C were derived from the vigilance condition described in the text and *Fig.* 1. No responses were required in the 'Resting' condition.

ponent peak latency and mean RT for the whole group gave a coefficient of $r = 0{\cdot}5$ indicating a positive correlation, significant at the 0·02 level. Correlations were also made between the amplitude of the negative wave following two odd numbers and RT on response trials during Stage 4. Only 4 *Ss* showed significant correlations, each of these 4 being negative.

The same general pattern of slow potential change was shown by all 27 *Ss* tested. Shifts were largest at the vertex but were widely distributed over the head and were visible in some degree at all electrodes except Fp_Z. Mean values of shifts are shown in *Fig*. 1. The distribution was bilaterally symmetrical.

A similar pattern of change was seen for the corresponding conditions during Stage 3. Amplitudes were slightly larger than those in Stage 4, for example $C_Z = -14\ \mu V$ compared with $-11\ \mu V$ in Stage 4.

Error scores for false or missed button presses were low, 9 *Ss* being error free. Where errors of omission were sufficient to permit separate averaging it was found that the pattern of slow potential change, which characterized the response trials, was absent.

The two patients showed the same patterns of slow potential change from the scalp. Corresponding changes could be observed from the cortical electrodes, being at their largest ($-13\ \mu V$) over the precentral gyrus and at their smallest, i.e. only just discernible, over prefrontal cortex. The electrode over the precentral gyrus showed an additional clear negative response associated with the button press made by the contralateral hand. The slow potential shifts could also be observed over post central gyrus although the amplitude was slightly lower (*see Fig*. 2).

Clear click evoked potentials were recorded during Stages 2, 5 and 7. The most prominent component had, at Stage 1, a mean latency of 109 msec, a mean peak amplitude of $-22\ \mu V$ and a mean peak-to-peak amplitude of $37\ \mu V$. No significant differences were found between amplitude or latency values at the beginning or end of the experiment or between the resting and high vigilance conditions.

DISCUSSION

The experiment provided a clear indication that CNV-like slow potential changes are not dependent upon well spaced presentations of paired stimuli. Shifts develop as the probability of having to make a response increases. Failure to note or act upon stimuli conveying information relevant to such probability results in failure of the potential shifts to develop properly. This finding augurs well for use of slow potentials to detect vigilance decrement in applied situations and as objective measures of at least some aspects of subjective state.

The brain systems on which the changes depend are still far from clear. However, the scalp distribution shows higher amplitude values over central and parietal areas than over frontal regions, a feature which was borne out by the cortical distribution seen in the two patients, where highest values were to be found precentrally and only very small values prefrontally. Among the questions which remain to be answered is the extent to which the cortical topography of these changes associated with assessment of stimulus significance and preparation for action overlaps, or is independent from, that of other activation states such as alpha rhythm desynchronization or that of sensory evoked potentials.

The consistency of the waveforms seen in the experiment suggest that the vigilance paradigm may well constitute a more reliable baseline condition for examining variations of slow potentials in experimental or clinical groups than the more conventional CNV paradigm.

The late positive component seen following the cut-off in the odd-odd-even (i.e. no response) condition can probably be equated with the phenomenon generally referred to as P300. This being the case it adds a further possible correlate for this wave, namely that of response inhibition.

Evoked potentials to the irrelevant clicks introduced in Stages 2, 5 and 7 did not vary significantly between states of high task involvement and low task involvement. This aspect of the study offered no support for the idea of a generalized change of excitability level discernible through changes in evoked potentials.

SUMMARY

Twenty-seven normal *Ss* displayed slow potential changes during a vigilance task. When required to respond to the occurrence of three odd numbers in a continuously changing single digit sequence, small negative slow potential shifts followed the occurrence of one odd number and larger shifts followed two odd numbers. Irrelevant clicks introduced during high vigilance conditions and resting control conditions elicited evoked potentials which were not significantly different from one another.

ACKNOWLEDGEMENTS

The research assistance of Mr P. Newton, the help of Mr P. Pocock with computer programs, and the general support of the W. Clement and Jessie V. Stone Foundation are gratefully acknowledged.

Further Studies of CNV, Stress and Sex using Discrimination Paradigms*

JOHN R. KNOTT and JON F. PETERS

University of Iowa, Iowa City, Iowa 52242, U.S.A.

Using a progressively more difficult discrimination RT paradigm, in which errors in discrimination were punished, CNVs from male and female *Ss* were compared. Females (N=12) showed a significant decline in CNV amplitude on response trials, and an increase on non-response trials, as discrimination became more difficult and punishment was increasingly given. Males (N=12) did not show significant changes under stress. The data indicate clear sex differences in effects of stress on CNVs which may confound CNV data when not taken into account. The data also indicate that CNV amplitude is not determined by 'response set' or 'expectancy' or 'template matching' alone.

* (Editors' note: For a fuller description of this work see: Knott, J. R. and Peters, J. F. (1974) Changes in CNV amplitude with progressive induction of stress as a function of sex. *Electroenceph. Clin. Neurophysiol.* **36,** 47–51.)

CNV and Post-response Negativity with Stressful Auditory Feedback

JON F. PETERS and JOHN R. KNOTT

University of Iowa, Iowa City, Iowa 52242, U.S.A.

Timsit *et al.* (1968, 1970) and Dongier *et al.* (1973) have described a prolongation of the CNV beyond S_2 which appears to be significantly related to certain psychiatric disorders. The phenomenon is reportedly uncommon in normals and, when present, disappears with retesting. Other authors have expressed the opinion that the rapidity with which the CNV returns to baseline may be related to the degree to which S_2 'resolves' the situation. Prolongation may be a reflection of the *S*'s attempt to evaluate his performance. Weinberg (1972, 1973) has shown that in a delayed feedback situation normal *Ss* develop a significant CNV in anticipation of a third stimulus regardless of whether it is informative or not. A recent experiment by Knott and Peters (1973a, b) provided an opportunity to investigate both the CNV and the prolonged CNV in two groups of normal *Ss* in which the feedback was both informative and stressful.

METHOD

Ss were 15 male and 14 female undergraduate students who had previously participated in a 'alpha conditioning' project but were naïve with respect to the CNV. The CNV paradigm employed both response (R) and non-response (NR) trials randomly intermixed in blocks of 40. A 5-min rest period was given between blocks, during which the instructions for the next block were read. The blocks of 40 trials were presented as follows:

Control (C_1)	S_1	1500 msec	S_2	R	
	$S_{\Delta 1}$	,,	S_2	NR	
(T)	S_1	,,	S_2	R	
	$S_{\Delta 1}$	,,	S_2	NR	(Squeal after wrong response)
Control (C_2)	S_1	,,	S_2	R	
	$S_{\Delta 1}$	,,	S_2	NR	

Tones 250 msec in duration, intensity 60 dB, with frequencies of 1500, 800 and 1478 Hz served as S_1, $S_{\Delta 1}$ and $S_{\Delta 2}$ respectively. S_2 was a single flash from a Grass PS-2 (intensity '1') 1·2 m in front of the *S*. The S_1–S_2 interval was 1500 msec with the i.t.i. varying between 10–25 sec. On the T trials, incorrect responses, i.e. pressing to S_2 after $S_{\Delta 2}$, or failing to press to S_2 within 1 sec following S_1, resulted in punishment, a 1 sec 95 dB mixed frequency 'squeal' being delivered by a loudspeaker placed 30 cm behind the *S*'s head.

CNVs were recorded using silver–silver chloride electrodes attached with collodion to vertex (C_Z); electrodes on both mastoids were linked as reference. Eye movements were monitored from electrodes placed above and below the eye. All electrode sites were vigorously abraded with Redux® (Picton and Hillyard, 1972). A Grass Model 7 with 7P1 DC pre-amplifiers was used to record CNV and eye movements, with on-line calibration (Emde, 1964). Data were stored on a PI 6100 tape recorder. Trials free of eye movements and other artefacts were averaged with a PDP–12, 12–15 trials being retrieved for each condition. CNV was measured as the peak negativity, visually determined within the S_1–S_2 interval from pre-S_1 baseline, and also as the mean post-response negativity in a 550 msec interval commencing 450 msec after S_2 (same baseline).

During collection of the electrophysiological data, *S* was comfortably seated in a darkened sound-damped chamber. *Ss* were told to fixate their gaze on a cross drawn on the PS-2 lamp. Warning stimuli were delivered from a speaker directly behind the lamp.

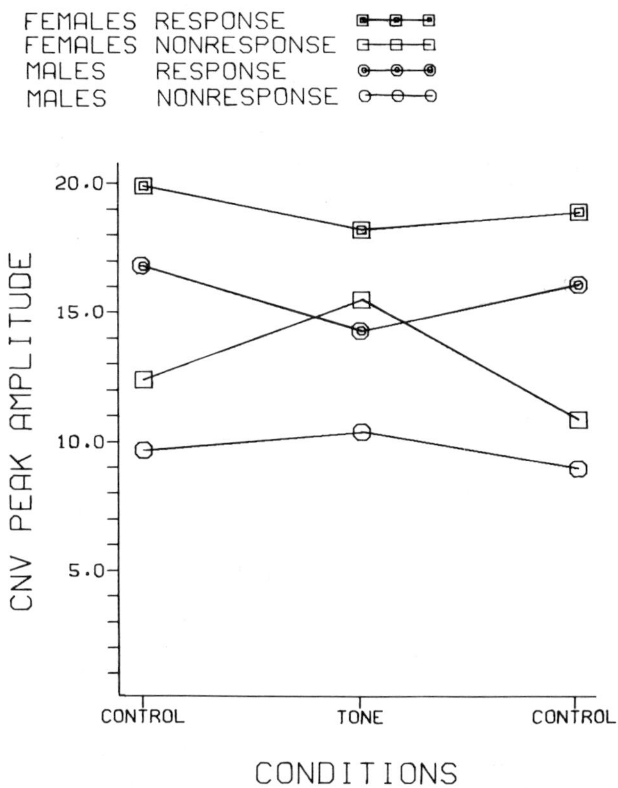

Fig. 1. Mean peak CNV amplitude as measured in the S_1–S_2 interval.

RESULTS

Fig. 1 shows the mean CNV amplitude measured in the S_1–S_2 interval. The data were examined separately for each group with related *t* tests. The females showed no significant amplitude changes within either the R or NR conditions, across conditions. However, the R and NR values showed significant differences in C_1 and C_2. The males on the other hand showed a significant depression in CNV amplitude from C_1 to T (punishment) and significant R and NR changes across all conditions. *Fig.* 2 shows the post S_2 amplitude data for both groups. The females showed a significant increase in post-S_2 negativity in the T condition compared to both C_1 and C_2 on both R and NR trials.

Again, R and NR differences were found only at C_1 and C_2. The males, however, showed a significant increase over C_1 and C_2 only on the NR condition. In addition no significant differences were found on the R and NR comparisons for the male groups on the C_1, T or C_2.

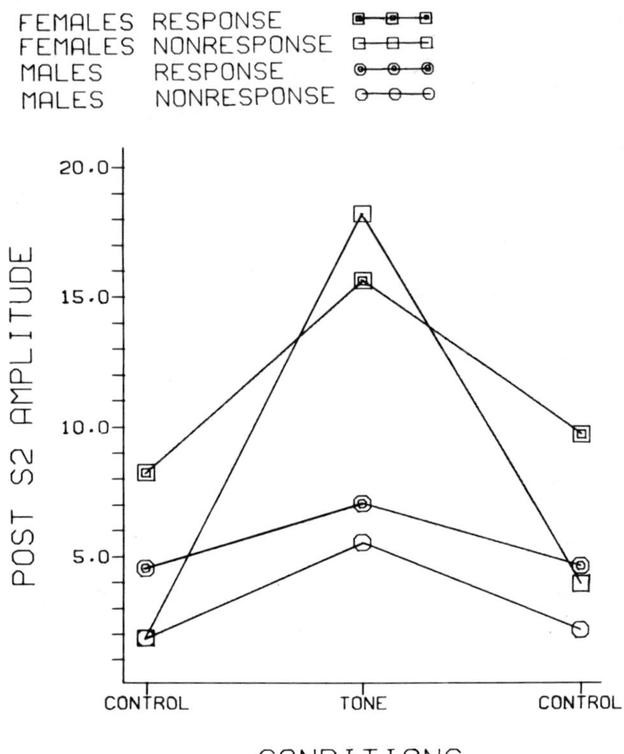

Fig. 2. Mean post S_2 amplitude measured over a 500 msec interval commencing 450 msec after S_2.

DISCUSSION

The data furnish some support for Weinberg's findings of increase in post-S_2 negativity under feedback conditions. There is also further evidence of sex differences in SP activity related to stress, which in this case continues beyond the immediate preparatory interval. However, the usual CNV difference in stress related to sex did not appear. This may be due to a sampling error, or a difference in paradigm.

SUMMARY

CNVs were elicited from 2 groups of *Ss* under feedback and non-feedback conditions. A significant increment in post S_2 negativity was seen on those trials in which feedback was expected, largely contributed to by female *Ss*.

ACKNOWLEDGEMENT

The authors are indebted to Maurice I. Mendel, Ph.D., Department of Otolaryngology and Maxillofacial Surgery, for use of the PDP–12.

Dependence of the CNV on the Probability of Occurrence of S_2

T. JÄRVILEHTO and S. MANTYSALO

Institute of Psychology, University of Helsinki, Finland

It has been suggested that the CNV is composed of at least two negative slow potentials, one of which is central-dominant and the other frontal-dominant (Järvilehto and Fruhstorfer, 1970). The former potential was proposed to reflect general readiness to perform a task, the latter uncertainty of the *S*. The present work was started in order to find out if the uncertainty of the *S* induced by varying the probability of occurrence of S_2 in an S_1–S_2–R paradigm has an effect on the configuration of the CNV.

METHOD

A standard S_1–S_2–R paradigm was used (2 auditory clicks, inter stimulus interval 1 sec). In a given experiment the probability of occurrence of S_2 was varied in 6 different blocks of trials, the probabilities being 0·1, 0·25, 0·5, 0·75, 0·85 and 1·00). The probability levels were randomized but the probability of occurrence of S_2 was reported to the *S* before each block of trials.

In each block 40 trials were given with a constant inter trial interval of 10 sec. The *S* had no prior information about the number of trials and he was instructed to react as fast as possible to S_2 by pressing a switch with his index finger.

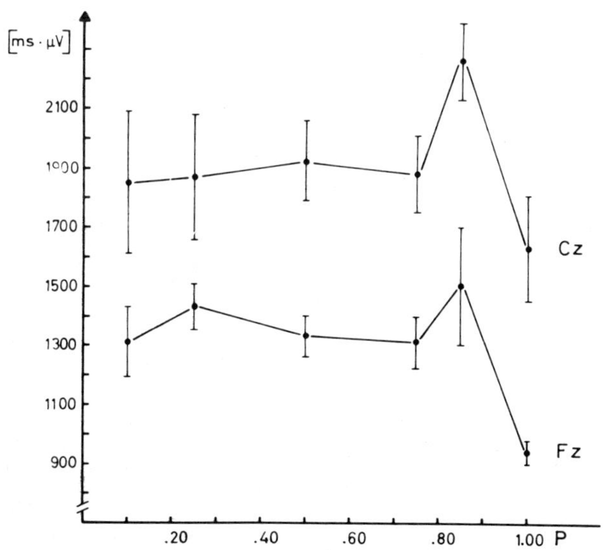

Fig. 1. Area of CNV at C_Z and F_Z as a function of probability of occurrence of S_2. Pooled data for four *Ss*. Bars give standard error of the means of the *Ss*.

The EEG was recorded at Cz and Fz and ocular movements from just above the right eye. The common reference electrode was at A_2 (the time constant of the recording system was 2 sec, the upper frequency limit 50 cps). The EEG during single trials was sampled off-line by a LINC computer (analysis period 2400 msec) and stored on digital tape with stimulus and reaction marks. The areas of single CNVs in both derivations were measured by a semi-automatic program between the point at which the rising negative slow potential crossed

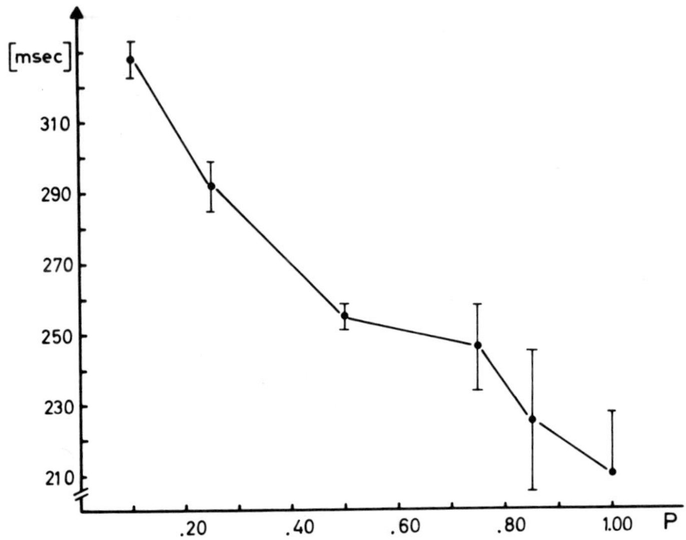

Fig. 2. RT as a function of probability of occurrence of S_2. Pooled data for four *Ss*. Bars give standard deviation of the means of the *Ss*.

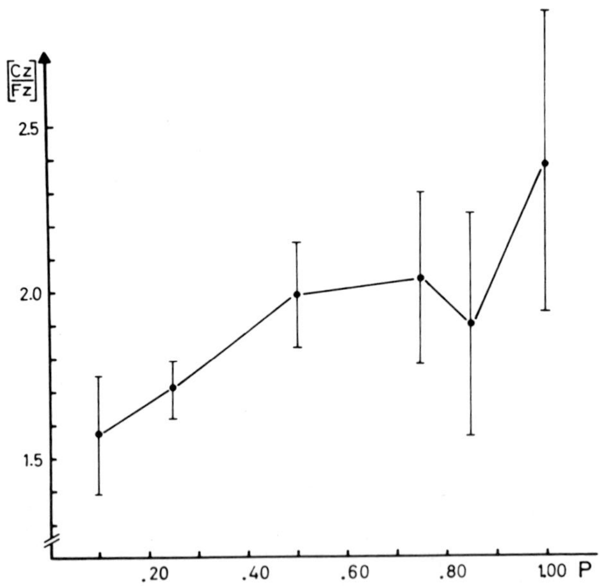

Fig. 3. Ratio between CNV areas at Cz and Fz as a function of probability of occurrence of S_2. Pooled data for four *Ss*. Bars give standard error between means of the *Ss*.

the baseline (determined as an average voltage during 250 msec before S_1) and the instant of occurrence of S_2. The potentials spoilt by eye movement, muscle, or drop-out artefacts were omitted. From single trial measurements the ratio between the CNV areas at C_Z and F_Z was calculated. Additionally, the RT of *Ss* to S_2 was measured.

RESULTS

The present results are based on preliminary data obtained from four *Ss*. The area of the CNV at C_Z and F_Z was clearly dependent on the probability of occurrence of S_2 (*Fig.* 1). The area was in both derivations smallest in blocks with an S_2 probability of 1·00. The maximal values were obtained when S_2 probability was 0·85. In contrast to the CNV, the RT of the *Ss* was a linearly decreasing function of the probability of occurrence of S_2 (*Fig.* 2). Similar dependence of the RT on the stimulus probability has been earlier reported by Näätänen (1972).

The C_Z/F_Z ratio was found to grow with increasing S_2 probability (*Fig.* 3), the frontal-dominant potential becoming smaller with respect to the central-dominant potential the higher the probability of occurrence of the imperative stimulus.

DISCUSSION

The present results confirm the findings of Karrer *et al.* (1973a) who found reduction in the CNV amplitude with an S_2 probability of 1·00 compared with a probability of 0·5. The present results further indicate that the changes in the probability of occurrence of S_2 do not produce correlated changes in the CNV and the RT. The change in the topography of the CNV as a function of the probability of occurrence of S_2 supports the notion of at least two different components of the CNV.

Contingent Negative Variation in a Verbal Learning Paradigm

T. W. BILLINGER, J. F. PETERS and J. R. KNOTT

University of Iowa, Iowa City, Iowa 52242, U.S.A.

Six *Ss* were studied in a paired associates paradigm, CNVs being measured at C_z, C_3 and C_4 (mastoid reference). A single trigram was followed 2·5 sec later by a tone, and 1·0 sec later the second, associated, trigram was given (via CCTV). *Ss* were instructed to respond to the tone by speaking the second trigram, if recalled. Stimuli were in blocks of 10 pairs, and up to 30 blocks given.

Although learning occurred, systematic CNV changes did not appear in all subjects and variability was extreme. CNVs rarely re-crossed the baseline (e.g. negativity was prolonged), so conventional measurement could not be made. The paradigm itself may account for this effect, and for the great variability, suggesting that further studies should be made using a revised experimental design.

(Editors' note: A later, redesigned study by these authors was presented to the American EEG Society at its 1975 meeting.)

Effects of Amphetamine on CNV and Behaviour in Man*

JOSEPH J. TECCE† and JONATHAN O. COLE‡

Laboratory of Neuropsychology, Boston State Hospital, and Department of Psychiatry, Tufts University School of Medicine

Twenty normal females were tested in a constant-foreperiod reaction-time situation (light-tone-key press) following the oral administration of dextro-amphetamine (10 mg) and placebo. In the first hour post-drug, a group of 13 individuals exhibited paradoxical drowsiness accompanied by lowered electrical brain activity (reduced CNV amplitude), whereas a group of 7 individuals showed behavioural alertness and increased CNV amplitude. During this period, 4 subjects verbalized feelings of sadness (e.g. 'the saddest I ever felt'). The majority of paradoxical individuals had Type B (slow rise time) baseline CNV shapes whereas the non-paradoxical group was made up of subjects with Type A (fast rise time) CNV shapes. Both paradoxical and non-paradoxical responder groups showed behavioural excitation in the second and third hours post-drug. In these late hours post-drug, there were reports of cognitive surges ('racing thoughts' and 'endless ideas'). We concluded that amphetamine is not a simple stimulant of the central nervous system but that it can also act as a depressant and that the CNV is a sensitive and accurate indicator of changes in psychological functioning produced by amphetamine.

* For a fuller account of this study *see* Tecce, J. J. and Cole, J. O. (1974) Amphetamine effects in man: Paradoxical drowsiness and lowered electrical brain activity (CNV). *Science* **185,** 451–453. Supported by USPHS research grants MH–19211 (JJT) and MH–16128 (JOC) from the National Institute of Mental Health.

† Research Scientist Development Awardee 1–K02–00016 of the National Institute of Mental Health.

‡ Psychiatrist to McLean Hospital.

CNV-Heart Rate Response Under Gradual Sleep Reduction

PAUL NAITOH and RAYMOND P. HILBERT

Naval Medical Neuropsychiatric Research Unit, San Diego, California 92152, U.S.A.

Lacey and Lacey (1970) and Lacey (1972) have shown that bradycardia occurs before the imperative signal (S_2) of a fixed foreperiod reaction time task. This is the same foreperiod that contains the contingent negative variation (CNV). Naitoh *et al.* (1973) reported that this anticipatory bradycardia remained constant after acute sleep loss (one night), and sleep stage deprivation (stage 4 or stage REM) for three nights. However, the CNV, during the same foreperiod, decreased significantly. Naitoh *et al.* (1973) suggested that cardiac slowing in anticipation of the imperative stimulus reflects 'not a degree of sustained attention, but rather a subjectively experienced task demand and the willingness to exert more effort in order to compensate for the detrimental effects of sleep debit.'

In the present study, 4 young adults (two couples) were observed over a 10-month period as they gradually reduced their sleep. This gradual sleep reduction permitted an experimental evaluation of the minimum hours of sleep which the subjects would tolerate (*see* Johnson and MacLeod, 1973). From these 4 subjects scores on the CNV, Wilkinson auditory vigilance, Wilkinson addition tasks (Wilkinson, 1970), subjective mood, and Sleep Logs were obtained on a scheduled basis.

These additional data allow us to explore further the psychological significance of the anticipatory bradycardia in the paradigm of the CNV.

METHOD

Four subjects (couple 1, S.S. and B.S.; couple 2, P.M. and J.P.) shortened their sleep by 30 min every *two* weeks, from their habitual 8 hours, to 6 hours. Then, sleep was cut by 30 min every *three* weeks until it was reduced to 5 hours. To give the subjects more time to adapt, at this point the sleep reduction rate was further slowed. The subjects cut their sleep by 30 min every *four* weeks until it was down to 4 hours per day. Each couple was given an active role in deciding when they could no longer tolerate the reduced sleep. When a couple decided that x hours of sleep per day were too short, they were asked to stay on x hours plus 30 min for 2 months. They then slept *ad lib.*

The subjects filled in Sleep Logs every morning. On selected nights, their sleep was electrophysiologically recorded.

The CNV, Wilkinson auditory vigilance tasks, and Wilkinson addition tasks were conducted just before the subjects changed their sleep duration. Couple 1 was always tested in the morning; couple 2 was always tested during the afternoon.

The CNV task was detection of a visual signal against a noisy background. The first warning stimulus (S_1) was a 50 msec presentation of an ellipsoid defined by dots on a TV monitor screen. S_1 was followed in 4 sec by S_2. S_2 was either the ellipsoid with added noise (dots), or the masking dots alone. S_2 was displayed on the TV monitor screen. If the subjects thought that S_2 contained the ellipsoid (the signal), they pressed a white response key as quickly as they could. If they thought that the S_2 was a noise alone, they pressed a red response key, but leisurely. The subjects were instructed not to look down on the response keys, and to keep their finger on the white response key before each trial. After the response to S_2, the subjects expressed their confidence in the response by pressing one of three keys: 'Not-sure-at-all,' 'So-so,' or 'Very sure.' Four seconds after S_2, the subjects were given knowledge of results (KR) by a 1-sec message flashed on the TV screen; 'OK' or 'NO GOOD'.

All of the stimulus presentations (including the KR) were controlled by a DEC PDP–12 computer. The program enabled us to alter, as required, the duration of the S_2 presentation (from 10 msec to 5+ sec), as well as the signal/noise ratio. The inter trial interval was 25 ± 5 sec. Sixty-four trials in four blocks of 16 trials each constituted the main test. The main test was preceded by an additional block of 16 'calibration' trials in which the subjects responded to S_2 by pressing the white response key as quickly as they could, regardless of the nature of the S_2.

The vertex EEG (referenced to tied mastoids), the vertical eye movements (of the right eye), and the electrocardiogram (EKG) were recorded with a Beckman Type R Dynograph on paper, as well as on a Hewlett-Packard FM instrumentation tape recorder, Model 3900. EEG activity and vertical eye movements were recorded in the DC mode with silver–silver chloride electrodes. A calibration signal of 25 μV was inserted in series with the vertex EEG on each trial 1 sec prior to the S_1 presentation.

To reduce skin potential artefacts, the mastoids and the vertex EEG scalp site were subjected to atropinization, using iontophoretic treatment (Lader and Montagu, 1962; Picton and Hillyard, 1972) with 3 per cent atropine sulphate for 20 min under 0·4 mA current strength.

The vertex EEG was analysed off-line by a Fabri-Tek Model 1072 after the trials with artefacts were eliminated from the analysis. The EKG analysis was performed off-line with a DEC PDP-12 after carefully editing out all trials with artefacts. The program printed out all heart rates from 3 sec prior to the S_1 to 9 sec after the S_1. Mean heart rates as well as variances were obtained at 0·25 sec intervals, starting 3 sec prior to S_1.

RESULTS

Sleep. Couple 1 could not tolerate a sleep duration of 5 hr per day. Couple 2 found 4·5 hr of sleep to be too short. So, the first couple stayed on 5·5 hr for 2 months and the second couple on 5 hr for 3 months, before they slept *ad lib*.

Sleep Logs suggested that all 4 subjects kept to the sleep reduction schedule.

There was no significant deterioration of the percent correct detection in the Wilkinson auditory vigilance task or the CNV visual signal detection task until the subjects approached the minimal sleep: 5·0 hr for couple 1 and 4·5 hr for couple 2. Even then, performance deterioration was not dramatic. The most consistent change was an increase in reported sleepiness.

Fig. 1 shows the CNV and the heart rate for subject J.P. during a session when he had slept only 4·5 hr a day for the preceding 4 weeks. The subject, J.P., reported that he could not tolerate 4·5 hr sleep. In *Fig*. 1, Hits refers to those trials where J.P. correctly detected the signal; CR refers to trials when J.P. made correct rejections; there was no signal. The CNVs were somewhat attenuated from the prior magnitudes of his baseline CNVs, and the baseline anticipatory bradycardia appeared also reduced.

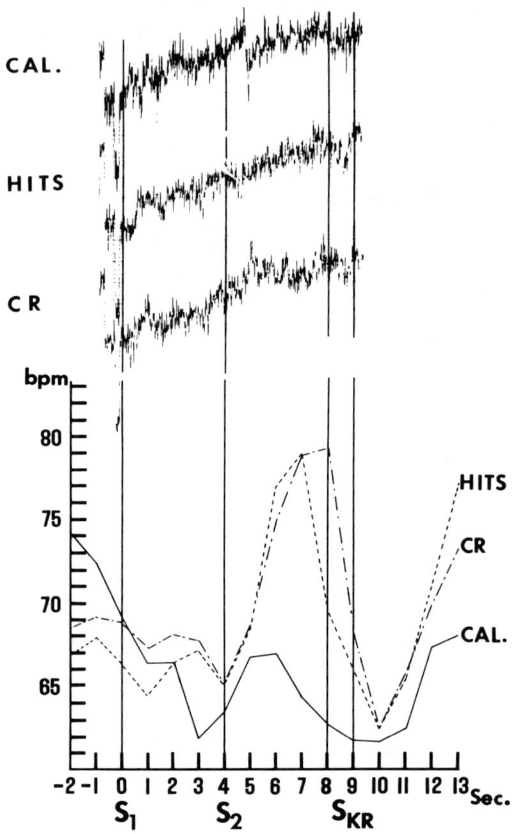

Fig. 1. The CNV and heart rate correlates of the subject, J.P. *See* text for details. CAL.= 'Calibration' trials. HITS=Correct detection of the ellipsoid. CR=Correct rejection. Averaged heart rate (the lower graph) is given in beats per min (bpm). S_{KR}=Knowledge of result.

Table 1 shows the anticipatory bradycardia for the 4 subjects under the response of correct detections (Hits), excepting the very first session. This cardiac slowing in anticipation to the S_2 was defined as the fastest heart beat during the S_1–S_2 interval minus the slowest heart beat observed near the time of the presentation of the S_2. The results suggest that anticipatory cardiac slowing accompanying the responses of Hits was attenuated during sleep reduction. For 3 subjects, the product-moment correlates between anticipatory bradycardia and the elapsed time since the start of the experiment were −0·92, −0·72 and −0·62. The fourth subject, B.S., had a correlation of near zero, 0·07. When 7 January was taken as the baseline, anticipatory bradycardia was attenuated to 5–10 per cent levels of significance during sleep reduction.

The expected cardiac slowing to S_3 (S_{KR}) was also observed, but it was difficult to tease this component out from the natural tendency of heart rate to decrease at this point in time (i.e. 4 sec after the S_2; *see* Naitoh *et al.*, 1973).

The relation of anticipatory bradycardia to the CNV magnitudes and reaction times is being analysed under 12 response contingencies (e.g. Hits with a high confidence, Hits with a medium confidence, *et seq.*).

Table 1. Magnitude of cardiac slowing in anticipation of S_2 (Trials of correct detection—Hits only)

Date	*Hours of Scheduled Sleep*	*Couple 1*		*Couple 2*	
		S.S.	*B.S.*	*P.M.*	*J.P.*
1/7/73	7·5–7·0	13·1*	4·4*	10·8*	28·0*
1/28/73	6·5	5·0	3·3	12·0	9·6
2/11/73	6·0	4·3	1·9	5·8	6·0
3/4/73	5·5	5·9	3·4	5·4	3·1
4/1/73	5·0	5·0	4·6	4·7	6·0
4/28/73	5·5 (No. 1)† 4·5 (No. 2)	3·0	6·6	2·1	3·3
6/3/73	5·5 (No. 1) 5·0 (No. 2)	3·8	1·8	3·7	4·4
7/1/73	*ad lib* (No. 1) 5·0 (No. 2)	—	—	0·7	7·5
7/29/73	*ad lib*	3·2	3·6	4·2	3·1

* Expressed in beats per min.

† 'No. 1' refers to the scheduled duration of sleep for couple 1 during the indicated period.

DISCUSSION

Cardiac correlates of psychological states have been extensively studied (Chase *et al.*, 1968; Connor and Lang, 1969; Lacey and Lacey, 1970; Jennings *et al.*, 1971; Lacey, 1972; Libby *et al.*, 1973). The present study confirms a previous report of Naitoh *et al.* (1973) that the CNV correlate of the cardiac slowing is a robust phenomenon observable under any CNV paradigms.

However, the cardiac slowing may reflect the learned memory sequence of timing, selective attention, or the willingness to perform (or all three of these factors).

The present study detected a decrease in anticipatory bradycardia after a sleep reduction régime which ordinarily produces few decrements (*see* Webb, 1969; Wilkinson, 1970).

ACKNOWLEDGEMENTS

The authors would like to thank Dr Ardie Lubin for critical reading of the manuscript and for editorial suggestions.

This work was supported in part by Department of the Navy, Bureau of Medicine and Surgery, and by a National Institute of Mental Health grant to Dr Gordon G. Globus, University of California, Irvine. The opinions and assertions contained herein are the private ones of the authors' and are not to be construed as official or as reflecting the views of the Navy Department.

The 'Form' of the CNV: the Relative Effect of Information in the Ready and Imperative Stimuli

H. WEINBERG, H. MICHELEWSKI and R. KOOPMAN

Simon Fraser University, Burnaby, British Columbia, Canada

Twenty normal subjects were tested in a CNV paradigm in which a warning signal (S_1) was followed after an i.s.i. of 1·5 sec by an imperative signal (S_2). The response, made to S_2, was contingent upon information presented in either S_1 or S_2. When the discriminative information determining the response occurred at S_1 the CNV assumed a form which was significantly different from that observed when it occurred at S_2. When the information was given at S_1 the CNV tended to show a ramp-like negative rise having its peak immediately prior to S_2. When the information was given at S_2 there tended to be a rapid early rise of the CNV followed by a slow decline in negativity. Both wave forms differed from that obtained in a 'standard' CNV stiuation where no discrimination was involved.

The findings suggest that the CNV may be the result of the interaction of two types of information processing. One type involves time estimation, which is positively related to CNV amplitude. The second type, which is inversely related to CNV magnitude, is information processing relative to the specific nature of response performance.

For a fuller account of this work see: Weinberg, H., Michelewski, H. and Koopman, R. (1976). The influence of discrimination on the form of the contingent negative variation. *Neuropsychologia* **14,** 87–95.

Slow Potential Correlates of Predictive Behaviour During a Complex Learning Task

L. J. LEIFER*, D. A. OTTO†, S. G. HART‡ and E. M. HUFF§

NASA—Ames Research Center, Moffett Field, California 94035, U.S.A.

Although event-related potentials (ERPs) are thought to be indices of cognitive processes, few efforts have been made to study ERPs associated with concept formation in a complex learning task. The present study independently examined the Readiness Potential (RP), Contingent Negative Variation (CNV), and P300 components of the evoked response in a complex probability learning task. The experiment was designed to test the hypothesis that the CNV varied directly with the expected probability of event occurrence (Walter *et al.*, 1964) and that the P300 varies inversely with the expected probability (information content) of event occurrence (Sutton *et al.*, 1965). The task was derived from studies investigating the joint influence of objective probabilities and sequential dependencies on learning (Huff, 1970; Hart and Huff, 1972).

METHOD

Nine adult *Ss* predicted which of 4 possible stimuli would occur next in a first-order Markov-structured sequence. In each trial, one stimulus had a high probability of occurrence (0·55) while other stimuli had equal low probabilities (0·15). Two stimuli had a high probability of repeating themselves (transition Class 1) while the other 2 stimuli had a high probability of alternating (transition Class 2). Each stimulus occurred 125 times during a 65-min session of 500 trials.

Stimuli were presented on a computer-driven oscilloscope. A question mark ('?') display (S_1) signalled subjects to predict the next event by pressing 1 of 4 microswitches within 1500 msec. Switch closure produced a display of the predicted stimulus (S_2). Two seconds later, the next stimulus (S_3) in the sequence was displayed for 200 msec. This was followed by an inter-trial interval of 2–6 sec which was determined on a uniform random basis (*Fig.* 1).

Monopolar recordings referred to linked ears were obtained at F_3, F_Z, F_4, C_3, C_Z, C_4, P_3, P_Z, P_4 with Beckman silver–silver chloride electrodes and an AMPEX DAS 100 set at a -3 dB bandpass of 0·1–50 Hz. Prior to signal averaging, EEG records were EOG compensated by weighted linear summation with supra-suborbital EOG. Mean and integral CNVs and RPs were computed relative to a 650-msec pretrial baseline. P300 peak-to-peak amplitude was computed relative to a 1-sec pre-S_3 baseline.

* Current address: Massachusetts Institute of Technology, Man-Vehicle Laboratory, 37–155, Cambridge, Massachusetts 02139, U.S.A.

† Current address: Environmental Protection Agency, University of North Carolina, Chapel Hill, North Carolina 27514, U.S.A.

‡ Current address: University of California, Berkeley, Department of Psychology, Berkeley, California 94720, U.S.A.

§ Current address: NASA—Ames Research Center, Man-Machine Integration Branch, Mail Stop 239-3, Moffett Field, California 94035, U.S.A.

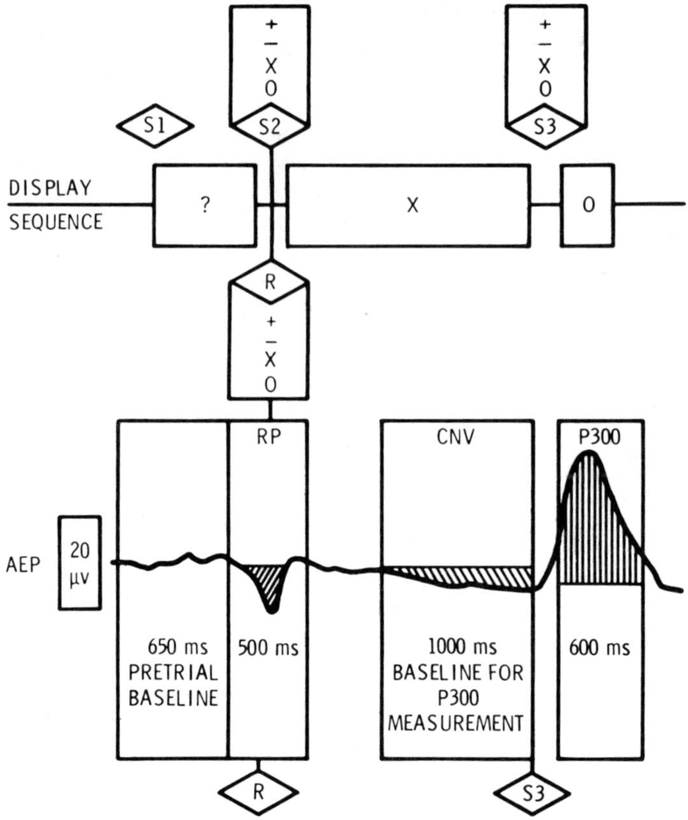

Fig. 1. Experimental paradigm and evoked potential measurement (schematic). (S1) = '?' imperative stimulus to predict next stimulus in sequence, (R) = response, indicating prediction which of 4 possible (+, –, ×, 0) stimuli would be presented next; (S2) = display of the predicted stimulus, presented immediately upon response; (S3) = feedback telling subject the next actual stimulus in the sequence.

RESULTS

Learning was measured as the proportion of astute predictions made per unit time. Astuteness, which is defined as prediction of the most probable successor to the outcome of the preceding trial, increased significantly across trials ($F=207{\cdot}5$, $df=2{\cdot}16$, $P<0{\cdot}001$). The level and rate of learning were different between transition classes ($F=121{\cdot}3$, $df=1{\cdot}8$, $P<0{\cdot}001$). Class 1 repetitions were learned relatively quickly. They were predicted as often as they occurred by the end of the first 50 trials, and over-predicted by 20 per cent at asymptote. Class 2 alternations were learned slowly and predicted less often than they occurred throughout the experiment. By asymptote, Class 2 transitions were predicted only 54 per cent of the time (*Fig*. 2a). Response latency was consistently shorter for Class 1 than Class 2 predictions until the final phase of the experiment (*Fig*. 2c) at which time they were predicted with equal facility.

The topographical distributions of mean RP, CNV, and P300 amplitude were assessed for 6 subjects by averaging 100 successive trials in each of 5 trial blocks (*see Table* 1, summary). CNV and P300 were maximal at P_Z and decreased progressively at central and frontal sites (Scheffe *post hoc* comparisons (Hays, 1963); F=80·3, F′=17·44, $P<0{\cdot}05$ and F=189·1, F′=17·44, $P<0{\cdot}05$, respectively). In contrast, RP was maximal at F_Z and diminished progressively at central and parietal sites (F=79·1, F′=17·44, $P<0{\cdot}05$). RP and P300 were bilaterally symmetrical in the coronal planes studied. CNV was larger in the right frontal region (F=20·2, F′=17·44, $P<0{\cdot}05$), but symmetrical otherwise.

Four-way analyses of variance were performed on selective averages of single trial C_Z data for 9 *Ss* according to the behavioural classifications: successive thirds of the experiment, trial blocks (TBK 1, 2, 3); correct *v.* wrong outcome (C, W); astute v. unastute prediction (A, UA); transition class (C1, C2) (*Table* 1).

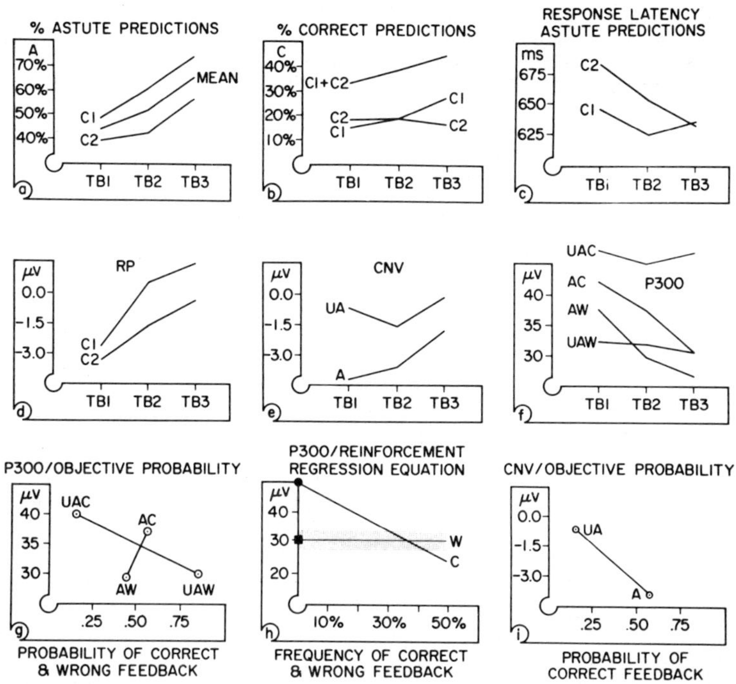

Fig. 2. Time, frequency, and reinforcement effects on behaviour, RP, CNV, P300. *a*, percentage of astute predictions across trial blocks, showing Class 1 and 2 differences which influenced RP, and the mean for C1 and C2 for comparison to CNV and P300; *b*, percentage of correct predictions across trial blocks showing Class 1 and 2 differences which failed to influence P300; *c*, response latency across trial blocks showing differences between Class 1 and 2; *d*, RP was significantly influenced by transition Class (C1, C2) across trial blocks; *e*, CNV differed significantly following astute (A) and unastute (UA) predictions over trial blocks; *f*, P300, across trial blocks was significantly influenced by astuteness and correctness, showing considerable habituation for astute correct (AC) and astute wrong (AW) trials, and minimal habituation for unastute correct (UAC) and unastute wrong (UAW) trials. P300 was consistently larger following correct than wrong feedback; *g*, P300 was inversely related to the objective probability of a correct outcome following unastute predictions (UAC, UAW), but directly related to the outcome of astute predictions (AC, AW); *h*, the regression equation for P300 and the frequency of occurrence of correct (C) and wrong (W) reinforcement shows a strong inverse relationship for correct outcomes, and no consistent relationship to wrong outcomes; *i*, CNV was inversely proportional to the objective probability of positive reinforcement to be expected from astute v. unastute predictions.

RP (*Fig.* 2d) was more negative for Class 2 than Class 1 (F=8·23, df=1·8, $P<0{\cdot}025$) and diminished significantly across trial blocks (F=5·11, df=2·16, $P<0{\cdot}025$). CNV (*Fig.* 2e) showed no class effect, was more negative following astute than unastute predictions (F=5·71, df=1·8, $P<0{\cdot}05$) and evidenced no significant change across trial blocks. P300 (*Fig.* 2f) was unrelated to transition class, larger following unastute than astute predictions (F=27·77, df=1·8, $P<0{\cdot}001$) and diminished over trial blocks (F=5·71, df=2·16, $P<0{\cdot}025$). There was significant interaction between astuteness and outcome (F=13·16, df=1·8, $P<0{\cdot}001$) and between astuteness and trial blocks (F=12·18, df=2·16, $P<0{\cdot}001$).

Table 1. *Analyses of Variance.* [1] trial blocks=successive thirds of experimental session (TB1=trials 1–166, TB2=trials 167–333, TB3=trials 334–500); [2] transition class= stimuli which tend to repeat themselves (C1), stimuli which tend to alternate (C2); [3] prediction astuteness=prediction of the stimulus with the highest objective probability (A), the converse, unastuteness (UA); [4] feedback outcome=stimulus which indicates next stimulus in the sequence: correct (C) and wrong (W); [5] topography=scalp electrode location.

	RP	*CNV*	*P300*
Trial Block [TB1, 2, 3]	TB1>TB2>TB3*	ns	Astute TB1>TB2>TB3‡ Unastute ns
Transition class [C1, C2]	C1>C2*	ns	ns
Prediction astuteness [A, UA]	ns	A>UA†	Class 1 UA>A‡ Class 2 ns
Feedback outcome [C, W]	na	na	C>W‡
Topography [F_3, F_z, F_4] [C_3, C_z, C_4] [P_4, P_z, P_4]	$F_z>C_z>P_z$‡	$P_3>C_3>F_3$‡ $P_z>C_z>F_z$‡ $F_4>F_3$*	$P_3>C_3>F_3$‡ $P_z>C_z>F_z$‡ $P_4>C_4>F_4$‡ $P_z>P_3$, P_4‡
RP	na	ns	ns
CNV	ns	na	r(TB)=−0·55† r(A)=−0·50‡ r(UA)=−0·33*
P300	ns	r(TB)=−0·55† r(A)=−0·50‡ r(UA)=−0·33*	na

na=not applicable.
ns=not significant.
*=rejection of the null hypothesis at $P<0{\cdot}05$.
†=rejection of the null hypothesis at $P<0{\cdot}01$.
‡=rejection of the null hypothesis at $P<0{\cdot}001$.

Neither RP nor CNV covaried significantly with behaviour occurrence frequency of any specific trial type. Contrastingly, a strong negative correlation was found between P300 and occurrence frequency, both across behavioural categories and within specific categories (*Table* 2). The most striking feature of this analysis is the strong correlation between P300 and frequency of occurrence of correct outcomes and lack of correlation between P300 and frequency of occurrence of wrong outcomes (*Fig*. 2h).

DISCUSSION

The discussion will be guided by the frequently made association between learning and habituation (Pribram, 1971). Habituation, a progressive decrementing of the neural response to specific stimuli or events, is both time and frequency dependent. Learning is similarly time and frequency dependent with the additional influence of positive or negative reinforcement. RP, CNV, and P300 will be reviewed as a function of these three variables.

Table 2. Regression Analysis, P300/Frequency of Specific Behaviours. Linear regression analyses of P300 amplitude and the frequency of occurrence (percentage) of specific behaviours (astute · correct, astute · wrong, unastute · correct and unastute · wrong) by trial block and transition class.

	Simple Linear Regression		*Analysis of Variance for Regression*		
	Intercept	*Slope*	*r*	*F*	*P*
P300/behaviour frequency all conditions	45·9 uv	−3·9 uv / 10%	−0·76	29·48(1,23)	0·001
P300/correct feedback	49·6	−5·1	−0·91	45·24(1,11)	0·001
P300/wrong feedback	31·4	—0·0	−0·0	0·72(1,11)	ns
P300/astute predictions	41·5	−2·7	−0·41	1·96(1,11)	ns
P300/unastute predictions	47·2	−3·9	−0·87	32.19(1,11)	0·001
P300/transition class 1	46·6	−4·4	−0·83	21·63(1,11)	0·001
P300/transition class 2	44·6	−3·3	−0·67	8·31(1,11)	0·05
P300/trial block 1	46·7	−2·9	−0·76	8·36(1,7)	0·05
P300/trial block 2	44·8	−3·7	−0·78	9·07(1,7)	0·05
P300/trial block 3	45·6	−4·9	−0·87	18·90(1,7)	0·01

ns=not significant.

RPs pooled across trials were somewhat less negative for astute than unastute predictions and significantly less negative for Class 1 than Class 2 predictions. Greater Class 2 negativity was associated with slower prediction of events which were most difficult to conceptualize. RP negativity decreased significantly across time as prediction accuracy and speed increased, but was independent of astuteness, correctness, and the occurrence frequency of any specific behaviour. These facts, as well as the anterior localization of RP and the contiguity of RP to a motor response, suggest that RP habituation may index motor learning, but not complex probability learning.

CNV showed no significant transition class effect, but was significantly more negative following astute than unastute predictions. When examined across time, CNV habituation was evident following astute predictions in which the expectancy of positive feedback might be expected to be increasing across trials. No habituation was evident following unastute predictions in which the occurrence of positive feedback was consistently very low (*Fig*. 2e). The relationship of CNV to event probabilities (*Fig*. 2i) is confounded by the lack of correspondence between CNV and occurrence frequency of specific behaviours. Thus, the hypothesis that CNV is proportional to event probability is both supported and contradicted. CNV is at first proportional to astuteness and the high

$P = 0{\cdot}55$ probability of positive reinforcement, and then diminishes as the frequency of astute behaviour increases.

The P300 wave evoked by the outcome stimulus (S_3) provided the strongest and most sensitive EEG index of predictive behaviour. S_3 conveyed two kinds of information to *Ss*: (1) direct feedback of correctness on the antecedent prediction and (2) indirect feedback of cumulative prediction astuteness. Across time, P300 varied significantly with both the correctness and astuteness dimensions, but not with transition class. The effect of correctness was much stronger than the effect of astuteness, consistent with the direct and indirect nature of the feedback.

The hypothesis that P300 varies inversely with event probability was partially confirmed if one assumes that the 'event' of interest is the occurrence of a correct trial outcome (positive feedback on *S*'s prediction). This is unlike most studies in which P300 has generally been measured in tasks where learning was minimized (e.g. Tueting *et al.*, 1971; Roth, 1973). For unastute trials, in which a minimum of learning could take place, and for which no P300 habituation was evident, a strong inverse relationship can be seen between P300 and the objective probabilities of correct and wrong outcomes (*Fig*. 2g). The probability of being unastute and wrong ($P = 0{\cdot}85$) was associated with a small P300.

For astute predictions, P300 pooled across trials varied directly with the objective probability of correct ($P = 0{\cdot}55$) and wrong ($P = 0{\cdot}45$) outcomes (*Fig*. 2g). Across trials, the marked habituation of P300 was inversely related to the increasing frequency of predictive astuteness and expectation of positive feedback (*Fig*. 2f). When P300 was compared to the occurrence frequency of positive and negative feedback, the inverse relationship was strong for correct outcomes and non-existent for wrong outcomes (*Fig*. 2h).

Assume that *Ss* engage in construction of an internal subjective model of the processes they are attempting to understand. Initially, with no correspondence between objective and subjective probabilities, there may be a multitude of models. A correct outcome acts to validate the model being tested, and in our study, P300 amplitude was highly correlated ($r = 0{\cdot}91$) with event frequency when the outcome acted as positive feedback. A wrong outcome neither confirms nor rejects the model being tested and P300 was found to be uncorrelated with event probability when the outcome constituted negative feedback ($r = 0{\cdot}01$). We may interpret this to mean then, that learning was enhanced following positive feedback as the model being tested received validation and habituation took place, and relatively unaffected by negative feedback, in which the model was neither confirmed nor rejected and no habituation was evident. Further, unastute predictions were rarely reinforced, and even more rarely supported the same internal model, whereas astute predictions were regularly reinforced, leading to selective adoption of the astute model. This model promoted increasing congruence between objective and perceived probabilities, and allowed *Ss* to optimize the opportunity for positive reinforcement.

SUMMARY

RP, CNV, and P300 were studied in a complex probability learning task. RP correlated with transition class, CNV with predictive astuteness, and P300 with astuteness, event probability and the occurrence frequency of positive reinforcement. P300 analyses are consistent with the conception that learning occurs mainly during positive reinforcement trials.

Auditory Evoked Potential and Lift/No-lift Reaction Time in Relation to Uncertainty

PATRICIA TUETING and SAMUEL SUTTON

Biometrics Research, New York State Department of Mental Hygiene and Department of Psychiatry, Columbia University, New York, U.S.A.

This experiment was in part motivated by the challenge presented by the recent data on Donchin *et al.* (1973b) which showed that requiring an RT response appeared to wipe out the difference in P_3 brought about by the presence or absence of uncertainty. This came as a surprise to us as P_3 is so sensitive to uncertainty in the guessing situation. In ten years of running *Ss* in various forms of the guessing situation we have never found a single *S* who on *any* occasion did not give a larger amplitude P_3 for the uncertain condition than for the certain condition. In fact, even guessing is not a necessary precondition, as both Ritter *et al.* (1968) and Roth (1973) have shown that simple unpredictability results in larger P_3s.

In our initial design, we used a press left *v.* press right choice response with the index finger—Donchin had used a right thumb *v.* left thumb button press. We noticed in our early pilot experiments that in the RT situation there indeed seemed to be only a small and inconsistent difference between the certain and uncertain condition. However, we were disturbed by the unusual and unfamiliar shapes of the wave forms in these experiments and this led us to substitute a lift/no-lift response.

METHOD

Electrodes at F_z, C_z, P_z and O_z with left ear lobe reference were used. Eight normal men (aged 17–30 years), who were paid for their services, participated as *Ss*. Evoked potentials to low- and high-pass clicks were averaged separately with a bandpass of 0·03–100 Hz (TC=5·3 sec). Three *Ss* were instructed to lift their index finger as quickly as possible to low click but to refrain from lifting to high click; the other five *Ss* lifted on high click, but refrained from lifting on the low click. The sequence of low and high RT stimuli was random with 50/50 probability. The RT stimulus, which was the third click in a trial, was preceded by a random delay intended to eliminate anticipation. Differential recovery effects were balanced by using four different trial sequences with the three clicks in a trial occupying different positions in each sequence.

Conditions of certainty and uncertainty were created by instruction and by the configuration of the two clicks preceding the reaction time click. Trials from two types of blocks in the certain condition were pooled for the analysis: for half of these blocks *Ss* were informed that the RT click would be of the same pitch as the first click in the trial, and for the remaining half *Ss* were instructed that the RT click would be of the same pitch as the second click in the trial. The uncertain condition was created by using the same random sequences and telling *S* that the two prior clicks were irrelevant in predicting the pitch of the RT stimulus. The certain and uncertain conditions were alternated by block within a session, but the certain condition always initiated a session. A small monetary payoff was given for an RT shorter than an RT on the last trial with the same stimulus sequence. There were two 45-minute sessions of four blocks each (one block=30 trials). Trials were 5·7–6·7 sec in length and the inter-trial interval was 5·5–6·5 sec.

RESULTS AND DISCUSSION

The essential results may be seen in *Fig.* 1. We have superimposed for each *S* the wave forms obtained at vertex separately for the lift trials and for the no-lift trials. The solid lines are the wave forms for the certain condition; the dashed lines are the wave forms for the uncertain condition. Negativity at the scalp electrodes (with respect to the ear lobe reference) is shown as an upward deflection.

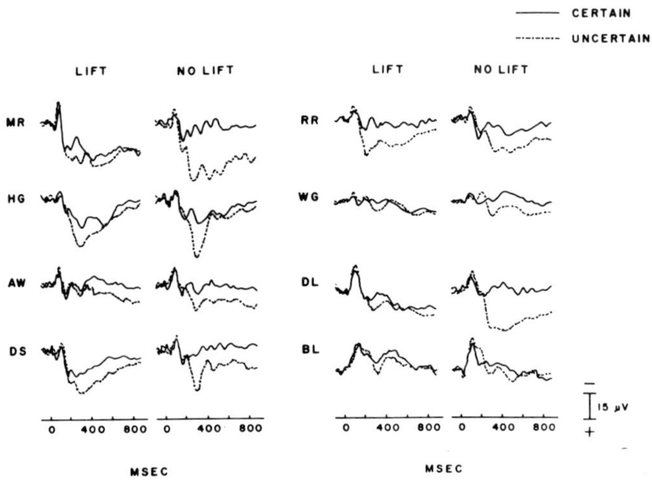

Fig. 1. Average evoked potentials for 8 *Ss* comparing uncertain and certain conditions for lift and no-lift trials. Vertex negativity with respect to left earlobe is up.

It is very clear that for both the lift and no-lift trials, P_3 in the uncertain condition is larger than in the certain condition. While the degree of difference between the uncertain and certain conditions varies from very large to very small, depending on the *S*, there is no exception to this generalization across the 16 pairs of wave forms. The differences between the uncertain condition and the certain condition are highly significant *both* for the no-lift and lift trials (t tests, $P<0{\cdot}01$).

Another finding for the vertex location, seen by inspection of this figure and supported, by the statistical results, is the degree of difference between the uncertain and certain conditions, which is greater for the no-lift trials than for the lift trials (t test, $P<0{\cdot}02$).

Somewhat less obvious by inspection, because of the way that we have superimposed the wave forms, is the fact that, within the uncertain condition, P_3 is larger for the no-lift trials than for the lift trials. This finding is in agreement with an earlier report by Karlin *et al.* (1970). However, this finding of a larger P_3 for the no-lift trials is not found in the certain condition. Although not statistically significant in these vertex data, the trend appears to be larger P_3 for the *lift* trials than for the no-lift trials.

We compared uncertain-certain differences in P_3 amplitude for both the lift and no-lift trials at frontal, vertex, parietal and occipital loci. The amplitude of P_3 was measured from a baseline defined by the median voltage in the 100-msec period preceding the RT stimulus. The uncertain *v*. certain differences are found to be statistically significant at all four loci (*Fig*. 2a). However, somewhat more interesting is the comparison of lift *v*. no-lift trials separately in the uncertain condition, and in the certain condition. It can be seen in *Fig*. 2b that in the uncertain condition the difference between lift and no-lift trials is largest at vertex and is also significantly larger at the frontal electrode. However, there is no significant difference in the uncertain condition between lift and no-lift trials at either parietal or occipital derivations.

In the certain condition, the relationship between lift and no-lift is reversed. First, at all electrode derivations, P_3 amplitude appears larger for the lift trials than for the no-lift trials. In this condition the pattern of significance is reversed; the difference is only significant at parietal and occipital derivations, in contrast with the uncertain condition where it is only significant at vertex and frontal derivations.

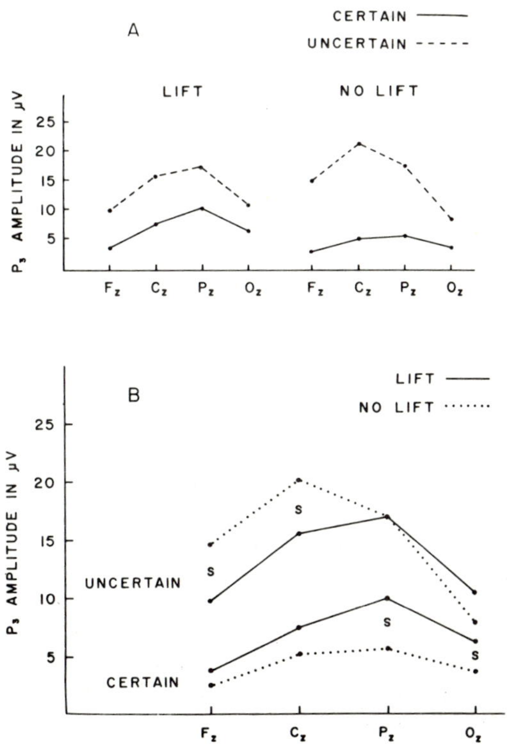

Fig. 2. A, P_3 amplitude averaged over 8 *Ss* for different scalp loci comparing uncertain and certain conditions for lift and no-lift trials. The average voltage in the 100 msec prior to stimulus onset was the baseline for P_3 measures. Fz, frontal; Cz, vertex; Pz, parietal; Oz, occipital. B, The same curves as in A above but superimposed to compare lift and no-lift trials. 'S' refers to statistical significance ($P<0{\cdot}05$).

The intriguing findings, that the certain condition P_3 is larger for lift than for no-lift, and that differences in the uncertain condition occur at the more forward electrodes whereas the differences in the certain condition occur at the more posterior electrodes, cannot be interpreted at this time. However, such findings tend to support the implication that two processes, rather than one process, are at work. We are forever cured of looking at P_3 at only the vertex derivation.

RT, as would be expected, is longer in the uncertain condition than in the certain condition (*t* test, $P<0{\cdot}01$). The average median reaction time was 326 msec for the uncertain condition and 260 msec for the certain condition. However, there are no significant correlations across *Ss* between RT and P_3 amplitude in either the uncertain or certain conditions. Furthermore, there is no significant correlation across *Ss* between the uncertain-certain differences in P_3 amplitude and the uncertain-certain differences in RT.

We do not consider our data to be necessarily in conflict with the findings of Donchin *et al.* (1973b) for the following reasons:

1. It may be that the task demand in a lift *v.* no-lift choice RT is different in some fundamental way from a press right/press left response. This possibility is supported by the finding of Karlin *et al.* (1970), as well as our own finding that no-lift trials give a consistently larger P_3 than lift trials. The fact that, at the same level of uncertainty, P_3 is larger for the no-lift than for the lift trials is itself an argument that P_3 is affected by the nature of the task over and above the uncertainty effect. Further evidence for the role of task demand is that in the certain condition the P_3 amplitude for lift trials is larger than for no-lift trials; this is opposite to the uncertain condition.
2. It may be that the degree to which *Ss* are pressed for rapid RT affects the degree to which the task demand becomes equated for the certain and uncertain conditions. While this argument, as a source of the difference between our findings and those of Donchin *et al.* (1973b), is not supported by our inability to find any relationship across *Ss* between RT and P_3, this is not conclusive. The comparisons between P_3 and RT would have to be made *within* individuals.

   We reinforced *Ss* for faster RT but there is no way of deciding whether our *Ss* were under the same pressure for high speed that they were in Donchin's experiment.
3. We have preliminary data which show no difference in P_3 as a function of the relative probability of occurrence of the two stimuli with the lift *v.* no-lift RT task. These results contrast sharply with the strong effect of probability in the guessing situation. We are inclined to believe that the lack of effect of probability in the RT task will be confirmed.
4. We have shown in our earlier work that P_3 occurs to uncertain, absent stimuli when the presence or absence of the stimulus is relevant to *S*'s task, but it does not occur to the absent stimulus when the presence or absence of the stimulus is irrelevant to *S*'s task (Sutton *et al.*, 1967).

   Given all of these considerations, we are sympathetic to the notion that P_3 is not uniquely sensitive to the uncertainty of the task but is also responsive to other aspects of the task demand. What are at issue are the specifics and fundamental logic of the crosscutting interactions.
5. As we had a random interval immediately prior to the choice stimulus, conditions were far from optimal for observing the CNV. Nevertheless, we did measure the CNV at the onset of the choice stimulus from a baseline

established 2 sec earlier—at a point in time when, in the certain condition, *Ss* know whether they will or will not be lifting. The CNV is on average 3 μV larger when *Ss* know they will be lifting to the choice stimulus than when they know they will not be lifting ($P<0{\cdot}05$). A parallel comparison with the uncertain condition would be of no value as in that condition *Ss* do not know prior to the choice stimulus whether or not they will be lifting. However, we compared the average CNV in the uncertain condition with the average CNV in the certain condition and found the former to be 5 μV larger. This parallels the findings by Friedman *et al.* (1973) of an average CNV which was 5 μV larger in the uncertain condition in a guessing situation.

We have argued elsewhere (Tueting and Sutton, 1973) that in some situations the CNV may influence P_3, while in others P_3 may vary quite independently of the CNV. In the present data both relationships are found. We showed that both the CNV and P_3 are larger in the uncertain condition than in the certain condition. Also, within the certain condition, both the CNV and P_3 are larger for lift trials than for no-lift trials. However, within the uncertain condition, in which the CNV is constant because the choice stimulus cannot be known in advance, P_3 is nevertheless larger for the no-lift trials than for the lift trials.

SUMMARY

Averaged evoked potentials and lift/no-lift RT were studied in relation to uncertainty. The results were: (1) Median RT was longer in the uncertain condition than in the certain condition. (2) P_3 amplitude was larger for the uncertain condition at F_Z, C_Z, P_Z and O_Z loci for both lift and no-lift trials. (3) In the uncertain condition P_3 amplitude in the no-lift trials had a more anterior distribution than P_3 in the lift trials. (4) In the certain condition P_3 amplitude in the lift trials was larger than in the no-lift trials. (5) CNV was larger in the uncertain condition than in the certain condition, and in the certain condition lift trials had larger CNV than no-lift trials.

ACKNOWLEDGEMENTS

We thank Marion Hartung, Robert Laupheimer, Raymond Simon, Robert Youdin and Joseph Zubin for assistance with various aspects of this work, and Dr M. Wallach, Director of Brooklyn State Hospital, for providing facilities for carrying out this research. This work was supported by Grants MH 14580 and MH 19812 from the National Institute of Mental Health, United States Public Health Service.

Covariation of the Magnitude of the CNV and P300 as a Function of the Subject's Task

EMANUEL DONCHIN, RAYMOND JOHNSON, jun.,
RONALD HERNING and MARTA KUTAS

Department of Psychology, University of Illinois, Champaign, Illinois, 61820, U.S.A.

This paper is concerned with the relationship between the CNV and P300. An opportunity to analyse jointly these two components of the human cortical evoked response was provided by an experiment, reported elsewhere (Donchin *et al.*, 1973b), designed to assess the effects of cognitive complexity on the amplitude of P300. The *Ss* in that experiment were instructed to predict the outcomes of series of Bornoulli trials (trials which can have one of two possible outcomes). In each experimental trial a warning click was followed after 1400 msec by a flash of light illuminating either the letter A or the letter B, the letter defining the outcome of the trial. *S* had to predict the trial's outcome prior to the presentation of the warning click. From one experimental series to another the rule which determined the sequence of outcomes was varied, thereby varying the degree to which *Ss* could utilize information from preceding trials in developing their predictions.

In the paper cited above we focused on P300 elicited by the flash and were able to demonstrate that its amplitude varies systematically with the sequence-generating rule. The experimental arrangement clearly allows a similar study of the effect of the same experimental variables on the CNV. Moreover, we can, by using these data, study conjoint changes in the CNV and in P300. Equally interesting is an analysis of the distribution of both components over the scalp, an analysis made possible by simultaneous recording from 8 electrode locations.

The interrelation between the CNV and P300 has been subject to considerable debate (Donchin and Cohen, 1969; Näätänen, 1969; Donchin and Smith, 1970; Donald and Goff, 1971). At issue is the degree to which the amplitude of the P300 component reflects differential, pre-stimulus, anticipatory processes as has been suggested by Näätänen (1969) and by Karlin (1970) rather than post-stimulus processing as proposed by others (Sutton *et al.*, 1967; Donchin and Cohen, 1967; Hillyard *et al.*, 1971). A different, though related, issue concerns the degree to which the P300 component interacts with the positive-going limb of the CNV. A detailed discussion of these issues appears elsewhere in this volume.

METHOD

The experimental procedures have been described in detail in our previous report. Briefly, 10 *Ss* were each presented with five series of trials, 64 trials per series. Each trial consisted of a warning click followed after 1400 msec by a

Supported by grant Number OEG–5–70–0030(508) from the U.S. Office of Education and grant number NSF–GB–30999 from the National Science Foundation.

flash illuminating the letter A, or the letter B, in an Iconix Tachistoscope. For stimulus parameters and instrumentation details see Donchin *et al.* (1973b).

The experimental series differed in the rule used to determine the sequence of outcomes. We used an *Alternating* series (ALT) in which the As and Bs alternated on successive trials. In the *Learn* series (LRN) *S* was handed prior to the series a sheet on which appeared a 9-element sequence of As and Bs. *S* was told that this sequence would be repeated in the coming series and that he was to memorise the sequence. In the *Pattern* (PAT) series a similar 9-element sequence was used, and *S* was so informed, except that he had to determine the elementary sequence from the presentations. In a series we called *Shafty* (SHF) *S* was again told that the series were generated by repeating an elementary sequence except that no such sequence was used, the series in fact being random. In the *Random* (RAN) series a random sequence of As and Bs was used, and *S* was so informed.

These five series were repeated in a counterbalanced order, under three different experimental régimes, each régime run on a different day. The following régimes were used:

1. *Guess* (GS) *S* was instructed to predict prior to the warning tone what the outcome of each trial would be. Proper care was taken to avoid contaminating the data by movement-generated noise in the EEG. *S* received a financial bonus for guessing correctly on at least 60 per cent of the trials.
2. *Reaction Time* (RT). In this set of sessions *Ss* were instructed to press one of two microswitches following the illumination of the letter. A right hand response followed the presentation of an A, a left hand response followed the presentation of a B. Both speed and accuracy were stressed; slow responses were indicated by repeating the flash; incorrect responses were indicated by a loud buzzer.
3. *Reaction Time with Catch* (RTC). These sessions were identical in all respects to the RT sessions except that catch trials were embedded in the series. On 10 per cent of the trials the warning click was not followed by a flash. *S* was abjured from responding in the absence of a stimulus. Catch trials did not disrupt the sequence generated in a specific series. The stimulus scheduled for the Catch trial was presented on the following trial.

Eight Beckman silver–silver chloride electrodes were referred to linked ear electrodes. Electrodes were placed at Fpz, Fz, T_3, C_3, Cz, C_4, T_4, O_2. EOGs were recorded from a pair of electrodes placed above, and to the right of, *S*'s left eye. The EEG was amplified using Brush amplifiers (13–4218–00) and recorded on FM magnetic tape (Hewlett-Packard 3955). The system's bandwidth was 0·01–30 Hz.

All processing was performed off-line using an IBM 1800 computer. All averages considered here were obtained by averaging together the data from all *Ss*. A detailed analysis of the data obtained from individual *Ss* is reported by Donchin *et al.* for the GS and the RTC condition. This analysis revealed a large measure of intersubject homogeneity which suggests that the overall averages used in this report are valid. We computed for each trace the area under the CNV and the P300 portions of the record. Integration was performed with respect to a base line defined as the Midmean of the 500 msec epoch immediately preceding the warning stimulus. (The Midmean, a statistic recently proposed by Tukey, is the arithmetic mean of the data falling into the interquartile range.)

RESULTS

In *Fig.* 1 we plot the area of the CNV and P300. Panels A and B present the areas for the Vertex electrode plotted as a function of the sequence generating rule (SGR) for each of the three experimental régimes. The other six panels plot the areas for all SGRs and experimental régimes (tasks) as a function of the location of the electrodes.

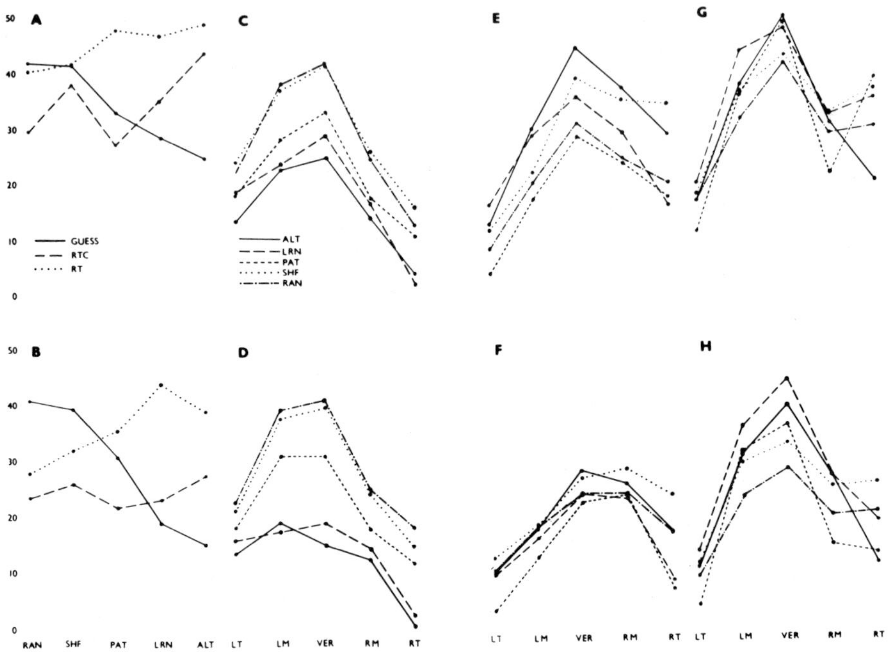

Fig. 1. The area of either the CNV or P300 at different locations on the scalp are plotted for different tasks and different series. Each point represents an area calculated by integration over a baseline defined as the midmean of the 500 msec preceding the warning stimulus.

Panels A and B plot the areas of the CNV and P300 respectively as a function of the sequence generating rules, for each of the three tasks (GS, RT and RTC) for the Vertex electrode.

The remaining panels present the distribution of CNV (C, E, G) and P300 (D, F, H) areas along the interaural line. The five lines in each panel represent the five sequence generating rules. Panels C, D represent the GS condition, E, F the RTC and G, H the RT conditions. Further details in text.

The data presented in Panel B, for the GS and the RTC task repeat in a condensed form the data presented in detail by Donchin *et al.* (1973b). The area of P300 is clearly a monotonic function of the SGR, from a minimum at ALT to a maximum at RAN. At the same time, when a motor response is required of the *S*, with catch trials included in the series, P300 area does not vary with the SGR. The values for the RAN and ALT series are virtually identical. When catch trials are eliminated from the series the relation between P300 and the SGR is different from that in the other two tasks. In the RT condition the area of P300 during an ALT series is larger than during the RAN series. In fact, P300 area for ALT during an RT condition is quite similar to the area of P300 for the RAN series during the GS régime. (A full report of these data, including the levels of statistical significance, are to be published elsewhere.)

Panel A presents the corresponding areas of the CNV. Interestingly the CNV areas vary as a function of the various experimental variables in much the same fashion as do the areas of P300. CNV area increases with the complexity of the SGR in the GS condition. It shows an inverse relationship in the RT condition and seems to fluctuate in the RTC condition. In general CNV areas obtained during the RT series are larger than CNV areas obtained in the other two régimes except for the RAN series, where the GS and RT areas converge.

The correlation between the CNV and the P300 areas is indicated by the behaviour of these two measures along the interaural line as shown by the remaining six panels of this figure. Two aspects of these plots are worthy of note. The distribution of the areas within any of these panels seems to be quite homogeneous. That is, while there are clear differences *between* the distributions in the different panels, within each panel the relative magnitude of the CNV at each of the five electrode locations is similar. In other words, the tasks (GS–RT–RTC) affect the distribution of the CNV and P300 along the interaural line. The SGRs appear to modulate the amplitude of this distribution without affecting its shape.

Of particular interest is the fact that, while in most cases the area recorded at the Vertex electrode is larger than that recorded at the more lateral positions, there is a substantial difference in the amplitude at the lateral electrodes between the data recorded at the GS and the two RT conditions. In the GS condition the left electrodes yield a larger area than the right electrodes. This relationship is reversed during the two RT conditions. It is important to note that the above statement is equally correct for the P300 areas. The distribution of P300 and the CNV, at least along the interaural line, seems positively correlated.

The effect of the SGR on the overall amplitude of both the CNV and P300 varies among the six panels. There is very little difference between the P300 areas recorded during the RTC condition. On the other hand for the CNV the areas are least variable during the RT condition. The areas of the CNV recorded during the RT condition are consistently higher than those obtained during the RTC and GS conditions.

DISCUSSION

The above results indicate that by varying *S*'s task it is possible to manipulate the distribution over the scalp of both the CNV and the P300 as well as the overall amplitude of these two components. Within each task, the magnitude of the components, though not their scalp distribution, is modulated by conditions under which the *S* is performing the task. In two tasks requiring a motor response we find right-hemisphere potentials exceeding the left-hemisphere potentials. In one task, not requiring a motor response, this relationship is reversed. However, in all tasks we find that the amplitude of the distribution can be shifted up or down as a function of the complexity of the sequence generating rule used to determine the succession of trial outcomes. In any event, P300 and CNV areas appear to be affected in a similar manner by all these manipulations.

We speculated about the implications of the effects of the sequence on P300 amplitude in our previous report (Donchin *et al.*, 1973b). For the present it is noteworthy that similar considerations might apply to the CNV. This of course revives the possibility that there is a strong relationship between these two components of the human evoked response. Needless to say, the demonstration of such a correlation does not imply a common causal source.

It is difficult to see why we should find the different tasks affecting the laterality of the CNV and P300. We are quite confident that this result is not due to any imbalances in our electrodes, amplifiers or recording system. That this is the case is attested by the fact that the evoked responses elicited by the warning click are virtually identical on the left and right hemispheres in all conditions. Had we changed in any way the parameters of our recording system between tasks, we should have seen a change in the relative amplitude of these evoked responses.

Why are the potentials recorded during the RT and RTC condition so different? The overall reduction in amplitudes during RTC is of course consistent with previous reports that diluting a series of RT trials leads to a reduction in CNV amplitude. Yet the dilution was only of the order of 10 per cent, far less than is required to obtain 'dilution' effects. It seems to us more likely that the differences in CNV amplitude are due to the relative usability of temporal information by the *S* in performing his task. Consider for example the ALT condition during the RT task. In this case the *S* can successfully perform the task by estimating the interval between the warning and the imperative stimuli. As he can well predict which hand will respond, the stimulus in itself is less critical. In the RTC series such temporal information is less relevant in that the *S* cannot be sure that he will indeed have to respond. Similar considerations can be applied to the other combinations of RT by series used. What we are proposing is that CNV amplitude is strongly affected by the degree to which the information about the interval between the warning and the imperative stimulus can be utilized by the *S* in optimizing his task performance.

Scalp Topography of the P_3 Wave in Different Auditory Decision Tasks

STEVEN A. HILLYARD, ERIC COURCHESNE,
HOWARD I. KRAUSZ and TERENCE W. PICTON

Department of Neurosciences, University of California, San Diego, La Jolla, California 92037, U.S.A.

Sutton and his colleagues initially described the late positive 'P_3' wave as an 'endogenous' potential that is emitted when task-relevant information is delivered that 'resolves' the subject's prior uncertainty (Sutton *et al.*, 1965; 1967). Subsequent studies have found that a P_3 or P_{300} wave occurs (after a latency of 250–450 msec) upon the delivery of information of several varieties, including specific 'target' signals that are being searched for, feedback cues that inform the *S* if a prior judgement or guess is correct or incorrect, and signals that command a prompt motor reaction in simple or disjunctive RT tasks (for review, see Squires *et al.*, 1973a). Other studies indicate that an infrequent, unexpected stimulus change may also elicit a P_3 wave (Ritter *et al.*, 1968; Roth *et al.*, 1973), although attending to such signals results in enhancement of the P_3 (Picton *et al.*, 1973b).

Numerous overlapping psychological constructs have been linked to the P_3 wave, including information delivery, uncertainty resolution, salience (Tueting and Sutton, 1973), orientation and cognitive evaluation (Ritter and Vaughan, 1969), reactive changes of non-specific state (Karlin, 1970), decision confidence (Hillyard *et al.*, 1971) and response set-selective attention (Hillyard *et al.*, 1973). The validity of such attempts to identify the P_3 wave with a particular psychological operation depends upon the unestablished proposition that 'the P_3' is a unitary physiological entity, arising from a single cerebral generator system, when it occurs in the diverse experimental contexts cited above. The present study investigates this 'unitary P_3' proposition by comparing the scalp topographies of P_3s obtained in six different experimental contexts.

A second aspect of this report is to examine whether the P_3 elicited upon detection of a threshold-level signal can be dissociated from the fall to baseline of the prior CNV, by comparing the scalp topographies of the two phenomena in D.C. recordings.

METHOD

Ten young adults participated in one experimental session lasting 4–6 hours for the 'mapping' of the P_3s, four serving in a later session in which both CNV's and P_3s were recorded in a signal detection task.

For the mapping of P_3 waves silver cup electrodes were attached with collodion to 15 scalp sites selected from the 10–20 system, and to the orbital ridges (vertical EOG). A non-cephalic reference electrode was used, which was the centre tap of a 50 kohm potentiometer, one end of which was placed at the

Supported by NASA Grant NGR 05–009–198.

sternoclavicular junction and the other on the back of the neck at the seventh cervical vertebra. Adjusting the potentiometer minimized pick-up of the EKG. Evoked potentials were recorded from all sites simultaneously with a Grass Model 6 polygraph (16-channel), with the TC set at the maximum (0·24 sec), and averaged on-line with a PDP–12 computer.

For the CNVs–P_3 comparison Beckman silver–silver chloride electrodes were applied to four midline scalp sites, left mastoid reference, and upper and lower orbital ridges (vertical EOG). The epidermis was scratched with a sterile needle to reduce electrode resistance well below 2 kohm. Event-related potentials were recorded with Grass 5P1 amplifiers (bandpass) D.C.—100 Hz), stored on FM magnetic tape, and averaged off-line with a Fabritek 1052 computer.

Auditory Decision Tasks

Ss reclined in an easy chair in a sound-attenuating chamber and fixated a spot on a screen that could be transilluminated with a Grass photostimulator. Auditory stimuli were delivered binaurally through earphones. The following tasks were performed in order, obtaining 30–40 evoked responses for averaging in each, with rest periods intervening. Then the entire sequence was replicated in reverse order as time permitted.

(1) *Passive Control.* Tone pips at 50 dB SL (1 kHz, 50 msec duration) were presented at a rate of 1–2 per sec while *S* fixated the point with no instructional set. (2) *Random Tones.* The same 50 dB tone pips were delivered at irregular, unpredictable intervals ranging between 5 and 20 sec. *S* had no instructional set. (3) *Omitted Stimulus.* The 50 dB tone pips were given regularly at the rate of 1 per sec. At irregular intervals (between 5 and 20 sec) one tone pip would be omitted from the sequence. *S* counted the number of omissions and reported it after the run. (4) *Threshold Signal Detection.* A warning flash was followed after 1·4 sec by a tone pip (1 kHz, 50 msec) on 50 per cent of the trials at random. A second flash appeared 2·5 sec after the warning, at which time *S* reported whether or not he had heard the tone signal. The intensity of the signal was reduced to the lowest level at which *S* could maintain 100 per cent correct performance. The i.t.i.s ranged from 6 to 15 sec. (5) *Correct Feedback.* On each trial a 1-sec burst of flashes was presented, either at a rate of 12 per sec or at a slightly higher rate adjusted to be barely discriminable (at a level of 80–90 per cent correct) from 12 per sec. *S* judged each train of flashes to be 'high' or 'low' in flicker rate and then received (after a variable delay of 2–4 sec) one of the two 50 dB tone pips, signifying that their choice was correct (1 kHz) or incorrect (2 kHz). 'Correct' tone pip trials only were averaged. (6) *Go, No-go. S* pressed a thumb-actuated switch as rapidly as possible to a tone pip of 1 kHz (50 dB), while not responding to randomly interspersed tone pips of 2 kHz. An equal number of each tone was presented at irregular intervals of 5–15 sec. 'Go' and 'No-go' conditions were averaged separately.

RESULTS

Topography of P_3 Wave

In every *S* the EP to the 50 dB tones contained an N_1 (80–110 msec) and a P_2 (150–190 msec) component. In most 'decision task' runs a later positive wave (P_3) could also be clearly identified, having a scalp distribution different from the N_1–P_2 complex and a latency ranging between 250 and 400 msec. Typical EP distributions are shown in *Fig.* 1. P_3s evoked upon detection of the threshold signal (HITS) and after the go and no-go signals appear largest at Cz, as were the N_1–P_2 components. However, the distribution of all three types of P_3 extended more laterally and posteriorly than that of N_1 or P_2.

Percentage amplitude distributions were calculated for each condition for each *S* and then averaged across all *Ss*. Only P_3s clearly discernible from P_2 and having a maximum value over 5·0 μV were included in the overall maps. P_3s elicited by the randomly presented tones fulfilled these requirements in only 5 *Ss*. They were not analysed in detail, but had a mean amplitude of 8·2 μV, mean latency of 302 msec and topography as follows: Fp_z (59 per cent), F_3 (77), F_z (97), F_4 (76), T_3 (65), C_3 (73), C_z (100), C_4 (86), T_4 (58), T_5 (61), P_3 (80), P_z (83), P_4 (72), T_6 (57), and O_z (59).

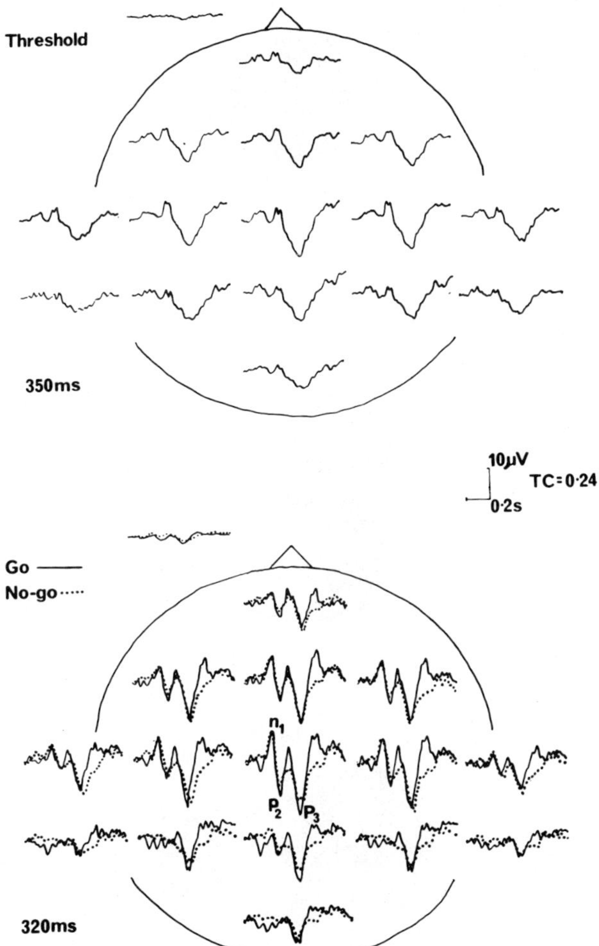

Fig. 1. Topographical representations of EPs during threshold signal detection (HIT trials only) and disjunctive RT task (go and no-go). The tracings on each 'head' are from sites Fp_z (top centre); F_3, F_z, F_4 (second row); T_3, C_3, C_z, C_4, T_4 (third row); T_5, P_3, P_z, P_4, T_6 (fourth row); and O_z (bottom). At left of nose is vertical EOG at a gain of half that of the calibration bar.

Amplitude distributions for P_2 and for two types of P_3s (detected threshold signal and omitted signal) are compared in *Fig.* 2. The two P_3s are virtually identical in topography and are much more widely distributed laterally and posteriorly than the P_2. The topographies of these two P_3s were reproducible within *Ss* and similar across *Ss*. Different *Ss* had maximum P_3 amplitudes at C_Z, at P_Z, or at both of these sites. P_3s obtained in all five experimental conditions shared the common characteristic of being significantly larger in relative amplitude across all ten sites of the lateral and posterior scalp than the P_2 component, on nearly every run in every *S*.

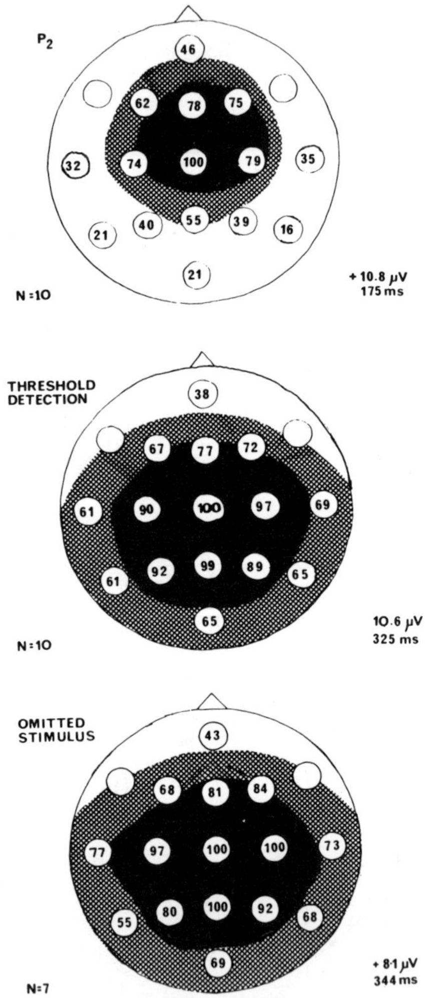

Fig. 2. Relative amplitudes at different scalp loci of the P_2 component evoked by the 50 dB tones and of the P_3s to threshold detection (HITs) and omitted stimuli (expressed in percentages of maximum). Electrode placements as in *Fig.* 1. Each map is a grand average of the percentage maps from N individual *Ss*. Black zone depicts regions of over 75 per cent in relative amplitude and hatched zone between 50 and 75 per cent.

Detailed comparisons among the various P_3 maps revealed some significant differences. In some *Ss* P_3s to both go and no-go signals had a more anterior scalp maximum than the threshold signal P_3 (largest at C_Z–P_Z). Some *Ss* had similar topographies for all five types of P_3s, others had either the go, no-go, or feedback P_3 shifted anteriorly relative to the others.

Topographies were compared statistically by calculating relative ratios of different P_3 amplitudes at each electrode site and performing *t* tests of dependent means upon them. These within-subjects comparisons indicated that the P_3 on no-go trials was relatively larger at F_Z and smaller at P_Z than the threshold-detect, omitted signal, and 'go' P_3s (all $P<0{\cdot}05$). Also, the lateral distribution of the omitted signal P_3 (at T_3 and T_4) exceeded that of the no-go and feedback P_3s ($P<0{\cdot}02$). It should be emphasized, however, that none of these differences was observed in every *S*.

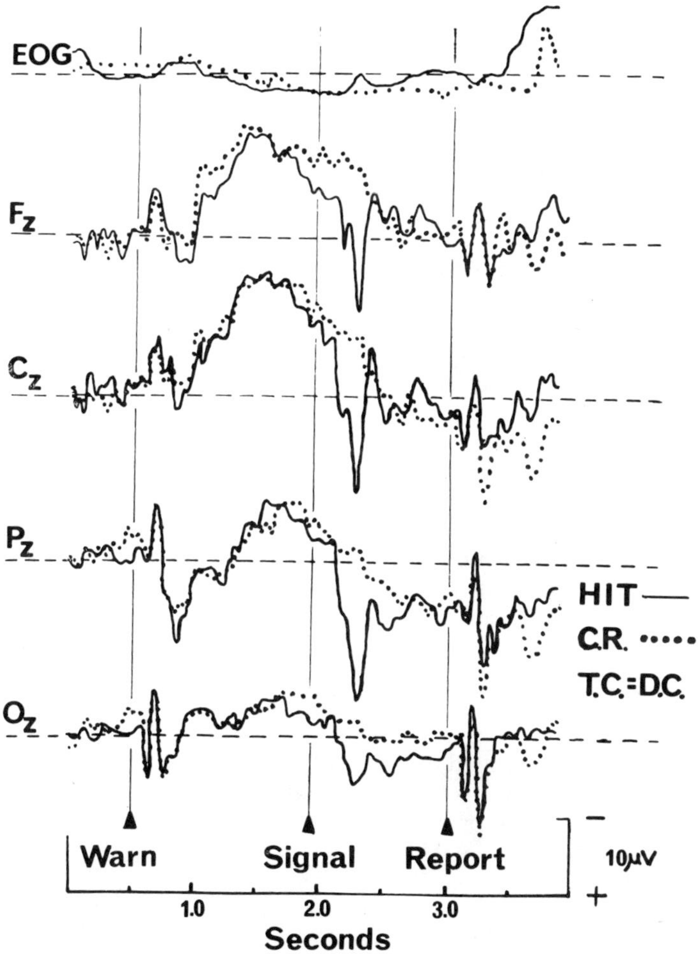

Fig. 3. CNV and P_3 at different scalp sites during threshold detection task. On HIT trials the signal was correctly detected and on C.R. (Correct Rejection) trials a non-signal was correctly reported.

CNV and P_3 distribution during Threshold Detection

The scalp distribution of the CNV prior to the threshold tone and of the P_3 upon a HIT (detection) are shown for one *S* in *Fig*. 3. The CNV was largest at C_Z and F_Z and small at P_Z and O_Z, while the P_3 (mean latency=360 msec) was largest at C_Z and P_Z; this pattern was found in all four *Ss*. The mean amplitude distribution of the CNV (measured over the 200 msec prior to the signal) is shown in *Table* 1. The distribution of the CNV is clearly anterior to that of the signal P_3 peak amplitude. A late positive wave (mean latency=340 msec), elicited after the warning flash, had a sharply posterior scalp distribution.

Table 1. Mean amplitudes (in μV) of event-related potentials at different scalp loci (N=4)

Component	F_Z	C_Z	P_Z	O_Z
CNV (*re* pre-warn baseline)	6·7	6·2	3·0	2·5
P_3 to signal (*re* pre-warn baseline)	6·5	10·4	13·4	6·3
P_3 to signal (*re* pre-signal CNV)	13·2	16·6	16·4	8·8
P_3 to warning flash (*re* pre-warn baseline)	3·4	6·4	11·2	5·5

DISCUSSION

All five types of P_3 were more widely distributed across the posterior and lateral scalp than the preceding P_2 component. All P_3s had relative amplitudes exceeding 50 per cent maximum at sites T_5, T_6 and O_Z, in contrast to the P_2 component, which was reduced to 16–21 per cent at those sites. Similar distribution of P_3s had been reported by Vaughan and Ritter (1970) and Picton *et al.* (1973b).

Despite similarities among the P_3 topographies, it cannot be concluded that they all arise from a common cerebral generator. In some *Ss* the no-go P_3 was clearly displaced anteriorly relative to the others. The data were not always extensive enough to certify that particular topographical differences are reliable and valid. There emerged, however, a statistically significant overall trend for the no-go P_3 to be situated more anteriorly than the threshold and omitted signal P_3s. These disparities, as well as others found in certain *Ss*, might be caused by the addition of more or less CNV fall-off, motor potential, or 'orientation response' (Loveless, p. 26) to an underlying 'unitary P_3'. Alternatively P_3s in the various tasks could arise from similarly disposed but distinctive brain systems.

The similarity of the omitted and threshold signal P_3s, within and between *Ss*, provides a strong basis for proposing that they represent identical brain events, associated with the detection of a specific 'target' signal among a set of non-signals (Hillyard *et al.*, 1973b). Whether or not all the types of P_3s fall under the rubric of 'target recognition', as Squires *et al.* (1973a) have proposed, or contain additional elements is as yet unresolved.

The CNV prior to the threshold auditory detection had a frontal amplitude maximum, like that reported by Järvilehto and Fruhstorfer (1970) for an auditory intensity discrimination. It is clear that the more posteriorly situated P_3 on the HITs is not simply a side effect of the abrupt termination or phasic modulation of this CNV. Their distinctive topographies suggest that a completely separate P_3 process is superimposed upon the CNV irrespective of whether or not it has an abrupt termination.

SUMMARY

Sixteen channel mappings of the scalp topography of the late positive P_3 wave were obtained during 5 different decision tasks. All types of P_3s had in common a very broad distribution across the posterior and lateral scalp. There were certain significant disparities in topographies among some of the P_3s, however, indicating that a simple 'unitary P_3' hypothesis may require modification. In addition, the P_3 accompanying detection of a threshold level tone was clearly dissociable topographically from the fall-to-baseline of the preceding CNV.

ACKNOWLEDGEMENT

We are grateful to Dr Reginald Bickford for permitting us to use his PDP–12 computer and 16–channel averaging facilities.

Relationship between CNV, its Resolution and the Evoked Response

ROBERT T. WILKINSON

Medical Research Council, Applied Psychology Unit (Annexe), Cambridge, U.K.

In this paper I will attempt to summarize and integrate the conclusions of three recent papers by myself and colleagues on the relationship between, on the one hand, the CNV and its resolution and, on the other, the N1–P2 wave and what has been called the P3 or P300 wave of the evoked response.

The first conclusion is that any CNV which is present at stimulus onset will be returned to baseline as a function of the relevance of that stimulus to the organism. Thus CNV resolution may be expected to vary from a maximum, when prior instructions or training establish the propensity for a highly motivated operant response to a particular stimulus at S2 of the CNV paradigm, to no CNV resolution at all, when the stimulus by habituation and instructions is of minimal importance to the organism and requires no operant response.

Another factor which must clearly influence the size of the CNV resolution is the amplitude of CNV at stimulus onset. With no CNV there can be no return.

The second conclusion pertains to the relationship of CNV resolution to N1–P2 and possibly P3 waves of the evoked response. Since CNV resolution may continue from 100 msec, or earlier, to 600 msec post-stimulus, its presence in the post-stimulus trace may coincide with that of N1–P2 of the evoked response and also possibly P3 (or P300). When this happens the peak-to-peak amplitude of N1–P2 will appear artefactually large, and so will the amplitude of P3 (or P300). Thus any tendency for these two components to reflect the degree to which a stimulus is relevant to the organism, as defined above, may be due wholly or in part to summation with a CNV resolution which is itself varying with (*a*) stimulus relevance and (*b*) prior CNV amplitude.

The evidence for the claim that CNV resolves selectively with stimulus relevance comes primarily from two studies, an early experiment by Näätänen (1967, Exp. III) and an experiment by Wilkinson and Lee (1972).

In *Figs*. 44–55 and 58a–f of the monograph by Näätänen (1967) recordings are presented of averaged EEG traces to relevant and irrelevant clicks alternating at 1-sec intervals. A bandwidth of 0·3–70 Hz allowed CNVs to appear. These slow potentials returned to baseline following a relevant stimulus, redeveloped usually before the onset of the irrelevant stimulus 1 sec later, and were again resolved by the relevant stimulus following 1 sec after that. Although only two *Ss*' records were presented, they suggest that the irrelevant stimuli failed to resolve any CNV that was present at their onset.

The experiment by Wilkinson and Lee (1972) differed from that of Näätänen in a number of important respects: Following a 'Ready' signal brief runs of about 25 tones were presented at rapid, quasi-random intervals, three tones appearing in random order. *Ss* counted one tone and ignored the other two, giving the total at the end of each run. As the run was short it was assumed that CNV would be present before each stimulus and, since the stimuli were unpredictable, of equal amplitude for both relevant (counted) and irrelevant (uncounted) tones. Nevertheless N1–P2 was larger following counted than uncounted tones. This difference did not correlate with performance. However, another measure did, presence of a long, positive-going wave extending from about 100 to 600 msec post-stimulus. As there was little sign of a transient P300 wave this slow positivity was taken to be CNV resolution occurring only to relevant stimuli as in Näätänen's study, and correlating with selective attention performance (accuracy of counting) for that reason.

Further correlational evidence relating such a slow positive going wave to both prior CNV and apparent effect of selective attention in N1–P2 and the P300 measure comes from an experiment by Wilkinson and Ashby (1973) which in part replicated and extended the Näätänen study.

A relevant and an irrelevant tone alternated at 1-sec intervals. 4-sec samples of EEG were averaged. These covered two relevant and two irrelevant stimuli and allowed the establishment of a reasonable baseline from which measures of CNV amplitude could be taken. CNVs developed before relevant but not before irrelevant stimuli and relevant/irrelevant differences in prior CNV correlated across *Ss* with corresponding differences in N1–P2 and the conventional measure of P300, positivity at 300 msec latency relative to stimulus onset of the EEG trace. However, the latter measure showed a high concordance (Kendall's $W = 0{\cdot}36$) with similar measures taken at 400 msec and 500 msec. This suggests that what was being measured was not a sharp P3 or P300 component of the evoked response but again a long, slow positivity extending from 300 msec to at least 500 msec. It seems most reasonable to identify this as the CNV resolution and as such positive only in relation to a negative level prevailing at stimulus onset due to prior CNV.

Thus it seems that prior CNV, CNV resolution, N1–P2 and the conventional P300 measure all correlate in reflecting stimulus relevance across *Ss*. Though causation is impossible to prove the implication is strong that the first two CNV influences may jointly determine, either in part or in whole, the apparent amplitudes of the two latter evoked response parameters as stimulus relevance is varied.

The third study by Wilkinson and Spence (1973) approaches the question from another aspect to ask what counts as stimulus relevance in the context of returning CNV to baseline: that an overt response be made, that a decision be made, that information be gained, or what?

A warning click (W) was followed by S1, a high or low tone, which was followed by S2, a high or a low click. If S1 was high a 2-choice response had to be made to S2, but if S1 was low the *S* was finished for that trial. Furthermore in half the trials a 2-choice response was required at S1, in half not. The CNV at S1 onset resolved identically to baseline (*a*) whether or not a motor response was required at S1, and (*b*) whether or not the tone at S1 indicated that continued expectancy was required in order to respond to S2. However, where such continued expectancy was required the CNV resolution gave way to a new CNV after a time. This produced a point of divergence between the continued and the noncontinued expectancy traces, the sharp change in the direction of the former occurring at latencies which varied from 250 to 500 msec and correlated with motor reaction time across *Ss*. It is suggested that CNV resolution occurs relatively early in processing with the first coarse identification of the stimulus as one of a general class of stimuli made relevant either by instructions or endogenously. CNV resolution proceeds until a second stage of finer processing is complete, when a further CNV may develop if that analysis counsels further expectancy. It is interesting that the sharp change in direction of the EEG at this point occurred at an average latency of 284 msec for responding trials and 354 msec for non-responding ones, the difference between the two being significant.

POTENTIAL CHANGES ASSOCIATED WITH MOTOR ACTION, REFLEX RESPONSES AND READINESS

Chairman's Opening Remarks

L. DEECKE

Sektion Neurophysiologie, Universität Ulm, German Federal Republic

Movement-related potentials (MRP) can occur in a number of conditions. They can precede conditioned movements as in the case of the contingent negative variation (CNV); they can precede voluntary, i.e. self-paced movements as in the case of the Bereitschaftspotential (BP)—also referred to as the readiness potential (RP) or N_1 component—and as pre-motion positivity (PMP) or P_1 component; finally they can be associated with the movement itself as in the case of the motor potential (MP) or N_2 component. After movement onset a potential complex occurs consisting of negative and positive components that might be termed reafferent potentials (RAP).

MRP can also be classified into very slow potentials and faster potentials. The two very slow potentials, CNV and BP, are negative shifts of the cortical DC potential. Slow negativity is a frequent phenomenon in the brain, for example it is found to be associated with arousal or alerting activity, recruiting and augmenting responses, and general facilitation processes.

It is tempting to relate negative potentials to excitation and positive potentials to inhibition of the majority of cortical synapses, although it can never be fully excluded that a surface-negative potential may indicate the sink of a positive source in deeper layers. However, let us assume in the case of the very slow potentials usually covering several lobes of both hemispheres that they are related to a general excitation in upper cortical layers (i.e. summated EPSP's in the dense superficial dendritic network), the propagation of which may also involve glial cells (Castel-Lucci and Goldring, 1970; Ransom and Goldring, 1973). We can then interpret the two very slow potentials, BP and CNV, as reflecting a general facilitation process, preactivating those brain regions which will be needed in the special experimental situation under study.

From this point of view it is, therefore, not surprising, that CNV and BP have different distributions on the scalp. The CNV is well developed frontally and tends to decline in the posterior direction, barely crossing the Rolandic line. The BP, on the contrary, has a parietal and precentral maximum and is absent, or can even be positive over frontal areas (*see* experimental demonstration of CNV–BP relationships, *Fig*. 2, p. 215). In the light of the above interpretation, we can say that in the experimental situation of the CNV the frontal lobes are active, whereas they are inactive or even inhibited in the BP situation. Obviously the brain cells of the frontal convexity are not needed for the initiation of a voluntary movement (more precisely: for the type of voluntary movements studied).

What does the frontal cortex do and what then is the difference between the CNV—and the BP—experimental situations? The only safe knowledge we have about frontal lobe function is that in animals delayed response tasks are no longer possible after destruction of the frontal convexity. It is not yet known—the experiment has to be done—whether frontal lobectomized monkeys would fail in the CNV task. However, contrary to the voluntary movement, which is initiated by the subject's internal cues, the CNV situation requires the coordination of internal cues with external stimuli. Since in the frontal convexity, the afferents from the tele-senses (eye and ear) and those from the motivational centres (limbic system, mediodorsal thalamus, Pandya and Kuypers, 1969; Kornhuber, 1974) converge, activity in this brain region preceding CNV—but not BP—movements can be understood.

The other difference in distribution of BP and CNV is over parietal areas. The BP is well pronounced there, whereas the CNV does not go much beyond the central sulcus (*Fig.* 2, p. 215). However, if one introduces visual cues into the experimental situation (as Marsh *et al.*, 1973, did with colour naming), the CNV occurs over parietal areas as well. Thus, it appears that the two very slow potentials—although perhaps only epiphenomena—are faithful indicators of brain activity (negative polarity). This means two things: (1) The differences between CNV and BP, which were stressed by some authors, may be only of an operational nature (i.e. depending on the experimental situation); and (2) the method of recording slow potentials from the scalp, employing multichannel recordings, can tell us where the action is, i.e. which cortical regions are active (excited) and which are inactive or inhibited in a certain experimental task. Planned modification of the experimental situation and delineation of the exact topographical distribution of slow potentials offers an important field for future investigation, although under the artificial circumstances of an experiment the ever-changing thalamocortical focus of activity of our daily life has to be subject to some degree of control to permit recording with our present methods.

The two faster potentials, PMP and MP, are probably the expression of movement-related cell activity changes (or average membrane potential changes in certain cortical regions) which are in closer time relation to the initiation of the movement itself. The PMP with its maximum over the sensory association areas of the parietal lobes may be resolution potential (positive), perhaps caused by the release of what may be called a motor command or a stored program of the 'know how' to make the movement. This assumption is tempting, since a patient with parietal lesions has apraxia, which means that he has lost knowledge of how to perform a certain skilled movement.

The last potential preceding movement onset—and the only one of the three being not bilaterally and widely distributed—is the motor potential. It can be solely recorded over the motor cortex contralateral to the movement. The temporal order of PMP (parietal cortex) and MP (motor cortex) suggests that they stem from subsequent stations in a nervous pathway. But not a direct path; direct cortico-cortical fibres from parietal to motor areas do not exist (Pandya and Kuypers, 1969). Instead, the cortical association areas strongly project to the cerebellum via the cortico-ponto-cerebellar path. The cerebellum in turn projects to the contralateral motor cortex via the cerebello-thalamo-cortical path. A recent theory (Kornhuber, 1971, 1974) thus assumes that the spatial program of the sensory association cortex is converted by the cerebellum (for slow movements by the basal ganglia) into a temporal code and then conveyed to the contralateral motor cortex. Data from single unit experiments in conditioned monkeys (Thach, 1970a, b for the cerebellum; DeLong, 1973, for the putamen; and Evarts, 1970 and 1972a for the ventrolateral thalamus and the motor cortex) are in good agreement with the onset times of PMP (90–80 msec) and MP (60–50 msec), (Deecke *et al.*, 1969).

After movement onset a negative-positive-negative-positive potential sequence is recorded over sensorimotor areas. This can also be recorded after passive movements of the same finger and hand (Kornhuber and Deecke, 1965) and closely resembles the somatosensory evoked response elicited by median nerve stimulation. Probably these potentials reflect reafferent activity from lower motor centres and/or peripheral kinesthetic receptors. They may, therefore, be tentatively termed reafferent potentials (RAP).

Co-chairman's Opening Remarks

D. PAPAKOSTOPOULOS*

Burden Neurological Institute, Bristol, U.K.

A substantial case could be made for asserting that what finally counts for survival is movement or motor control. Historically, the study of movement has been at the root of most major developments in the psychophysiology of behaviour. From Galvani's animal electricity sprang the discovery of membrane and action potentials. From movements elicited by cortical stimulation by Fritz and Hitzig, the cortex of the hemispheres became the focus of research for understanding mental activities (Brazier, 1961). From the variability in RT of the astronomers have developed key concepts such as attention, set (Boring, 1950), psychological refractory period, channel capacity and parallel processing (Welford, 1968). Event-related potential changes in man present a new and potentially important contribution, which has come at a time when the basic ideas of movement organization, which predominated during the first half of this century, are under drastic revision. There has been a shift in emphasis from particulars of neural organization to the examination of whole structural systems and a consequent need for molar electrophysiological data from man.

A motor action, whether reflex or otherwise, involves displacement of a load with a velocity, force and direction, carried out by co-operative contraction of thousands of individual muscle fibre units. An action starts, is supported and finishes in a background referred to as 'posture', provided by co-operation of other muscle groups.

The starting position and the progress of this spatio-temporal coordination is directly or indirectly signalled to the central nervous system. This feedback operation constitutes the sensory aspect of movement and is a fundamental factor in the general notion of the motor sensory system. It is not clear in what circumstances this feedback is necessary, as some complex activities can be initiated and executed without it (Konishi, 1963; Taub and Berman, 1968).

Such findings are damaging to the explanation of movement as interlocked reflexes. 'Reflexology' pre-supposes an input control which, it would appear, does not always exist. Nevertheless, most movement terminology originated from such a scheme. The result has been an oversimplified classification of movements as reflex, spontaneous or voluntary.

Nystagmus after thermal stimulation of the labyrinth is a *reflex* induced movement of the eyeball. The rapid eye movements of the paradoxical phase of sleep could be called *spontaneous* and a shift of gaze could be classed as a *voluntary* movement. The output, the nerves and the motor centres used during these very different behaviours are the same. Nystagmus can be induced by brain damage, in which condition it is stimulus independent. Similarly, rapid eye movements can be stimulus independent and stimulation or ablation of appropriate brain structures can abolish or induce deviation of the eyes, i.e. change of

*Supported by a grant from the Wellcome Trust.

gaze. One must conclude that, although activation of the motor-sensory system can be induced in various ways, the system has its own internal organization. Some movements are apparently pre-programmed, but this pre-organization is not rigid; it can be biased at various levels by other systems such as sensory, cognitive, motivational and emotional systems, which are concerned with the integration of the organism with its environment.

The concept of feedback can therefore be broadened to include not only information about the state of the executing organ, the muscle, but that from the neuronal centres organizing the muscular action. The former tends to be referred to as response, the latter as internal feedback. Knowledge of the results of the action is variously referred to as 'learning feedback', 'reinforcement' or 'extrinsic feedback' (Evarts, 1971).

A much favoured and fruitful research strategy with regard to the motor system has been the 'isolational' approach (Granit, 1970; Matthews, 1972), which has concentrated upon the activity of single muscular or neuronal units. Such an approach is not possible in man, or even in alert animals. In human experimentation movement, posture, intention, feedback and affect are inextricably interwoven.

Although it is possible to elicit and record monosynaptic reflexes in humans Papakostopoulos and Cooper (1973b and p. 114) have shown that they can

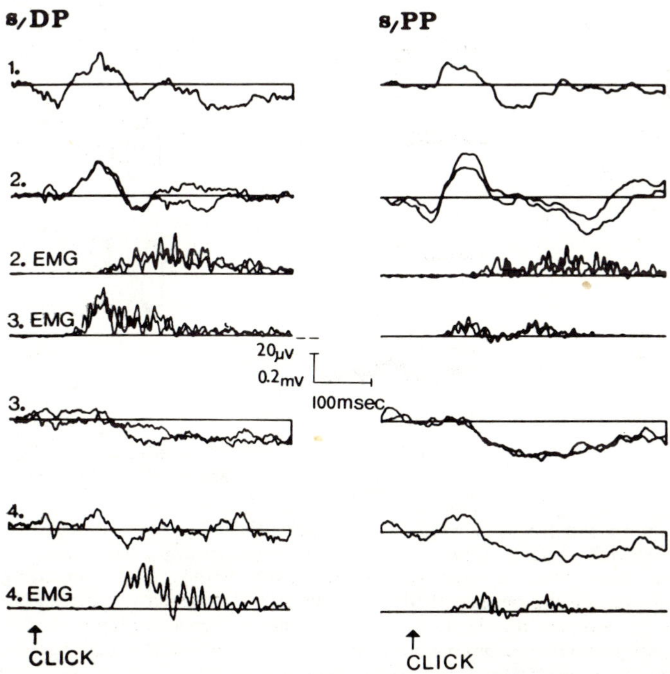

Fig. 1. Vertex auditory evoked responses and forearm EMG in 2 *Ss* (1.) without instructions, (2. and 2. EMG) with instructions for simple RT experiment, and (3. and 3. EMG) with instructions for foreperiod RT experiment. The warning tone preceded the click in 3. and 3. EMG by 1 sec. Note the faster EMG onset and peak for the warned condition (3. EMG in comparison with 2. EMG). The evoked responses on the same occasion (3.) have almost disappeared compared with the unwarned situation (2.). 4. and 4. EMG are selected trials, of the warned condition, with long RT. Note the re-appearance of the evoked responses (N=16).

no longer be regarded as quite such fundamental and autonomous elements. Such examples indicate a need for caution in experimental strategy and in the interpretation of results.

We must be more precise in our ideas about measurement. When we speak of the latency of a simple RT as being 200 msec, this is an oversimplification and could mislead. Usually what is meant is that 200 msec is the time taken by a whole series of procedures which terminate in the closure of a circuit by a microswitch. Peak EMG activity might well be 140 msec for the same event or,

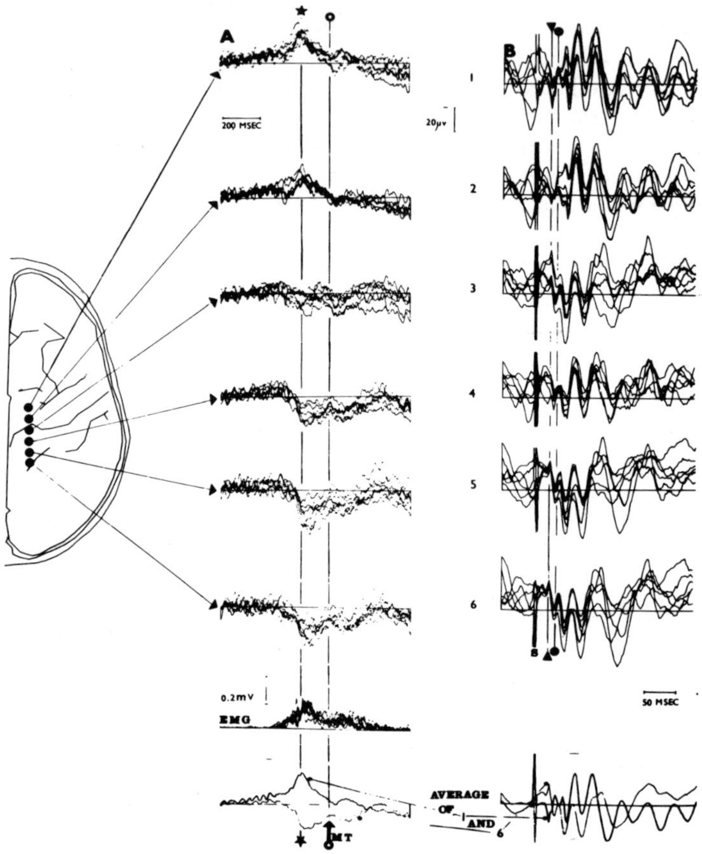

Fig. 2. Superimposed traces (N=16) of slow potentials (*left*) and somatosensory evoked responses (*right*) to R. median–nerve stimulus at the wrist from 6 cortical electrodes over the left hemisphere. Note that the negative peak (indicated with asterisk) preceding movement (MT) does not precede the onset of EMG (rectified) from the active muscle. Note also that this negative peak reverses in polarity in postcentral electrodes as do the early components of the somatosensory evoked responses (indicated with triangle and dot). Bottom traces are averages of activity at electrodes 1 and 6 (N=96).

if one chose to measure the onset of EMG activity, the latency might be nearer 100 msec. Using this last method we have measured RTs of 80 msec for auditory signals and Marsden *et al.* (1972) have values of 50 msec for somatosensory stimuli.

Thus a behavioural measure such as RT can be a misleading oversimplification when divorced from the underlying physiology. But physiology is only one aspect of the problem of motor control, another is the end or context of the action. RTs to an auditory stimulus are shortened if preceded by a warning signal, although the auditory stimulus itself and the act remain the same. Nevertheless, the main negative component of the evoked response to the stimulus diminishes significantly (*Fig.* 1). This reduction tends to disappear if the foreperiod changes fail to produce a shortening of RT (*Fig.* 1, 4. and 4.EMG). It is clear that motor-sensory integration cannot be divorced from its context or its final outcome. Adequate controls and monitoring are possible in man but a more rigorous approach is essential.

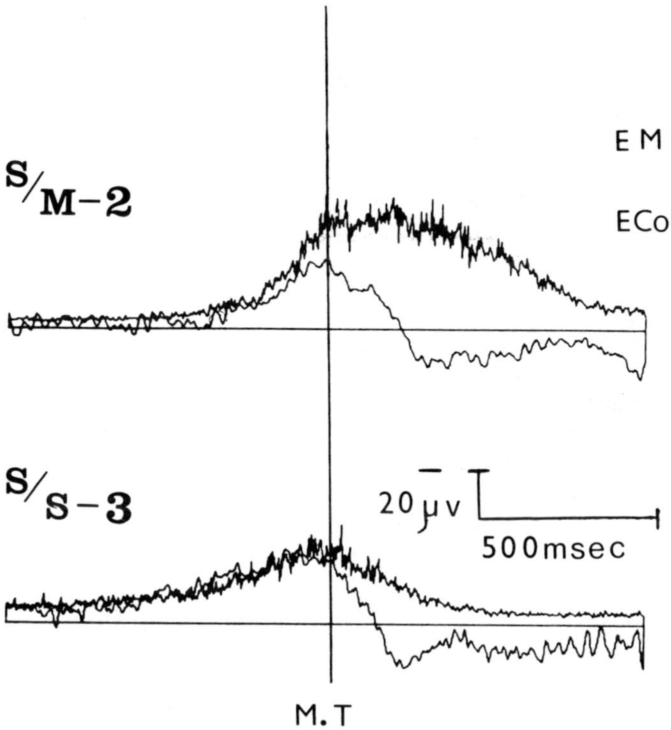

Fig. 3. Cortical activity from the right precentral gyrus preceding and following movement (MT) of the left index finger, for 2 *Ss* (M–2 and S–3). Note that the main cortical negativity and EMG are in parallel up to the point of movement; thereafter they diverge.

Before speaking of motor potentials we should define whether they reflect the cause, the result, or correlated—but irrelevant—changes of the movements. If they reflect the cause of movement, then which aspect: its initiation, its timing, its force, its control in space or its postural support? If these potentials are the result of movement as indicated by the recordings of Gerbrandt *et al.* (1973) from scalp and by our cortical recordings (*Fig.* 2), in what sense can we understand them? Do they originate from activity in muscle receptors, joint receptors or only skin receptors, as our subject presses the button? Do they possibly have little to do with the movement in progress, but rather with some aspect of its future or its past? Do they reflect cortical, subcortical or cerebellar aspects of the movement, afferent or reafferent? In this last context, even to speak of reafferent activities is not enough. *Fig.* 3 shows that the peak of the precentral negativity coincides with the peak of the EMG in a self-paced index finger flexion, but that the EMG activity continues while the cortical negativity changes to positivity. Which then is the reafferent potential? Or are both perhaps reafferent potentials but from different structures?

There remains another question to be answered, namely that of afferent or efferent transformations (Phillips, 1969; Brooks and Stoney, 1971; Asanuma and Rosen, 1972; Porter, 1972). How does the motor cortex translate its input and how do the subcortical structures translate the commands received?

There has already been a marked shift of emphasis in motor control research from static and isolated events to the dynamic and complex organization of movement in alert animals or man. There has also been a major change in the field of sensorimotor integration. A re-evaluation is in progress of the whole concept of motor activity and its place in understanding brain function in both physiological and psychological terms. Evarts (1973) recently expressed it as follows: 'Indeed, it seems possible that understanding of the human nervous system, even its most complex intellectual functions, may be enriched if the operation of the brain is analyzed in terms of its motor output rather than in terms of its sensory input.'

This echoes Sperry's remarks some 20 years earlier: 'Instead of regarding motor activity as being subsidiary, that is, something to carry out, serve, and satisfy the demands of the higher centers, we reverse this tendency and look upon the mental activity as only a means to an end, where the end is better regulation of overt response. Cerebration, essentially, serves to bring into motor behavior additional refinement, increased direction toward distant, future goals, and greater over-all adaptiveness and survival value. The evolutionary increase in man's capacity for perception, feeling, ideation, imagination, and the like, may be regarded, not so much as an end in itself, as something that has enabled us to behave, to act, more wisely and efficiently.'

Bereitschaftspotential Preceding Voluntary Slow and Rapid Hand Movements

W. BECKER, K. IWASE, R. JÜRGENS and H. H. KORNHUBER

Sektion Neurophysiologie, Universität Ulm, German Federal Republic

Recently a new theory of the neural organization of voluntary movements has been presented by Kornhuber (1971). The theory postulates different neural schemes for rapid (ballistic) movements which are thought to be preprogrammed, and for slow (smooth) movements which are ascribed to a ramp generator. The present study was undertaken to find out whether the differences postulated by this theory are reflected in the brain potentials preceding slow and rapid hand movements.

METHOD

Ss were seated in an electrically shielded room and held the grip on the end of a rod of 10-cm length with their fingers, the elbow resting on a table for support. They were required to push the rod into a 10-cm tube at regular intervals and to vary the speed of these movements randomly so as to achieve rapid ballistic movements as well as smooth slow ones. All movements had to be of approximately 10-cm length and were purely voluntary. *S's* head was supported on a chin rest and a forehead holder to prevent head movements synchronous with the hand movements. In addition, *Ss* were trained to relax all muscles but those involved in pushing the rod.

The EEG was recorded from positions C_z C_3' and C_4' (1 cm anterior to C_3 and C_4 respectively) by means of Beckman adhesive electrodes. The linked ears served as a reference. The EEG was amplified on a Siemens Mingograph having a 1·6 or 5·0 sec time constant, and was stored on magnetic tape together with the position of the rod, the maximum speed reached during the movement, various control data, such as EOG and EMG, and a motion-triggered pulse. The latter was elicited as the rod crossed the first millimetre of its total course. With the time resolution of the subsequent averaging (20 msec/point) no delay could be detected between the start of the rod movement and the trigger pulse, no matter whether slow or rapid movements were considered. Artefacts and erroneous triggering were rejected on-line.

The procedure for data analysis was essentially that of time-reversed averaging (Kornhuber and Deecke, 1964) using a BIOMAC Model 1000 averager. An additional electronic device categorized the movements as to speed and gated only those into the averager which corresponded to a preset range. The range of slow movements was set to be below 10 cm/sec and the fast range was above 110 cm/sec. The sweep time was 5·12 or 10·24 sec and a total of 32 or 64 sweeps were averaged. Ten *Ss* have been investigated at this time.

RESULTS

In all cases, a clear Bereitschaftspotential (BP) was observed (*Fig.* 1). For ballistic movements, it starts about 0·8 sec prior to the onset of the movement whereas for slow movements it precedes the movement by 1·3 sec (*Fig.* 2A). This latency difference between slow and rapid movements was highly significant at all 3 positions tested (*t* test for paired values, $P \leqslant 0{\cdot}01$). No significant latency differences were observed between different leads.

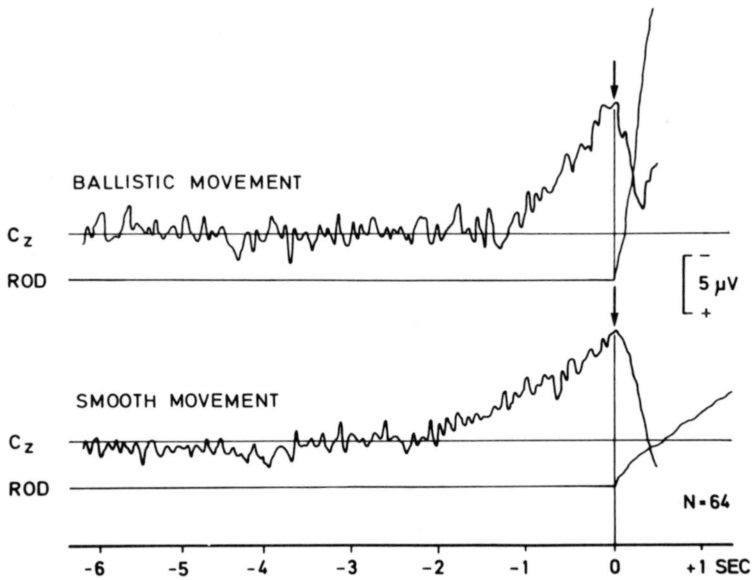

Fig. 1. Brain potentials preceding rapid ballistic movements (upper plot) and slow smooth movements (lower plot) in the same subject. Upper trace of pair shows average potentials at the vertex (TC=5 sec, 64 sweeps), lower shows rod positions. In both cases there is a clear Bereitschaftspotential.

The amplitude of the Bereitschaftspotential was read from its maximum shortly before or coincidently with the start of the movement and was referred to a baseline representing the average of a 2-sec interval preceding the movement by 2·5 sec. The maximum amplitude for both slow and fast movements was at the vertex electrode. Preceding slow movements, the amplitude of the BP was generally higher than when preceding rapid movements (*Fig.* 2A). This amplitude difference was found to be significant in recordings from the precentral regions contralateral to the moving hand (Wilcoxon test for ranged matched pairs, $P \leqslant 0{\cdot}05$). A similar, but insignificant difference is observed in recordings from the vertex and from the ipsilateral precentral region.

Since the latency difference suggests that it might take more time to prepare for a smooth movement than for a rapid one, it was felt desirable to compare the latency of smooth and rapid movements upon a command signal. In order to do so, *Ss* were given a short auditory stimulus at irregular intervals using the same set-up as for the previously described experiment. *Ss* were required to react as fast as possible to these command stimuli by pushing the rod into the tube. Again, the speed of pushing was to be varied randomly from movement to movement, and RTs, as measured from the onset of the stimulus to the start of the movement, were averaged for ballistic and smooth movements separately. In agreement with the longer brain potential, RT to auditory stimuli was significantly longer (paired *t*-test, $P \leqslant 0{\cdot}01$) than for fast movements (*Fig.* 2B).

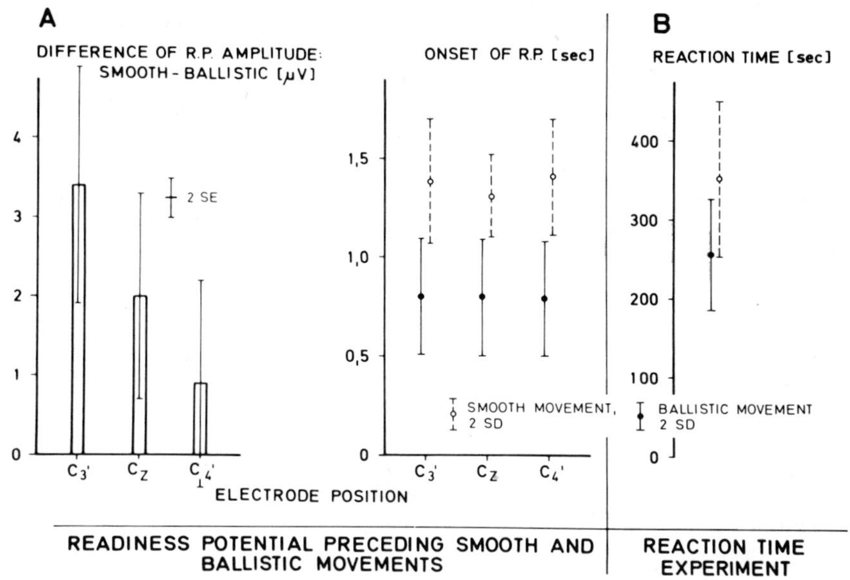

Fig. 2. A, Bereitschaftspotential (=readiness potential, RP) preceding slow smooth and fast ballistic movements. Average from 10 subjects. *Left:* Difference of RP-amplitude preceding smooth movements minus amplitude preceding ballistic movements at different electrode positions. Vertical lines inside the bars indicate standard error. C_3' and C_4' are 1 cm anterior to C_3 and C_4 as defined by the 10–20 system. *Middle:* Onset of RP preceding smooth movements (open circles) compared to onset for ballistic movements (filled circles). Vertical lines give standard deviation. B, Reaction time to auditory command signals for the execution of smooth movements (open circles) and of ballistic movements (filled circles). Average of 5 subjects. Vertical lines give standard deviation. The latency is measured between the onset of the stimulus and the instant when the rod has crossed the first millimetre of its course.

DISCUSSION

The start of the Bereitschaftspotential about 0·8 sec prior to rapid ballistic movements is consistent with the results of Deecke *et al.* (1969) for rapid finger flexions and of Becker *et al.* (1972) for rapid (saccadic) eye movements. The fact that the negative potential shift begins some 0·5 sec earlier if the movement is slow and smooth suggests that it takes more time to prepare for voluntary smooth movements than for ballistic jerks. This idea receives some support from our RT experiments. The RT observed for rapid hand movements is within the range of values reported in the literature. The increase of latency by about 100 msec in the case of smooth movements indicates that the initiation of such a movement is probably more complex than the preparation for the jerk, even if the amplitude of the jerk is to be fairly accurately preprogrammed, as was the case in our experiments. Another fact that points to this conclusion is the larger amplitude of the BP in the case of slow movements. The maximum difference in BP amplitudes is found over the precentral area contralateral to the moving hand. Whether this difference is due to the BP solely or whether it is caused also by different motor potentials (as defined by Deecke *et al.*, 1969) cannot yet be answered since the present study was not aimed at investigating the details of potential changes immediately before the onset of the movement (pre-motion-positivity and motor-potential). Such an investigation would require an extensive survey of EMG latencies to insure that the start of the rod movement has no variable delay with respect to the earliest change of EMG activity which could distort average potentials of short duration. Our results are consistent with the idea that the neural organization of voluntary slow movements is different from that of fast movements. It would be interesting to carry out a similar experiment using a CNV paradigm. On the basis of the results reported here, we would predict a higher CNV amplitude if the conditioning stimulus is to be responded to by a smooth slow movement than if a ballistic movement has to be performed.

SUMMARY

The Bereitschaftspotential (BP) preceding voluntary self-paced slow and rapid movements of the hand was investigated. Preceding slow smooth movements the BP starts 0·5 sec earlier and tends to have larger amplitudes than for rapid ballistic movements. If the movements are not self-paced, but elicited by an auditory signal, the RT to the signal is 100 msec longer for slow movements than for rapid ones. It is suggested that these differences reflect a different central organization of slow and rapid voluntary movements.

EEG [illegible]gation of Hemispheric Asymmetries Precedin[illegible] [illegible]peech: the R-wave

B. GRÖZINGER, H. H. KORNHUBER and J. KRIEBEL

Sektion Neurophysiologie, Universität Ulm, German Federal Republic

In our investigations of brain potentials preceding speech, with EEG scalp recording we detected respiration-correlated potentials, with frequencies of about 0·25 Hz (Grözinger *et al.*, 1972, 1973). As far as we know this is the first study of respiration-correlated slow brain potentials using averaging techniques. In routine EEG tracings for clinical purposes these potentials have remained undetected because of the fast (0·3 sec) time constant used. In recordings with longer TCs they were generally assumed to be artefacts.

METHOD

From a total of 31 normal human *Ss* in 93 separate experiments, the EEG, EOG, phonogram (throat microphone) and nasal respiration were recorded preceding speech, phonation and articulatory movements. Recordings were made from the vertex and from bilaterally symmetrical electrodes which were placed so as to overlie the speech area on the dominant hemisphere. The trigger signal was either the onset of EMG of the orbicularis oris muscle for speech and articulatory movement (in contrast to McAdam and Whittaker, 1971) or the onset of humming as detected by throat microphone.

The movement-related EEG potentials were analysed by time-reversed averaging (Kornhuber and Deecke, 1965) using a DEC PDP-12 digital computer (DECUS PROGRAM No. 12–1, modified by K. Murata). All measurements were referred to a 'normalized' baseline. Head and lead movements were effectively eliminated via a special head holder or a bite board. Segments of recording containing artefacts such as eye movements were rejected during analysis. Handedness was established by test and history.

RESULTS

In contrast to the relatively simple forms of potentials preceding limb movements, more complicated patterns prior to speech were discovered. In particular a periodic slow potential was discovered, which exhibited a high degree of correlation with respiration and which has therefore been termed the R-wave.

Fig. 1 shows typical R-waves. In the left part of the diagram the *S* performed articulatory movements and in the right part he hummed without articulatory movements. The sweep time was 8 sec and the vertical lines define the onset of activity. Definite lateral asymmetries can be seen in this example; before articulatory movements the potentials are more pronounced over the right hemisphere and before humming over the left. This hemisphere difference of the R-waves parallels an alpha and theta asymmetry in the original EEG recordings. In the lower half of *Fig.* 1 the less pronounced R-waves correspond to an almost complete block of alpha and theta waves in the original tracings, although contralaterally these frequencies are clearly seen. Bearing in mind the well known correlation between vigilance and reduced alpha-amplitude, one might hypothesise an inverse correlation between vigilance and the R-wave. For this *S* (right-handed) one could conclude that, prior to articulatory movements, the left hemisphere is more active and, prior to humming, the right. These results, probably, correspond to a lateralisation of these activities. This finding was reproduced for this *S* in two experiments.

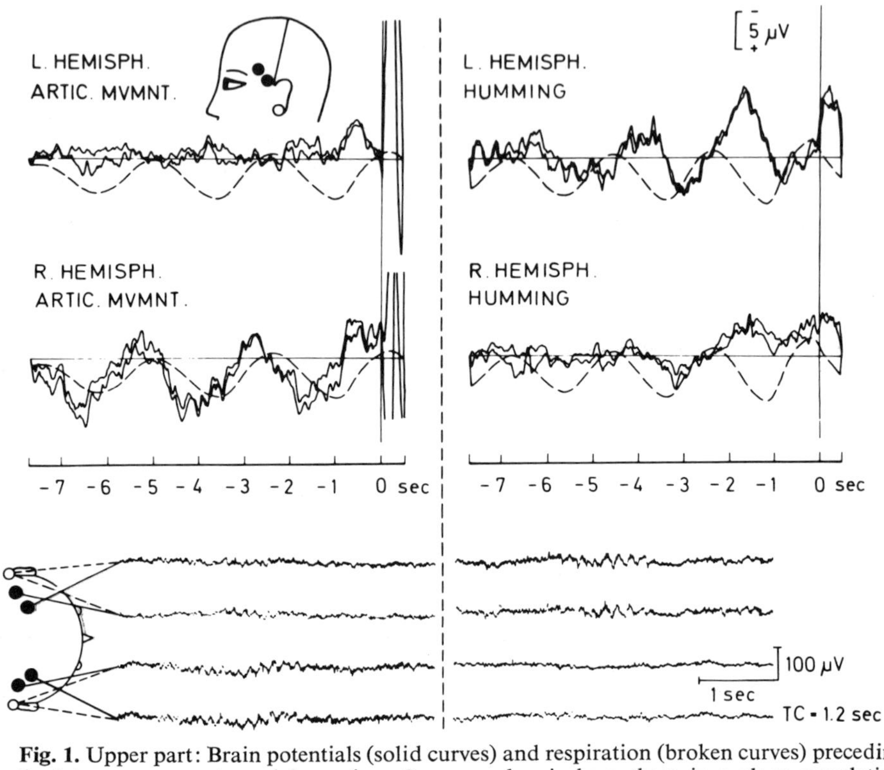

Fig. 1. Upper part: Brain potentials (solid curves) and respiration (broken curves) preceding articulatory movements and humming. Over one hemisphere there is a clear correlation between the average respiratory curve and the wave-like brain potentials (R-wave) whereas the opposite hemisphere shows almost no correlation. This hemispheric asymmetry may reverse depending on the task to be performed by the subject. In this example, the R-wave appears over the right hemisphere preceding articulatory movements, and over the left hemisphere prior to humming. Average of 250 trials, TC=1·2 sec. Lower part: Primary EEG trace from the same subject. There are marked hemispheric asymmetries of the α- and θ-amplitudes which correspond to that of the R-wave: The α- and θ-rhythms disappear over that hemisphere where no R-waves are observed.

The amplitude of the R-waves is often so large as to mask the other brain potentials prior to the voluntary activities (as in *Fig.* 1), making measurements of amplitude and latency impossible. In *Fig.* 2A, another example of bilaterally asymmetrical R-waves is depicted. In this case the sweep time was 4 sec. The test word was 'Lotte'. The *S* was right-handed. A comparison of corresponding electrode positions over both hemispheres shows pronounced R-waves over the left side. The right hemisphere, in addition to a smaller R-wave, exhibits a slow negative potential commencing about 1·8 sec before movement (upper diagram of *Fig.* 2A) which could be a BP. Over the left hemisphere the R-wave may mask an additional positive potential near the onset of EMG. Positive and negative potentials similar to the BP were also found in other experiments. The differences between the potentials can even lead to a polarity reversal between the hemispheres in some *Ss*. In contrast to *Fig.* 1 the results in *Fig.* 2A are not in good agreement with hemispheric dominance. In *Fig.* 2B, brain potentials prior to the test word 'Puppe' are shown. As is typical for most of our experiments, the vertex potential had the greatest amplitude. About 350 msec before movement, a sharp additional negative potential appears over the vertex. The other electrodes recorded laterally asymmetrical R-waves. High amplitudes are only seen over the right hemisphere. The most frontal electrode registered an additional slow positive potential.

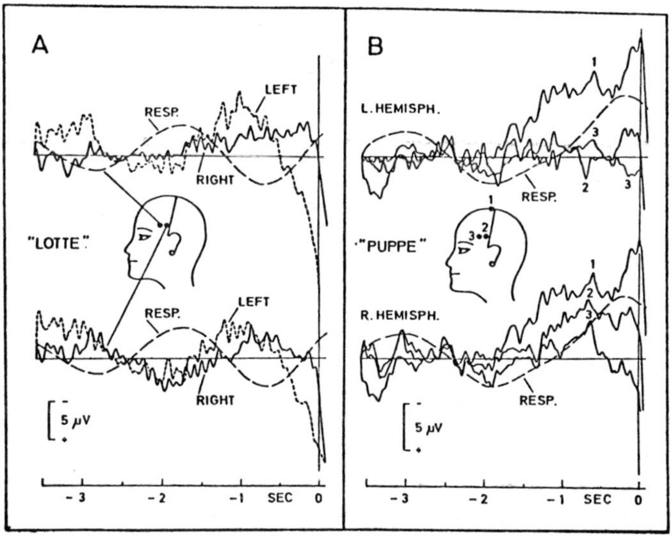

Fig. 2. Brain potentials and respiration preceding the test words 'Lotte' (A) and 'Puppe' (B). On both figures respiratory curves are given by broken lines. A, Solid curves indicate potentials from right hemisphere, dotted curves correspond to left hemisphere. Note small negative potentials at the more anterior electrode on the right hemisphere and pronounced R-wave at the posterior electrode on both hemispheres. B, Upper part is for left hemisphere, lower for right hemisphere. (The vertex potential is identical for upper and lower plot.) Electrode positions indicated by corresponding numbers. Note hemispheric asymmetry of R-wave at positions 2 and 3: no R-wave at the left hemisphere.

DISCUSSION

Preceding voluntary movements of the limbs in man, Deecke *et al.* (1969) have reported three characteristic brain potentials:

1. The almost bilaterally symmetrical negative Bereitschaftspotential (BP).
2. The bilaterally asymmetrical pre-motion positivity (PMP).
3. The unilateral motor potential (MP).

These potentials show a fairly good inter- and intra-individual constancy, in comparison to those preceding speech, and are meaningfully interpreted as bioelectrical correlates of neuronal motor events. These authors (Kornhuber *et al.*, 1965) did not find slow wave patterns. The reason may be that the onset of speech, in contrast to limb movements, is correlated with respiration. In limb movements the phase relationships are independent of each other and therefore the average amplitude of the R-wave actually decreases. The time courses of brain potentials prior to the onset of speech show quite definite interindividual variability and good reproducibility for a given *S*. We find, for example, positive and negative potentials similar to BPs with latencies of up to 2 sec, but with distinct hemispherical amplitude differences, which might correspond to the dominance of one hemisphere. There are not as yet sufficient criteria to identify the more active hemisphere in every case. The upper curves in *Fig.* 2A, for example, show a negativity similar to the BP only over the right hemisphere. There is no obviously comparable potential over the contralateral hemisphere and this could be interpreted as indicating the right hemisphere to be more active. The *S* in this case was right-handed, and we cannot offer a satisfactory explanation. Moreover, the recognition of the already mentioned positive or negative potentials is complicated by superimposed R-waves.

Because speech and respiration are coordinated in a complex way and speech involves cortical activities, one may assume that the rhythm of respiration during speech is influenced by cortical impulses. If respiration during speech is under cortical control, we might ask if the R-wave represents a bioelectrical correlate of the cortical respiratory coordination. In this context, the topography of the R-wave is of interest. The most pronounced R-waves are often recorded from the vertex. Over Broca's area and corresponding contralateral regions the waves are often asymmetrical although large in amplitude. Temporally, parietally and occipitally, they are much less pronounced and normally symmetrical.

That mechanical artefacts such as head- or lead-movements could be the cause of wave-like potentials was ruled out by detailed control experiments (Grözinger *et al.*, 1973).

The GSR can also be ruled out as a source of the R-wave on the basis of control experiments and experiments with cats. Nevertheless, the GSR is an undesired superimposition in some cases (Grözinger *et al.*, 1973). Respiratory pressure changes of the CSF are unlikely sources. It is more likely that the R-wave is at least partially dependent on cortical neuronal activities caused by both afferent and efferent impulses. The coordination of speech and respiration has already been mentioned and respiration naturally activates the mechanoreceptors and their cortical projections.

Another possible mechanism might be small changes in P_{CO_2} of tissues in the cortical region caused by respiration. That changes in P_{CO_2} lead to changes of cortical DC-potentials is well known from investigations by Caspers *et al.* (1963). The above mentioned interpretations do not as yet explain the observed laterality of the R-wave (*see Fig.* 1). The correlation of R-waves and asymmetries of alpha-waves in the original tracings might correspond with the functional laterality. The possible reason for the hemispheric differences of the R-wave amplitude itself may be found in different neural activities in the dominant and non-dominant hemisphere during speech and similar activities. The different neural activities may again result in local cortical changes of metabolism and blood supply as demonstrated by Lassen and Ingvar (1972).

The variety of slow potentials with positive and negative time courses or even with polarity reversal between the hemispheres are in accordance with the complexity of speech and require further investigation. The positive and negative time courses preceding speech, phonation and articulatory movements are cortical preparation processes similar to the BP before voluntary limb movements. From our results it is still questionable whether the nomenclature of the brain potentials preceding limb movements can be used for speech as well.

SUMMARY

1. Preceding speech, articulatory movements, and phonation, slow waves (R-wave) correlated with respiration were found by time-reversed averaging of chronologically stored EEG recordings, triggering from the onset of EMG (orbicularis oris muscle).

2. R-waves often show marked hemispheric asymmetries in their amplitude. This points to a corresponding difference in correlation between respiration and cortical preparatory processes. We suggest that these hemispherical differences might be a bioelectric correlate of hemispheric dominance.

3. In addition we found slow brain potentials of both negative and positive polarity (referred to the mastoids). Compared to the time course of potentials preceding voluntary limb movements (BP, PMP, MP) these potentials are more complex. They start as early as 2 sec prior to the investigated event and show marked hemispheric asymmetries like the R-wave.

4. Although extremely variable between *Ss*, the slow potentials exhibit only small changes in the same *S* in different experiments.

ACKNOWLEDGEMENT

The authors thank Mrs U. Becker for preparing the diagrams.

Preliminary Observations on the Effects of Response Parameters on Pre-response Potentials

E. DONCHIN and M. KUTAS

Department of Psychology, University of Illinois, Champaign, Illinois, U.S.A.

This is an interim report of an investigation into the effects which response parameters may have on the slow pre-response cortical negativity which is sometimes labelled the Readiness Potential (RP). It seemed to us that if these potentials reflect preparatory activity by the motor control centres, in anticipation of the execution of a response, then the characteristics of the anticipated response should affect the amplitude, or the cortical distribution of the potentials. It is also possible that these potentials are CNVs recorded over intervals defined by intra-subject signals rather than by external stimuli. It has been demonstrated that CNVs can be recorded in the absence of a motor response (i.e. Donchin *et al.*, 1972); it is difficult therefore to view CNVs as solely reflecting motor preparedness. Our intention has been to conduct a factorial experiment varying both the cognitive and the motor parameters of the task, and thus to determine the effects of such variations on pre-response potentials. At this time only part of the data has been collected and analysed.

METHOD

This investigation had its genesis in an attempt to derive information relevant to the interpretation of the RP from data collected in a different investigation (Donchin *et al.*, 1973b). In a study focused primarily on P300 we presented *Ss* with sequences of choice RT trials. *S* was to press a button with either his right or left thumb, depending on the outcome of the trial. Outcomes were presented in different sequences, in some series the outcome being completely predictable, in others completely unpredictable. If the outcome is predictable the responding hand is also predictable. A 'motor' interpretation of the pre-response potentials should predict a different cortical distribution of pre-response potentials during predictable and unpredictable series. We reasoned that when *S* knows the hand he is to respond with preparatory activity should appear unilaterally. When both hands may have to respond a more symmetrical distribution should be expected. This however was not the case. The distribution of the two potentials along the interaural line was the same for predictable and unpredictable series (*see* Donchin *et al.*, p. 76). However, the microswitch depressed by *Ss* required a force of about 450 grams for a full depression. It was possible (Otto, personal communication) that the 'motor' characteristics of these potentials may be brought out by requiring a more forceful response. It is in the attempt to resolve this issue that we developed response-devices with which the parameters of the response could be adequately measured. Two different devices were constructed and will be discussed separately.

This study is supported by a grant from the University of Illinois Research Board, as well as by NSF Grant 6B–30999.

Series I: Work and the Slow Potential (squeezing a spring)
Response device: Two padded vertical poles protruded from the subject's arm rest. The subject held his palm against the proximal pole and his fingers around the more distal pole. The two poles were attached to a heavy steel spring. The distal pole could be moved towards the proximal pole thereby depressing the spring. A microswitch was installed in such a way that it could be tripped by the moving vertical pole. A successful squeeze was defined by the tripping of the microswitch. By varying the position of the microswitch with respect to the moving pole the force with which *S* was required to squeeze could be varied. Of course, force requirements in this device were confounded with the displacement requirements. Thus, strictly, we were varying the *work* required of the *S* rather than the *force* which he had to generate.

Ten *Ss* were paid for participating in the experiment. Beckman cup electrodes (silver–silver chloride) were placed at C_Z, O_1, C_4, C_3, T_4, T_3. *Ss* were instructed to squeeze on the bars until they heard the audible click of the microswitch. An intersqueeze interval of 3–4 sec was maintained. Each *S* generated three 50 trial series with each of three different work levels. In addition, they each pressed 50 times with each thumb on the button press device used in our previous experiments.

RESULTS

The results obtained with this primitive response gadget were rather inconclusive. They are discussed here mainly because of the results we obtained with a more sophisticated device. We note here two aspects of the data. The records for C_3 and C_4 were superimposed and the degree of asymmetry determined by visual inspection. When *S* responded with the right hand we found an asymmetry in 1 out of 9* *Ss* for the button press, 4 out of 9 for the lowest work level, 4 out of 8 for the medium work level and 5 out of 7 for the hardest level. Similar results were obtained for the responses with the left hand. The potentials were larger at the contralateral hemisphere. The work level requirement had a clear effect on the amplitude of the components. The harder, and more displaced, the squeeze, the larger were the amplitudes of the potentials recorded at C_3 and C_4. However, the data showed considerable intersubject variability and it would be difficult to accept these as more than preliminary results.

There were many evident difficulties in the technique described above. The force level requirements could not be accurately set and calibrated. The force requirements were confounded with the displacement as well as with the time required by *S* to respond. We have therefore developed a system without these drawbacks.

Series II: The Electronic Dynamometer
Response device: Two metal hand grips were each connected to an LVDT force transducer (Daytronic Model 152 A). These transducers provide a precise and high resolution measure of the force applied to them. The maximal displacement of the transducer is 0·001 in and the displacement is independent of the applied force. The transducer provides a continuous output proportional to the applied force and is energised electronically. It provides a continuous display of

---

*Data from subjects generating excessive eye movements were discarded. Hence the unequal number of subjects.

its output as well as allowing a rapid setting of force limits which activate a Schmitt trigger. One of the problems we encountered in the use of the previous grip device derives from the fact that *Ss* differ widely in the ease with which they generate different force levels. We now define force levels not in terms of their absolute level but rather as percentages of the maximal force level a *S* can generate at the beginning of a session. Having determined for each *S* his maximal level he was then tested at 25 per cent, 50 per cent and 75 per cent of this level.

Ss (12, all male, 7 right handed) sat in a comfortable chair viewing an illuminated circle in a Tachistoscope. The circle was superimposed on an illuminated square which was always on. Whenever *S* reached the force level set for a given series the illuminated circle disappeared. Thus from *S*'s point of view his task was to operate a switch extinguishing the circle. (An equivalent series was subsequently run in which no such feedback was provided.) Scalp electrodes were placed at C_Z, C_3, C_4 as well as 2 and 4 cm lateral, and 4 cm anterior to either side of the vertex. Recording was referential to linked ear electrodes. In addition the EOG and EMGs from either the left or the right arm were recorded. All data were recorded with Brush amplifiers (band pass 0·1–30 Hz) and analysed by an IBM 1800 computer. Trial series of 50–100 trials each were obtained for each hand, for each of three force levels.

RESULTS

All conclusions at the present are based on visual inspection of superimposed records averaged for each *S* and obtained from right-handed *Ss* only. With these restrictions the results may be very simply summarized. We find that the amplitude over the left motor cortex of the potentials (all components) are larger than the corresponding potentials obtained over the right motor area, when *S* responds *with his right hand.* This asymmetry between the hemispheres disappears when *S* responds with his left hand. The above result appears at all force levels.

In general, we find that the force level has no effect on the amplitude of the potentials, the amplitudes at all three levels used, for each of the right-handed *Ss* is identical. We also note that the presence or absence of the feedback stimulus at the termination of the movement has no immediately obvious effect on the potentials.

DISCUSSION

As the data from left-handed *Ss* has not yet been analysed it is not possible to tell if the asymmetry observed for the right hand response is characteristic of right hand responses in general or is associated with the response of the dominant hemisphere. Either way, the result will have important implications for the interpretation of these slow potentials. Equally puzzling is the fact that force level fails to affect the potentials, particularly since work-levels did affect the potentials in our first series of *Ss*. Due to the crudeness of the initial response device the early data must be accepted with caution. A systematic investigation of the effects of displacement and force pre-response potentials is required. Finally, we have still to reach the stage of our study in which the force level, and the work level, are examined in a CNV producing paradigm, namely when *S* operates the dynamometer in response to an imperative stimulus, preceded by a warning stimulus.

Somatosensory Evoked Potentials, Attention and Voluntary Self-Paced Movements

P. HAZEMANN and F. LILLE

Laboratoire de Physiologie (Pr. J. Scherrer). CHU-Pitié-Salpêtrière, 91 Bd de l'Hôpital, Paris 13, France

Previous work (Lille, 1973) demonstrated attenuation of the somatosensory evoked potential (SEP) during a variety of discrimination tasks (visual or auditory) involving motor responses. In these tasks *Ss* were required to discriminate between two signals of weak intensity responding with the appropriate left- or right-handed bar press. During this time somesthetic stimulations were delivered and the corresponding EP studied. It was observed that the amplitude of the EP decreased during the discrimination and that a similar decrease appeared when *Ss* performed simple repetitive movements without discrimination. To distinguish between the effects of discriminatory and motor involvement *Ss* also performed the discrimination task without a motor response, but were required to report the number of stimuli occurring of either type. Again the EP decreased during the discrimination task.

To define further the effect of a motor response the present study was designed to identify the temporal changes in the SEP associated with a self-paced motor task.

METHOD

Seven normal *Ss* received median nerve stimulation at approximately 2-sec intervals during seven 5-min trial periods. In each of the trial periods *Ss* performed a repetitive (self-paced) bar-pressing task at intervals exceeding 4 sec. These movements were performed ipsilateral to the nerve stimulus. Control SEPs were recorded for 5 min before the experiment began, and again for 5 min after the experiment. Five-minute periods of inactivity without stimulation separated the test periods.

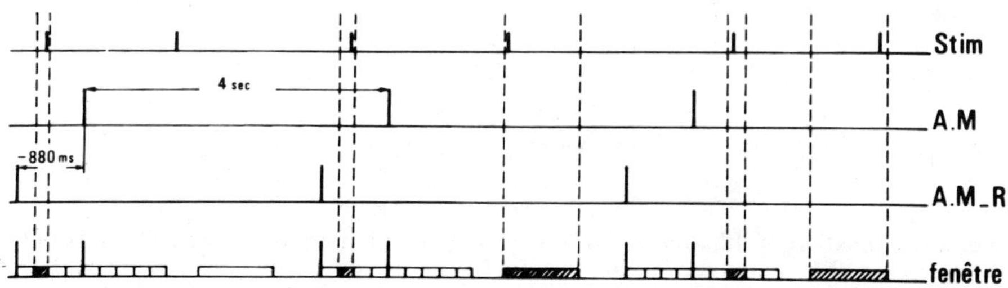

Fig. 1. Experimental paradigm. Stim: nerve stimulus marker. A.M: bar press contact closure. A.M_R: contact closure re-recorded with a lead of 880 msec. fenêtre: represents the successive 220 msec time windows which permitted the averaging of SEPs according to their time of occurrence relative to movement. Only the 'nerve stimulus marker' arriving within a specific window was taken into account for the SEP average corresponding to that epoch.

The EEG, which was recorded from a scalp electrode over the somatosensory area contralateral to the nerve stimulus referred to a linked-ear reference, was

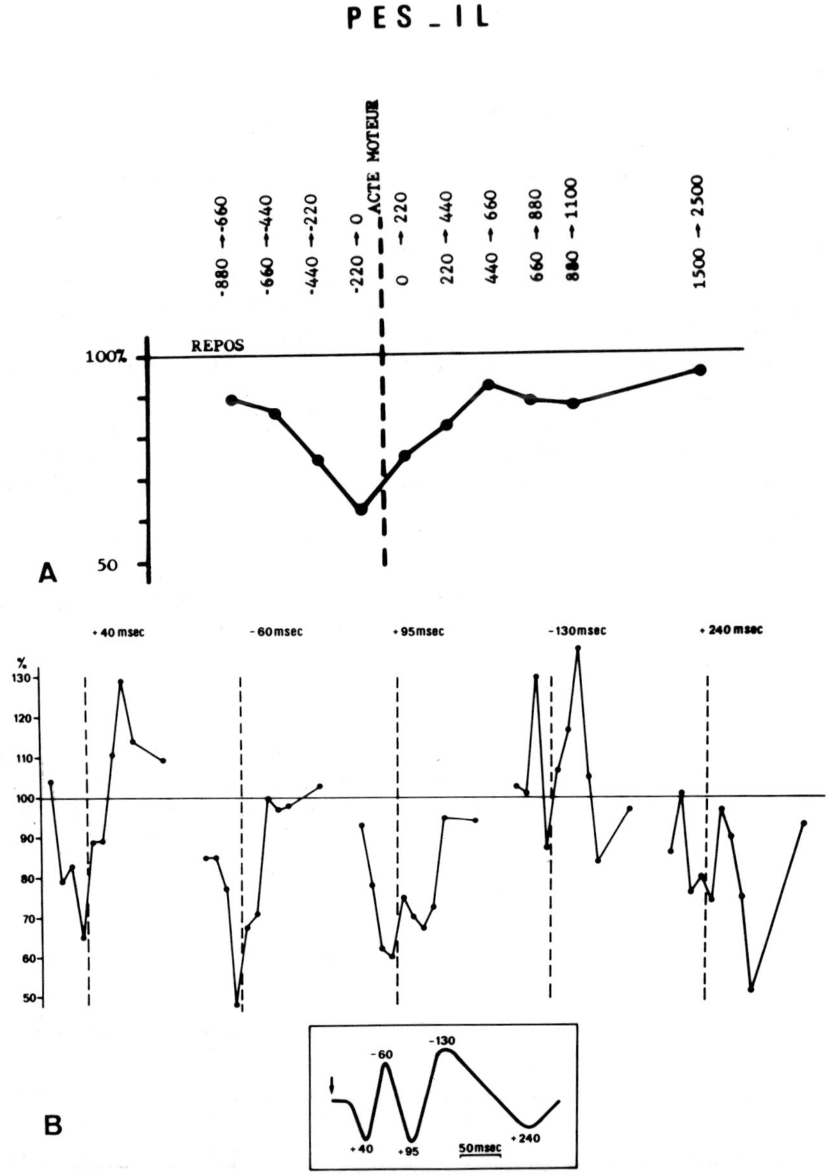

Fig. 2. Attenuation of the averaged SEP relative to a voluntary motor act. The horizontal line represents SEP amplitude during rest (mean of the pre- and post-experimental rest sequences). The vertical line represents the time of bar-press contact closure. The successive points represent the SEP amplitude in the successive periods of 220 msec. The values are expressed in percentage of the rest value. A, Modulation of the total SEP amplitude (total amplitude: summation of successive peak/peak amplitudes for SEP components averaged within 300 msec of the nerve stimulus). B, Modulation of SEP component amplitudes for peaks occurring at 40, 60, 95, 130 and 240 msec after nerve stimulation.

recorded with two time-constants; a time-constant of 0·3 sec was used for averaging the SEP and a time constant of 2·0 sec was used for averaging readiness and motor potentials.

Contact closure for the bar press was recorded; then, during a backward replay, it was re-recorded to derive a time marker preceding actual closure by 880 msec. This marker and the nerve stimulus marker were used to derive the following averages (*Fig*. 1).

SEPs (N=40) were selectively averaged according to their temporal relationship to bar pressing. Nine averages were constructed for stimuli occurring within successive 220-msec epochs, 4 preceding and 5 following the bar press. A tenth average was computed for stimuli occurring between 1·5 and 2·5 sec after the bar press.

Average SEPs were measured by summation of successive peak/peak amplitudes occurring within 300 msec of the trigger. These scores are presented in *Fig*. 2A with respect to the average SEP derived from the pre- and post-test periods. Individual component amplitudes are shown in *Fig*. 2B.

RESULTS

Summation of component changes clearly demonstrates that the SEP was attenuated by the temporal proximity of a voluntary motor act. This attenuation was greatest in the 220-msec epoch just preceding switch closure.

Modulation of the individual components was generally similar to those of the summated score. Only the late (+240 msec) component failed to reach its minimum just prior to switch closure, while the −130 msec component was the only component which was regularly augmented by the motor activity.

DISCUSSION

Mechanisms responsible for modulation of the SEP in our experiment are open to speculation. Coquery *et al.* (1972) studied SEPs and motor activity, but obtained somewhat different results. They found an increase in the SEP prior to the onset of electromyographic activity and SEP diminution following contraction when the motor act and stimulus were ipsilateral. When the motor act and stimulus were contralateral the SEP increase persisted during and following the contraction. We have never observed an enhancement of the SEP related to the motor act.

Broughton *et al.* (1949) observed a decrease in the SEP during fist clenching and Dinges and Klingaman (1972) observed an enhancement of the VEP concurrent with 'induced muscle tension'. Karlin *et al.* (1970) observed a variety of AEP variations but could not attribute them unequivocally to the occurrence of a motor response. It may be that attentional orientation to the motor act with concomitant distraction to somatosensory perception should be taken into account. Thus, the *S* who was required to bar press rhythmically, might have paid sufficient attention to the task to attenuate the SEP in somewhat the same fashion as has been demonstrated for the discrimination task alone. Passive induced motor activity might overcome this problem. In any case, SEP attenuation prior to movement strongly suggests an effect specific to motor activation.

Brain, Spinal Cord and Autonomic Changes before, during and after a Planned Motor Action in Man

D. PAPAKOSTOPOULOS* and R. COOPER

Burden Neurological Institute, Bristol, U.K.

The CNV appears as the electroencephalographic index of the foreperiod in a simple reaction time experiment. During the same period particular and repeatable changes in heart rate, blood pressure, respiration, electrodermal responses and pupil diameter occur (Lacey and Lacey, 1970; Papakostopoulos and McCallum, 1973; Papakostopoulos, 1973). In addition there is an increase of the excitability of the spinal monosynaptic reflex even for muscles not involved in the planned motor action (Papakostopoulos and Cooper, 1973b). In this work the monosynaptic reflexes were measured a few seconds before and after the motor action to the imperative stimulus. More recently measurements have been taken during the motor action itself and at various times shortly after. By so doing it is now possible to describe the temporal pattern of spinal monosynaptic reflex excitability before, during and after planned motor action and the concomitant changes in brain potentials and autonomic function.

A further series of experiments has been completed which determine the changes when the muscle used for the motor action is the same one from which the spinal monosynaptic reflex is obtained.

METHOD

The method used was similar to the one already described (Papakostopoulos and Cooper, 1973b) except that the subjects were in the sitting position and more variables were stored in digital form.

Briefly, monosynaptic reflexes (H reflex) were electrically elicited at various times during and after the 2-sec foreperiod of a simple RT experiment with a click as S_1, a tone as S_2 and an average inter-trial interval of 29 ± 11 sec. The H reflex was elicited in a pseudo-random order 0·2, 1, 1·5, 1·8 and 1·9 sec after S_1 and 0·2, 0·5, 2, 5 and 10 sec after S_2 by a 0·5 msec electrical pulse applied to the tibial nerve in the popliteal fossa.

Six sequences each of 10 trials were presented when *S* was responding to S_2 by pressing a button with his left thumb (Press Hand—PH—condition) and 5 sequences when he was making an isometric plantar flexion with his left foot (PF condition). Either of these actions caused the cessation of the tone (S_2). The occurrence of the stimuli for eliciting the CNV and the H reflex was controlled by the PDP-12 computer which was also sampling the data at appropriate rates and displaying some of the results immediately. RT was measured and printed out on-line. Before initiating a trigger pulse the computer measured and stored the R to R interval for 5 heart beats. The fifth beat initiated the stimulus sequence and S_1 appeared 1 sec later. This procedure was used to establish the EEG, EMG and heart rate baseline before S_1. The rectified EMG of the arm flexors, gastro-

*Supported by a grant from the Wellcome Trust.

cnemius and tibial anterior muscles from surface electrodes were sampled at 1 msec intervals until 1 sec after S_2. Respiration curves and heart beat intervals continued to be stored for 6 and 10 sec respectively after S_2. Electrodermal resistance was also monitored. In addition to the H reflex, which is the electromyographic sign of the muscle contraction, the resultant mechanical twitch was measured with a force transducer. The evoked response to the same electrical stimulus was recorded from two electrodes, one 2 cm behind the vertex and the other 2 cm to its right.

For the force and the evoked potentials the total sampling time was 500 msec; for the H reflex 50 msec. The CNV, with 4-sec collection time, was recorded from the vertex, C_3 and C_4 electrode positions all referred to linked electrodes on each mastoid process. A nasion+2 cm electrode referred to the same reference was used to monitor eye movements and also served for compensation of the vertex channel (McCallum and Walter, 1968). A modified Elema-Schönander 16-channel machine acted as the main amplification system. For the EEG, respiration and GSR a time constant of 5 sec was used. Special silver–silver chloride electrodes, selected to have less than 1 mv potential between any pair in distilled water, were used. Separate amplifiers with 10 kHz bandwidth were used for recording the H reflex.

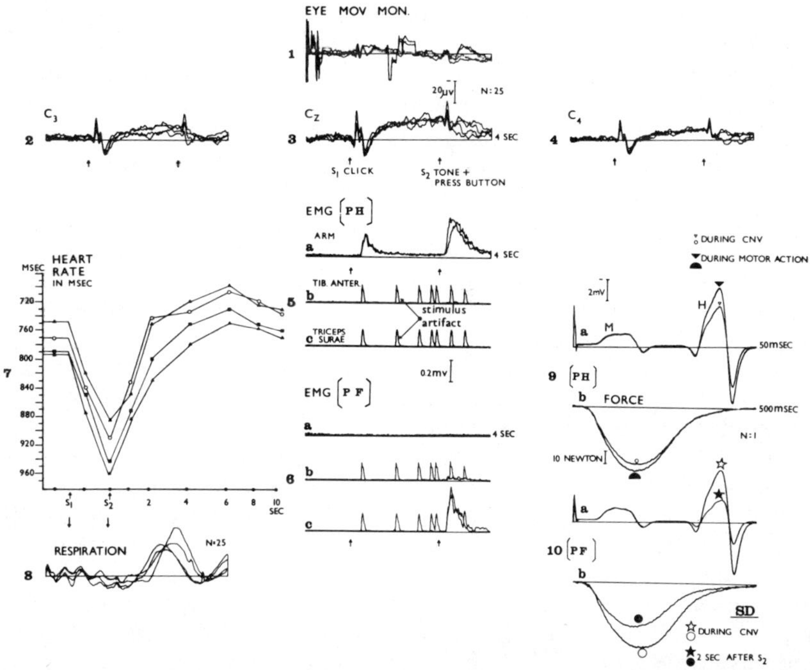

Fig. 1. Superimposed averages for eye movements (1), the CNV (2, 3 and 4) and the EMG (5 and 6), for the Press Hand (PH) and Press Foot (PF) situations. Trials with the 'H' stimulus present during the foreperiod and trials where it is presented later are averaged separately. At 7 and 8 the corresponding respiration and heart rate are shown. At 9 and 10 appear two examples of the 'H' reflex and the corresponding force changes during PH and PF. At 9 comparison of H and force during the foreperiod and during motor action, at 10 and H reflex during the foreperiod and 2 sec after. The average of 5 heart intervals before S_1 have been taken as base line and the average duration (N=6) of the heart interval at the indicated points in time shows the change for 10 sec after S_2 (open circles and triangles PH situation; full circles and squares PF situation).

Nine female and 3 male normal *Ss* were tested; paid volunteers with a mean age of 24 years.

All data were stored as individual trials and further analysed off-line.

RESULTS

The average vertex CNVs for the two situations (press hand PH and press foot PF) were practically identical. In addition there were no differences in the CNVs recorded on trials when the H reflex eliciting stimuli were presented during the interstimulus interval and those recorded on trials when the H reflex was elicited outside this interval. These are shown for a representative *S* in *Fig.* 1.3 where the corresponding average vertex CNVs for the four situations are superimposed. Simultaneously recorded activity from an electrode 2 cm above the nasion, monitoring eye movements, is shown in *Fig.* 1.1. Traces 2 and 4 of *Fig.* 1 show the activity from electrodes C_3 and C_4. The variability of the CNV recorded from the area ipsilateral to the active limb (C_3) is substantially higher than that recorded from the contralateral area (C_4).

In all situations a pronounced bradycardia developed during the foreperiod, the lowest rate occurring at the time of the motor action to S_2 (*Fig.* 1.7).

Respiration during the same period lacks any particular pattern across *Ss* although the *S* of *Fig.* 1 showed a tendency for inspiration to occur about 3 sec after S_2 (*Fig.* 1.8).

The behaviour of the electromyographic activity during the period of CNV development varies according to the muscle group and the *S*. During the PH situation, in the case of the *S* shown in *Fig.* 1, the EMG of the arm flexors increased during the foreperiod (*Fig.* 1.5, a) but the leg muscles (tibialis anterior, *Fig.* 1.5, b, and triceps surae, *Fig.* 1.5, c) were silent. During the PF situation there was no increase of the EMG of the active muscle (*Fig.* 1.6, c) and the other muscle groups were silent during the foreperiod. During the motor act a small EMG activity developed in the antagonist muscle (*Fig.* 1.6, b) but both agonist and antagonist myographic activity (*Fig.* 1.6, c) ceased 600–700 msec after S_2.

The amplitude of the H response elicited during the foreperiod was significantly greater than that elicited during the intertrial interval for both PH and PF conditions. The H reflex elicited during the motor action was significantly higher than that elicited during the foreperiod, *Fig.* 1.9, a, as was the force developed by the muscle twitch, *Fig.* 1.9, b. This behaviour is similar in both PH and PF situations across *Ss* as can be seen in *Fig.* 2, where data of 8 *Ss* are presented. In A and C the individual *S* H reflex data for PH and PF situations respectively are superimposed; each point plotted is the mean of 6 H reflexes in the PH condition and 5 in the PF condition. *Fig.* 2, B (dots) is the average across *Ss* of the data in A with the corresponding force (stars). *Fig.* 2, D (stars) shows the average across *Ss* of the H reflex data of the individuals shown in C (condition PF); the average of the PH situation is superimposed for comparison.

In the PF condition (*Fig.* 2, C) there is a small (10 per cent) decrease in the amplitude of the H reflex immediately prior to the motor action. During the motor action itself the amplitude increases by more than 30 per cent for the PF condition and 20 per cent for the PH condition. This increase in amplitude of the H reflex is inversely correlated with the *S*'s RT. Both amplitude values then show a decrease; in the PH condition the H reflex falls to 85 per cent which is about the intertrial level, whilst in the PF condition it falls to 40 per cent of the foreperiod level (*Fig.* 2, D). [This can also be seen in *Fig.* 1.10, a, where the averaged H reflexes for one *S* are presented. The developed force follows the same pattern (*Fig*. 1.10, b).] This inhibition in the PF condition continues for at least 2 sec after S_2 which is 1 sec after the cessation of EMG activity of the agonist and antagonist muscles. The average heart rate for the group is shown in *Fig*. 2. The deceleration peaking at S_2 is clear. The mean respiration for the group and its standard deviation are displayed underneath; both are virtually flat lines.

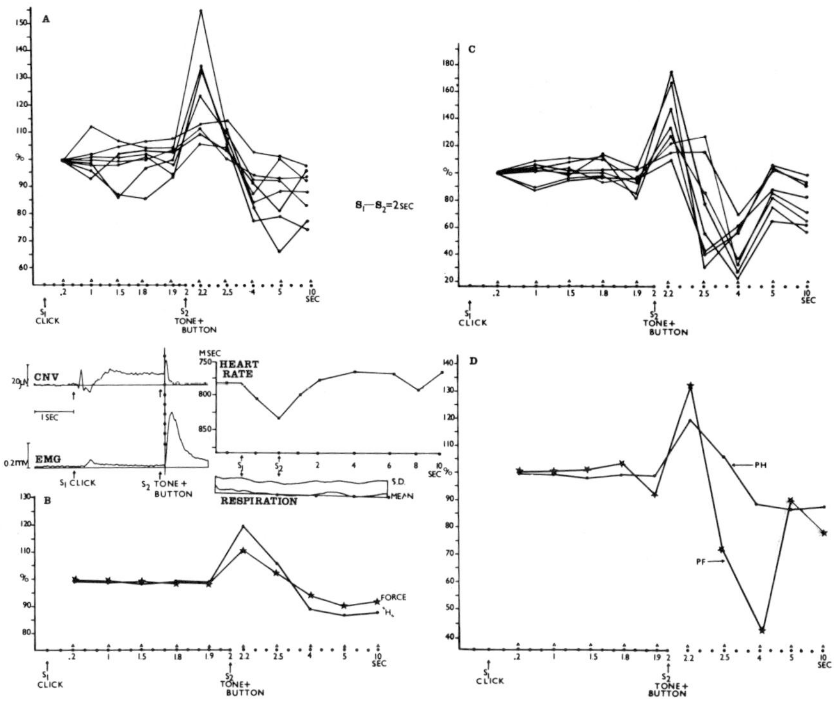

Fig. 2. The 'H' reflex amplitude normalized to the one elicited 200 msec after S_1 (100 per cent). Data from 8 *Ss* are superimposed at A for the Press Hand (PH) and at C for the Press Foot (PF) situation. At B the average of the data at A and the force (starred line) developed by the twitch. At D the average of the data at C (starred line) and the 'H' curve of B superimposed to facilitate comparison. Between A and B are shown the average of all *Ss* for the CNV, the corresponding EMG activity of the arm muscles, the average heart rate in msec and the average and standard deviation of the respiration.

The average vertex CNV for this group is 12 μv and it cuts-off about 160 msec after S_2, *Fig*. 2. The average forearm EMG responses for the whole group in the PH situation is displayed underneath the CNV and shows that the beginning of the EMG response is as early as 90 msec after S_2.

DISCUSSION

A number of systems of the human organism participate in the execution of a planned motor action. The data presented show that this involvement is specific in pattern and monotonically repeatable across *Ss*.

Cortical negativity (CNV), heart rate deceleration, and increase in spinal monosynaptic reflex excitability during the foreperiod was observed in all *Ss* tested. Because of the difference in time course for the several variables it has been suggested that, though independent, there is a structure within the brain stem, probably the reticular formation, that is the common controller (Papakostopoulos and Cooper, 1973b). Recordings from brain stem structures of CNVs with shorter latencies than those recorded on the cortex (McCallum *et al.*, 1973) give direct support to this hypothesis.

Although CNVs in the *Ss* tested terminated about 160 msec after S_2, the peak EMG associated with the motor action had a delay of about 200 msec. H reflexes elicited at this point in time (200 msec after S_2, when it might be deduced that reticular activation had ceased, present the highest values recorded. This is in agreement with recent observations from other laboratories, without EEG recordings (Requin, 1969; Gottlieb *et al.*, 1970; Coquery and Coulmance, 1971; Pierrot-Deseilligny, 1971).

The amplitude of the H reflex during this period of time is inversely correlated with the RT of the *S* to the imperative stimulus (S_2). Involvement of the pyramidal system for this second increase is suggested. Such an involvement could be accomplished through coactivation of α and γ motoneurones (Granit, 1970). Hagbarth and Vallbo (1968) have shown it to happen in man during voluntary contraction.

The reflex inhibition observed after the motor action could be the result of many segmental mechanisms (Eccles, 1967a), but the Golgi tension receptors are the favourite candidates. Nevertheless, the inhibitory activity of these pathways is mediated through interneurones and, as Granit observed, is not a necessary but a facultative one (Granit, 1970). In other words it can be reversed by supraspinal actions upon the interneurones (Matthews, 1972). It is these considerations and particularly the prolonged duration of the inhibition that favours supraspinal participation perhaps of the cerebellum (Eccles, 1967b, 1973). The P300 component of the evoked response has been proposed as a sign of central inhibitory mechanisms (Papakostopoulos and Crow, p. 201); its relationship with inhibition of the spinal monosynaptic reflexes needs further experimentation.

The behaviour of respiration does not affect the time course of the CNV, the heart rate or the spinal motoneurone excitability during the CNV or during motor action, which means that a general postulate of motor and autonomic inhibition during the foreperiod (Obrist *et al.*, 1970) or later cannot explain the phenomena.

It seems that the various biopotentials reflect various aspects of the cognitive and operational actions of the nervous system. Some reveal the particular spatio-temporal neuronal organization involved in carrying out a specific task and some the general principles of task conception and execution. The principles of integration of the nervous system will be found by the simultaneous registration of the changes in the various systems.

SUMMARY

Simultaneous multichannel recordings of brain activity and the activity of various motor and autonomic systems, from 12 *Ss*, have been analysed. The data have permitted inference about the spatio-temporal participation of the pyramidal, extrapyramidal and cerebellar systems before, during and after a planned motor action in man.

Whether neurological and psychiatric states and the effect of various drugs can be assessed by the above methods is under investigation.

ACKNOWLEDGEMENTS

It is a pleasure to acknowledge the help of Mr W. J. Warren, Mr P. V. Pocock, Mr P. Newton and Mr R. Jefferies. Dr Delwaide and M. J. Delbovier supplied us with drawings of the apparatus they have developed for holding the leg during the stimulation. The Wellcome Trust and the W. Clement and Jessie V. Stone Foundation gave financial assistance.

Slow Potentials Evoked by Involuntary and Voluntary Movement, Unconditioned and Conditioned Sensory Stimulation

G. LELORD, F. LAFFONT, Ph. JUSSEAUME and N. BRUNEAU

Laboratoire d'Explorations Fonctionelles Psychopathologiques Hôpital Bretonneau; Laboratoire de Neurophysiologie du Conditionnement Faculté de Médécine (Tours), France

Experiments were made in man in which *S* was asked to carry out a movement after a stimulus or in which a reflex movement was provoked by specific stimulation. S_1 was a click, S_2 either a flash, after which the *S* made the movement, or external stimulation (percussion of the Achilles tendon or direct electrical stimulation) which elicited the reflex movement. The i.s.i. was 700 msec.

Movement, voluntary or involuntary, evoked a slow cortical wave predominating over the anterior region of the scalp. The characteristics of this wave were very similar irrespective of whether the elicited response was clenching the fist, the ankle reflex, or the pronouncing or thinking of a word.

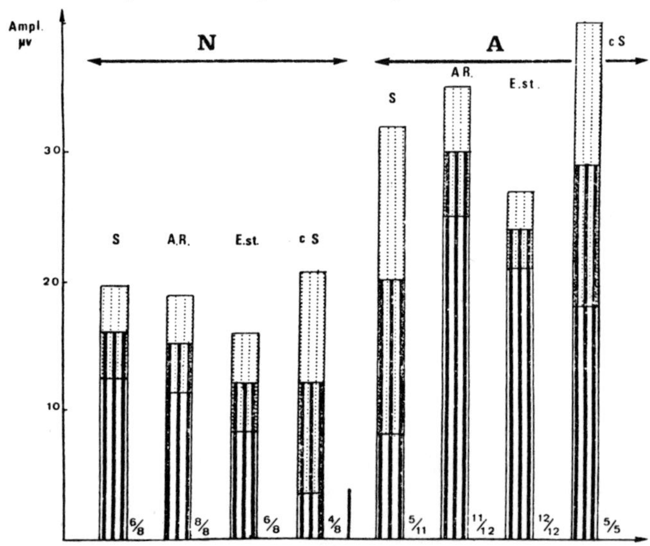

Fig. 1. Mean amplitude of slow negative waves evoked by reflex movement. N, 8 normal children; A, 12 autistic children. S, sound alone; A.R., Achilles reflex; E. st., electrical stimulation; c S, conditioned sound. The values at the bottom of each column indicate the number of children in whom a slow wave could be recorded.

Note: For a fuller account of the adult studies reported, *see:* Lelord, G., Laffont, F., Sauvage, D. and Jusseaume, P. (1973). Etude chez l'homme d'activités évoquées lentes succedant un mouvement volontaire et au langage articule (Evoked slow activities in man following voluntary movement and articulated speech.) *Electroenceph. Clin. Neurophysiol.* **35**, 113–124.

These slow waves can be conditioned. In this case the anticipatory phenomenon which appeared after S_1 resembled a CNV.

Important individual variations have been observed. In autistic children slow waves are large and diffuse and their conditioning is evident, they can be evoked be sensory stimulation and they resemble slow waves recorded in normal newborn infants. In normal children they are smaller, more localised and their conditioning is more irregular (*Fig*. 1).

The method used for the study of involuntary movement has the advantage of universal application in that it is not dependent upon the voluntary participation of the subject.

Some Relationships between CNV, P300, and Task Demands

GAIL R. MARSH, LEONARD W. POON and LARRY W. THOMPSON

Center for the Study of Aging and Human Development and Department of Psychiatry, Duke University Medical Center, Durham, N.C. 27710, U.S.A.

The relatively slow electrocortical phenomena CNV and P300 have been found to be related to various psychological/cognitive demands of the testing situation rather than to the physical dimensions of the stimuli involved (Tecce, 1972). Apparently with the correct combination of tasks both variables may be enhanced (Donchin and Smith, 1970), but with proper instructional set to disregard certain stimuli they can be dissociated (Donald and Goff, 1971). The two phenomena may, however, share the common thread of sensitivity to arousal and attention (Tecce, 1972).

Most recently attention has been focused on the topography of the CNV and P300 in an attempt to provide a further understanding of their functional correlates (Borda, 1970; Järvilehto and Fruhstorfer, 1970; Donchin *et al.*, 1972a). The results have shown topographical shifts of the CNV in an anterior or posterior direction with changes in 'cognitive set' with eye movement artefact eliminated as a factor. This raised several theoretical issues: (1) would the CNV appear in parietal or temporal cortex if the appropriate task was imposed, (2) would the CNV and P300 be highly correlated when recorded from these 'association' areas, and (3) would verbal-spatial tasks evoke left-right hemisphere differences in CNV and P300 reflecting the unequal involvement of the hemispheres in the respective tasks.

METHOD

Experiment 1

Subjects. Seventeen *Ss* (8 males and 9 females) from Duke University participated. All were naïve with regard to electrophysiological experimentation.

Procedure. Each *S* was instructed that he was in a binary-choice gambling situation in which on each trial he was to predict if a red or green light was likely to be presented. A tone at the start of each trial signalled the subject to make a prediction and place a bet of from 0 to 5 cents on the outcome. Shortly thereafter a yellow warning light (S_1) warned that the answer was to be given in 1·5 sec by the flash of a red or green light (S_2). Following S_2 the subject verbally reported the amount won or lost. He was instructed to win as much money as possible and that he could keep his winnings.

An initial habituation set of alternating red and green lights was given, followed immediately by a more complex repeating pattern of G, G, R, G, R. This was continued until the subject correctly guessed one complete 5 element pattern. Four iterations of this pattern were then given as overtraining.

Forty disjunctive RT trials were then administered using the same S_1–S_2 stimuli with reaction required to the green, but not to the red light. Red and green appeared in a random sequence and the subject was so instructed. The response was speaking 'yeah' as fast as possible.

Apparatus. EEG recordings were obtained from F_Z, C_Z, and P_Z via Beckman silver–silver chloride electrodes referenced to the linked mastoids. The epidermis was pierced and electrode impedance kept below 2000 ohm. The amplifying chain on each channel was a Grass P17 DC high impedance probe amplifier led to a Grass 7P1 DC pre-amplifier and driver amplifier modified to provide an 8 sec time constant. The high frequency filter was set at 35 Hz.

Silver disk electrodes placed above and below the left eye and at the outer canthus of each eye provided information on eye movements. Any eye movements or blinks during a trial period caused that trial to be excluded from further data analysis.

All data analysis was performed off-line using a general purpose computer interfaced with an A/D converter sampling FM tape recorded data at 50 samples/sec. The amplitude of the CNV was measured between the last 200 msec of activity preceding S_2 and a baseline of 600 msec of activity preceding S_1. Amplitude of P300 was measured from 200 msec of prestimulus baseline to the most positive point 200–400 msec after S_2.

Experiment 2

Subjects. There were 11 male and 10 female volunteer *Ss* in group 1 and 11 males and 8 females in group 2. All were students at Duke University.

Procedure. Group 1 *Ss* had bilateral placement of non-polarizing electrodes on T_3 and T_4, and in the parietal area at a point halfway between the mastoid and a point on the midline 2 cm posterior to C_Z. Each *S* fixated a small point on a screen placed 7 ft in front of him. Words and slanted lines were tachistoscopically flashed into either the right or left half of the screen on a level with the fixation point. The fixation point lighted green (S_1) to warn of a verbal trial and red to indicate the start of a spatial trial. The flashed word or line (S_2) followed S_1 by 1·4 sec. At 1·4 sec after S_2 the fixation point was turned off (S_3) signalling *S* to speak the word or point out the line on an answer slide. Forty-eight word and 48 line trials were randomly presented.

Apparatus. The same equipment was used in this experiment as in Experiment 1.

RESULTS

Experiment 1

When *Ss* were engaged in learning a sequence of coloured lights they showed the largest P300 over P_Z, next largest at C_Z, and smallest at F_Z. Also, the P300 was larger during the acquisition phase than during the overtraining phase. The amplitude for P300 was significantly larger for those stimuli which *S* guessed wrong than for those guessed correctly ($P<0{\cdot}05$), which may serve as a partial explanation of the difference between acquisition and overtraining.

Interestingly the CNV results did not always reflect the same differences as those seen in the P300 data. When the CNVs were compared between sites the largest was found at P_Z and the smallest at F_Z, which was comparable to the P300 results. However a comparison between the acquisition and overlearning phases showed no differences in CNV amplitude. Comparing the distribution of potential across the head in the acquisition phase with that obtained during the disjunctive RT task showed significant differences ($P<0{\cdot}007$). The acquisition phase had the largest potential posteriorly with a steady decrement anteriorly while the RT task had the largest CNV at C_Z with both P_Z and F_Z smaller in amplitude.

Experiment 2

Using a task of perceiving tachistoscopically presented words or slanted lines allowed investigation of left-right hemisphere differences as well as verbal-spatial differences at each recording site. Comparisons between verbal and spatial trials showed that they were significantly different ($P<0{\cdot}01$), with most of the differences being limited to the right hemisphere. When the CNV data were gathered using only trials on which *S* had responded correctly there was a significant difference for left-right hemisphere ($P<0{\cdot}03$). However, when all trials were used the left-right hemisphere difference failed to reach statistical significance ($P<0{\cdot}06$). The CNV from the parietal area was significantly larger than that from the temporal area which often tended to be positive.

Another group of *Ss* tested under the same set of circumstances with frontal electrodes substituted for the temporal ones showed essentially the same results: (1) larger potentials in the parietal region, (2) differences largely in the right hemisphere, and (3) left-right hemisphere differences only when restricted to the correct answer trials.

DISCUSSION

Much of the work in CNV correlates has not emerged from a strong theoretical background. To some extent the same claim could be made for research in P300 correlates. However there exist models from other areas that might be used with some utility. Borrowing from the animal literature, a model has been developed for a brain mechanism controlling attention in the cat and monkey (Gerbrandt *et al.*, 1970; Pribram, 1969, 1970). Specifically, frontal cortical areas were seen to influence the animal toward sampling his environment rapidly but with the sampling restricted to a small area of interest. The temporal-parietal area was seen as an opposite influence, moving the animal toward sampling less often from a widened field from which information was extracted and integrated. The majority of CNV experiments were done using one of the standard RT formats, seemingly biasing *Ss* into the type of set that was hypothesized to be induced by frontal cortex activity. Thus, if the CNV is a sign of increased activity, one would expect to see CNVs over the frontal areas during the period from the warning signal to the imperative stimulus. Correspondingly, when turned toward more cognitive tasks the parietal areas should show more CNV activity. Our results from the first experiment give partial support to these speculations since greater activity was seen in the parietal area during problem solving than during overtraining. However, changing to an RT task did not elicit the greatest amount of activity from the frontal regions, but rather from C_z. This may be due to the addition of a readiness component in some areas (Otto and Leifer, 1972a, b), or that even the small cognitive component of whether to respond or not might excite some posterior activity.

The size of the CNVs recorded in these experiments ranged from -1 to $-9\ \mu V$. These are much smaller than those reported at an earlier time in the literature and with other tasks. This may be due to removal of eye movement artefacts, but it may well also reflect other ill-understood processes. Following the report of positive shifts in the monkey when performing a key-release response (Donchin and Smith, 1970), Otto and Leifer (1972a) showed positive shifts in man with a key-release response and negative shifts with a key-press response. That the shifts reported here were of opposite polarity in different portions of the cortex could be a sign of such a different process, or merely the summated output of a differently convoluted set of sulci and gyri. The notion of diminished amplitude due to some sort of distraction (Tecce and Scheff, 1969) would not seem applicable here since correlated changes in CNV and P300 would be expected. Such correlations were not obtained.

To the extent that the CNV and P300 may be related to the arousal level of the *S* they may co-vary. However, to the extent that one reflects an expectancy or set with regard to impinging stimuli while the other is a reaction to the presentation of a stimulus they should be quite different. The present results, where arousal level changed very little, seem to indicate that even though they may have similar topographic distributions during cognitive tasks, these two electrocortical phenomena reflect quite different processes.

DC Potential Consequences of Induced Muscle Tension: Effects on Contingent Negative Variation*

CHARLES REBERT, ROBERT BERRY and JAMES MERLO

Stanford Research Institute and Stanford University, Menlo Park and Stanford, California, U.S.A.

Recently Otto and Leifer (1972a, b) demonstrated that when a motor response is made to the warning stimulus (WS) in a CNV paradigm, the CNV is greatly reduced and positive potentials are generated at several recording sites, maximally in the parietal region. Similar findings were reported by McAdam and Rubin (1971) and Hamilton *et al.* (1972). When the RT task was initiated by the *S*, a readiness potential was observed before the WS and was followed by a positive potential (McAdam and Rubin), and the CNV was smaller with task initiation by *Ss* than by the *E* (Hamilton *et al.*). McCallum and Papakostopoulos (1972), however, showed that sustained button pressing had no effect on the CNV, i.e. when the press was not made specifically at the onset of the WS.

In the foregoing tasks, the motor activity has been nonstrenuous, and as yet, the effect of sustained *effort* on the CNV has not been examined. The *anticipation* of exerting greater muscular effort enhances CNV amplitude (Rebert *et al.*, 1967), so it might be expected that induced muscle tension (IMT) would also enhance the CNV. The induction of muscle tension has often been used as a means of increasing general arousal levels, and has been found (within limits) to enhance performance on a variety of tasks (Eason *et al.*, 1964a) and to influence visually evoked potentials (Eason *et al.*, 1964b) and the motor potential (Dinges and Klingaman, 1972). The effects of IMT are usually attributed to its activating influence on the reticular system. The fact that the CNV has often been thought to reflect general arousal (Irwin *et al.*, 1966) also suggests that some interaction between CNV amplitude and IMT might occur.

METHODS

Thirteen male undergraduate students from Stanford University were used as *Ss*. DC recordings were made from C_Z and P_Z referentially to linked mastoids, and from above and below the right eye with Grass 5P–1 preamplifiers. Data were stored on magnetic tape (Ampex SP300) and analysed off line with a Linc–8 computer. Stimulus presentations were programmed with Iconix time base and logic set. The WS was a 100-msec duration, 1 kHz, tone burst that preceded by 1 sec the imperative stimulus (IS, which was a 1-sec train of 10/sec flashes from a Grass PS–2 photostimulator). *S*'s response terminated the IS. The PS–2 lamp was placed against the outside window of the shielded chamber and was viewed by the *S* through a 15-cm square aperture at eye level approximately 2 m from the *S*.

Five seconds before onset of the WS, a small green lamp (usually a 'fixating stimulus', FS) situated a few cm to the right of the square aperture was lighted

* Supported by Grant NS 08248 from the National Institute of Neurological Diseases and Stroke, USPHS.

for 300 msec. The FS indicated to the *S* to pull, with the left hand, a stirrup arrangement attached to a weight carrier, which was loaded with 0, 15, 30, 40 or 50 lb. The 40 and 50 lb weight conditions were begun later in the experiment after it was suggested that the 30 lb condition was not overly difficult; only 9 *Ss* were run in those conditions. Initially 50 lb was used, but proved excessive for the cable arrangement and was reduced to 40 lb. The highest weight condition will be referred to below as the 45 lb condition.

Weight presentations of 0, 15 and 30 lb were counter-balanced across *Ss*, but because the 45 lb condition was introduced later, it was always presented last. Trials were continued until 12 of them were presented without error (E) or excessive eye movement. No particular condition was associated with an abundance of repeated trials.

Slow potential (SP) data were analysed by digitizing 8-sec epochs and CNVs by digitizing 2·5 sec epochs. SPs induced by pulling weights were hand scored by measuring mm deflections from a pre-FS, 500 msec baseline period at an 'initial' (SP_i) and 'terminal' (SP_t) period, then converted to μV (*see Fig.* 1). In the final analysis, SP_i and SP_t were averaged to reduce variability. CNVs were scored by computer, using the average level of activity over the 250 msec before IS with respect to the baseline defined as the average level prior to WS.

RESULTS

A typical response at the vertex obtained with the 8-sec epoch is shown in *Fig.* 1a. When the FS went on and the weight was pulled, a negative shift occurred that usually declined somewhat before onset of the CNV trial. A clear CNV is evident following a −25 μV calibration pulse. The average SP deviation (combined SP_i and SP_t) for various weight conditions are shown in *Fig.* 2a and summarized in *Table* 1. At the vertex, a negative SP is produced even by the 0 weight condition (pulling the carrier only), and the SP increases in negativity with increasing weight. Using one-tailed *t*-tests, the differences between 15 and 0 lb, and 30 and 0 lb were significant ($d_{15-0}=4{\cdot}7$ μV, $t=1{\cdot}99$, $df=11$, $P<0{\cdot}05$; $d_{30-0}=6{\cdot}9$ μV, $t=4{\cdot}2$, $df=12$, $P<0{\cdot}001$).

Table 1. Descriptive statistics for slow potential shifts produced by muscle tension

		Weight (lb)			
Placement	*stat.*	0	15	30	45
	n	13	12	13	9
Vertex	$\bar{x}$	−2·0	−7·0	−8·9	−8·1
	σ	4·7	4·9	5·6	9·2
	n	13	12	13	8
Parietal	$\bar{x}$	3·1	0·8	−2·2	−0·7
	σ	5·8	6·8	6·7	9·0

Comparison of SP_i and SP_t separately with the baseline (i.e. 0 μV) for the 0 lb condition at the vertex showed a significant 'initial' negative SP from pulling the carrier ($d_{0_i}=-4{\cdot}3$ μV, $t=2{\cdot}9$, $df=12$, $P<0{\cdot}01$) but no significant 'terminal' SP, i.e. just before the CNV trial ($d_{0_t}=-0{\cdot}4$ μV). For the combined score, there was no significant effect of the 0 lb condition ($d_0=-2{\cdot}0$ μV).

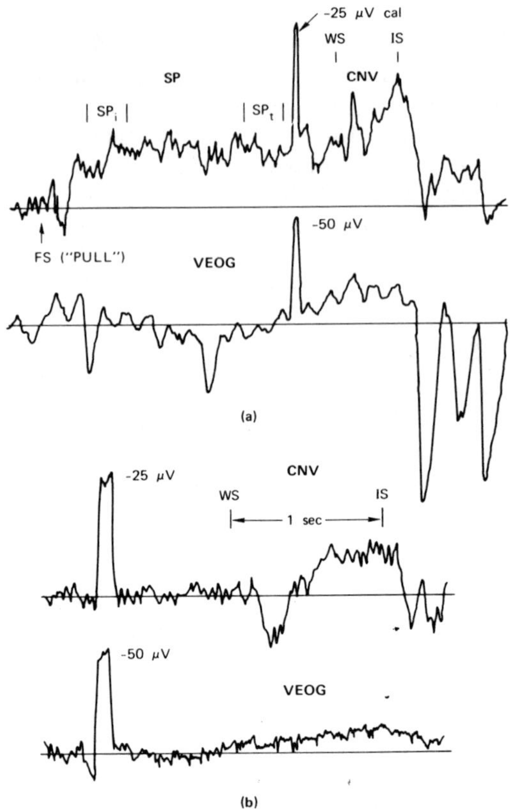

Fig. 1. A, *Upper*: Slow potential (SP) change and CNV at the vertex in one *S* induced by the pulling of weight. *Lower:* Vertical electro-oculogram (VEOG) associated with the SP and CNV above.
B, Typical CNV (upper) and VEOG (lower) from another *S*.

Results were opposite in the parietal region. There was only a small (d_{0_i}= 1·3 μV), nonsignificant 'initial' positive SP_i in the 0 lb condition, whereas the 'terminal' positive SP_t was significant (d_{0_t}=4·9 μV, t=1·9, df=12, P<0·05). The combined SP_i and SP_t score was also significant (d_0=3·1 μV, t=1·9, df=12, P<0·05). The difference between 15 and 0 lb did not reach significance in the parietal region (d_{15-0}=3·1 μV), but the 0 and 30 lb conditions did differ (d_{30-0}=5·3 μV, t=3·2, df=12, P<0·005), with the SP becoming more negative as the weight increased.

Data from *Ss* given 40 lb (n=5) were combined with those from *Ss* receiving 50 lb (n=4) as there were too few *Ss* to make meaningful comparisons between those weights. At the highest weights, there was no significant change from the 30 lb condition for either the vertex or parietal placements.

The vertex and parietal regions differed significantly at all weight conditions (e.g., d_0=5·13, t=3·7, df=12, P<0·005), the parietal region being the more positive.

Eye movements for the combined SP_i and SP_t measurements across weight conditions were 0·9, 0·6, 7·9 and 2·9 μV respectively. The variances averaged

13·1 μV, considerably higher than those for CNV ($\bar{x}=7{\cdot}5$ μV). Since (1) VEOG amplitudes were small (and only approximately 25 per cent of EOG amplitude is reflected at the vertex), (2) the pattern of change over weight conditions was not similar to the CNV profile, and (3) VEOG variances were high, it is concluded that eye movements were not responsible for the SP shifts. The SPs also did not appear to result from galvanic skin responses as their latency was short, and they declined rapidly during the holding period when only the carrier was pulled.

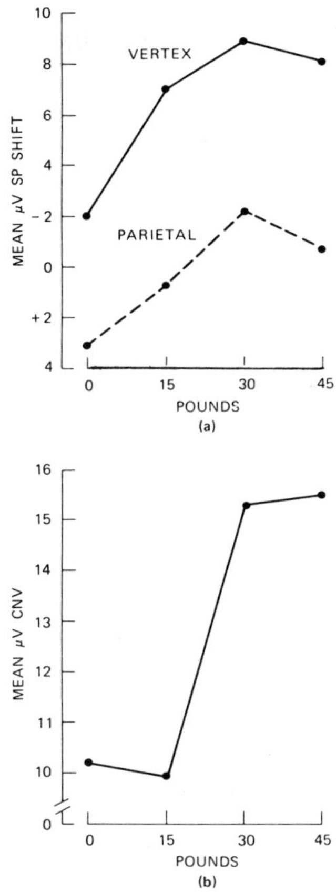

Fig. 2. A, Mean SP shifts at the vertex and parietal regions associated with pulling different amounts of weight.
B, Mean CNV amplitude as a function of amount of weight pulled.

A typical CNV and associated vertical EOG are shown in *Fig.* 1b. Summary statistics for CNV amplitudes at the vertex under the various weight conditions are included in *Table* 2 and plotted in *Fig.* 2b. Parietal CNVs were extremely small or absent so were not subject to analysis. There were no differences between CNVs associated with the 0 or 15 lb conditions, but at 30 lb a large increase occurred, which remained the same at the 45 lb condition ($d_{30-0}=5{\cdot}04$, $t=1{\cdot}84$, $df=12$, $P<0{\cdot}05$; $d_{30-15}=5{\cdot}54$, $t=2{\cdot}7$, $df=12$, $P<0{\cdot}01$). A rank order correlation between the SP and CNV summed across *Ss* and weights was marginally significant (rho$=0{\cdot}47$, n$=13$, $P\approx0{\cdot}06$).

RT was influenced only at 15 lb. Mean RTs in msec for 0–45 lb were 201·5, 190·8, 199·6, and 200·7, respectively. Eleven of the 13 *Ss* had shorter RTs at 15 lb than at 0 lb ($d=10{\cdot}7$ msec, $t=2{\cdot}4$, $df=12$, $P<0{\cdot}025$), and 10 were less at 15 lb than at 30 lb ($d=8{\cdot}8$ msec, $0{\cdot}05<P<0{\cdot}10$). For scores summed across *Ss* and weights, there was no significant correlation between the SP and RT (rho=0·09) nor between CNV and RT (rho=0·13).

Table 2. Descriptive statistics for CNV amplitudes at various levels of induced muscle tension

	Weight (lb)			
stat.	0	15	30	45
n	13	13	13	13
$\bar{x}$	−10·2	−9·7	−15·3	−15·5
σ	6·1	5·4	8·9	9·5

DISCUSSION

Results associated with the pulling of weights are consistent with those of Otto and Leifer (1972b) in showing the occurrence of a positive SP in parietal regions with motor activity. Further, they indicate that the parietal region (as well as the vertex) becomes more negative when the response involves greater exertion. The results also show, as an extension of the study by McCallum and Papakostopoulos (1972), that sustained motor activity does influence the CNV if it involves sufficient effort.

The SP responses evoked by IMT appeared to be a reflection of cortical responsiveness to increased afferent input, mediated probably by some aspect of the nonspecific arousal system. Negative SPs are known to occur in parts of the cortex during increased arousal (Caspers, 1963), and electrical stimulation of nonspecific nuclei also produces cortical negativity (Brookhart *et al.*, 1958). Thus, it appears that an objective measure of arousal induced by muscle tension might be available in the scalp-recorded SP. The findings that IMT above a certain level influenced the CNV, and that a very nearly significant correlation of SP and CNV amplitude occurred, support the notion that the CNV is associated with activity in some portion of a nonspecific arousal mechanism (but perhaps not the brainstem reticular system—*see* Rebert, 1972c, and p. 191). Because no increment in SP amplitude was produced by the 45 lb condition, it is possible that some limit exists in the extent to which the cortical surface can normally be shifted. Knott and Irwin (1967) suggested that cortical saturation could account for low-amplitude CNVs observed in highly anxious *Ss* under stress. However, if the cortex were saturated by high muscular exertion in this study, the CNV should have *declined* in the 30 and 45 lb conditions. The data are somewhat anomalous, however, in that no further increase in CNV occurred above the 30 lb condition. The lack of increase in RTs in the 45 lb condition suggests that the absence of further CNV increase at that point was *not* due to distracting effects of pulling that much weight (a viable alternative explanation of the anxiety-stress effect). In this context, these data imply a limit to cortical negativity induced by muscle tension, but show no suppression of CNV at those limits. Thus, the SP and CNV may reflect separate but interacting cerebral processes.

SUMMARY

Thirteen undergraduate male *Ss* were studied to determine whether or not induced muscle tension (IMT) had an influence on scalp-recorded slow potential (SP) changes and on the CNV. IMT resulted in negative SPs at the vertex which increased in amplitude with greater tension. In the parietal region a low effort motor response produced a positive SP which increased in negativity as tension level increased. CNV amplitude was augmented by moderate and high tension levels. The data suggest that an objective measure of IMT-produced arousal is available in the scalp-recorded SP, and that the SP and CNV are mediated by separate but interacting cerebral mechanisms.

Post-warning Signal Positivity in Relation to Development, RT Performance, and Warning Signal Compounding

R. KARRER and J. IVINS

Illinois State Pediatric Institute, Chicago, Illinois, U.S.A.

Walter (1966) was the first to report event-related slow potentials (the CNV) in children. The CNV of children was lower in amplitude and could not be detected in children below about 3–4 years of age using the standard RT task. It tended to reach adult characteristics in the late teens. Consequently, the CNV was considered a 'sign of mature, orderly, sensorimotor conditional association' (Walter *et al.*, 1971). Since Walter's early work, developmental studies using the standard RT task have shown the CNV to change in amplitude and topography with age (Cohen, 1973a). At ages 6–12 the CNV tended to have a higher amplitude on the posterior scalp, while at ages 12–18 it tended to higher amplitude on the frontal scalp. Others have had variable success in obtaining artefact free CNV recordings from young children (Low *et al.*, 1966; Small, 1971; Low and Stoilen, 1973). That event-related slow potentials (SP) may be recorded reliably as early as two years by using novel, attention-getting paradigms has been demonstrated by Gullickson (1973b).

Manipulation of stimulus and response variables and task requirements in children should lead to important information on the relation between SP phenomena and known developmental changes in motor and cognitive behaviour. The study to be reported manipulated the modality of the warning signal (WS) in an attempt to increase its arousal or alerting effect. A bi-modal WS should decrease response latency and increase CNV amplitude more than a unimodal WS in the child. Young children should also have slower and smaller CNVs than older children if the CNV is developmentally dependent.

METHOD

Ss were 26 normal school children from the public schools of Bordentown, New Jersey. One group of 13 ranged in age from 13 to 18 years ($\bar{x}=15{\cdot}1$) while a younger group of 13 ranged in age from 9 to 12 ($\bar{x}=11{\cdot}1$).

A standard fixed foreperiod (1 sec) RT task was utilized. The WS (S_1) was either an auditory (600 Hz, 80 dB tone) or tactile (vibration to non-dominant wrist) stimulation, or the two combined. S_2 was a 1-in diameter white light (0·8 ft L) embedded in the back wall of a carrel 18 in from the *S's* eyes. Twenty trials of each WS condition were presented in random order (60 trials per session). The *S* was instructed to respond as fast as possible to S_2 onset by pressing a telegraph key with his dominant hand. He was given practice trials until he understood the task thoroughly. Trials in which the *S* anticipated S_2 or had obvious movement artefacts in the EEG were repeated.

Steady potentials were determined from EEG recordings of midline frontal (F_Z) and vertex (C_Z) sites referred to left mastoid. Vertical eye movement was monitored with electrodes in the centre of the superior and inferior orbital ridges of the left eye. These data were recorded DC on a Grass (Model 5) polygraph with outputs to an Ampex (Model FR 1300) FM tape recorder.

EEG epochs of 5·2 sec were averaged over trials in each WS condition. The interval between the WS and 200 msec after S_2 (1200 msec) was divided into two 600-msec segments. From the first segment, peak *post-WS positivity* (PWSP) was determined, while peak *CNV* negativity was determined from the second segment.

RESULTS

Older children had faster RTs than younger children and the compound WS condition gave faster RTs than either single modality WS.

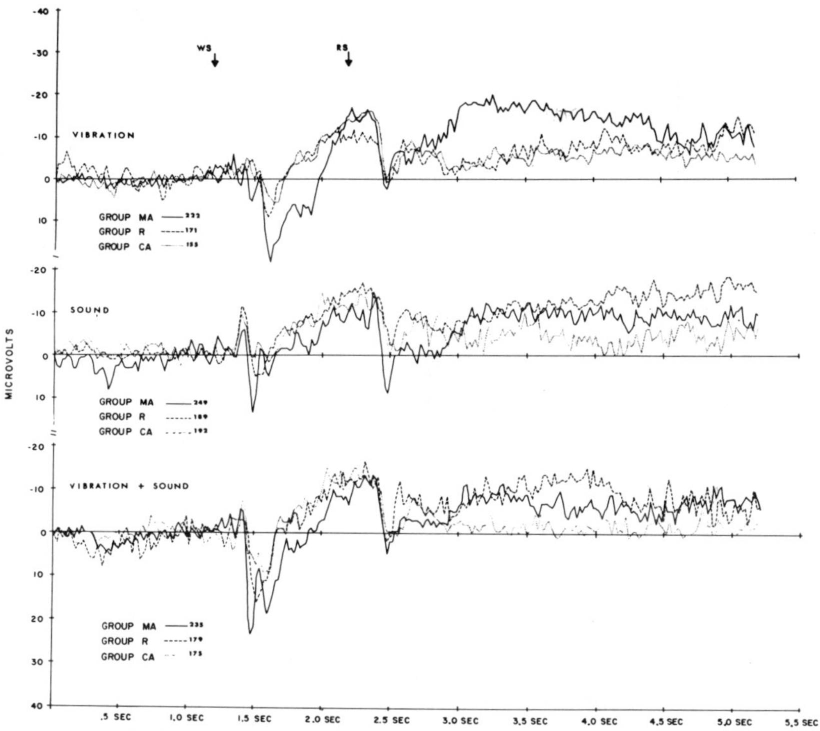

Fig. 1. Group averages of vertex activity for each WS condition.

The group averages of the SP data may be seen in *Fig.* 1 for vertex. There were consistent oscillations following the WS. Four positive peaks are discriminable with the largest and last at about 300–500 msec. These oscillations seldom crossed baseline in the frontal lead. There were significant differences in the PWSP measure between groups and between WS conditions: Young children had larger PWSP than the older children for vertex activity but the same trends were not significant in frontal areas. The amplitude of this PWSP was largest in the bimodal condition and smallest in the sound condition for both vertex and frontal regions. The PWSP, therefore, paralleled the WS manipulation of RT (*Fig.* 2). The larger the peak positivity after the WS in frontal and vertex, the faster the RT.

Correlations between PWSP and RT indicated this was a weak relationship existing primarily in the young child. The only significant between-*S* correlation for the frontal area was in the bimodal condition of the young children (r= 0·564). For vertex, 8 of 9 non-significant correlations also indicated this trend. Fast RT young children had larger PWSP than slow RT children. There were no significant intra-individual correlations of PWSP and RT, but the sign of the correlations in the young children was consistently positive ($\bar{r}$=0·34).

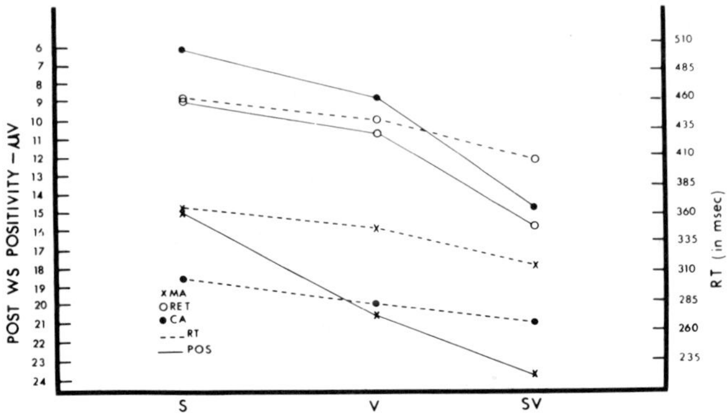

Fig. 2. Vertex PWSP as a function of warning signal condition.

Contrary to expectation, there were no group differences in CNV amplitude and the sound WS condition gave larger CNVs than the vibration or compound WS. The CNV and PWSP were minimally correlated ($\bar{r}$=0·385), indicating that peak activity early (PWSP) and late (CNV) in the preparatory interval reflect relatively independent processes.

DISCUSSION

The bimodel WS produced the best performance and the largest PWSP. There was a clear developmental trend for RT and PWSP amplitude to decrease with age. The relation of PWSP to RT may also be developmentally linked.

In our task an evoked potential to vibration may be responsible for the PWSP effects. The analysis was concerned only with SP activity rather than EP components. This could explain the smaller positivity in the sound condition, but not the almost additive effect of the bimodal stimuli (*see Fig*. 1). The PWSP developmental trend may reflect the same processes that produce larger visual evoked potentials in young children (Dustman and Beck, 1969).

The PWSP probably also has a P300 component that is seen in evoked potentials to visual and auditory stimuli. The P300 has been interpreted as reflecting selective attention (Donchin and Cohen, 1967). Paul and Sutton (1972) suggest that 'P300 amplitude is increased whenever the "salience" of a stimulus is enhanced'.

In the present study, the larger PWSP in young children may be interpreted within the P300 conceptualization as reflecting greater conscious (cognitive) effort to perform the task. While this greater cognitive effort does not seem tenable if considered solely as greater attention or stimulus salience, it does seem tenable if considered as effort to control irrelevant activity and greater uncertainty in responding. Lack of motor differentiation is known to result in 'motor overflow' in non-responding muscles in young children of the age used here (Cohen *et al.*, 1967; *see also* Stern and Plapp, 1969, for discussion of this issue). The PWSP, then, may reflect inhibitory processes that facilitate a motor response set. In this case, the young normals would require greater inhibition of activity irrelevant to the motor key press because of still relatively immature differentiation of motor skill performance. These inhibitory processes could take the form of voluntarily increasing conscious effort to respond, as well as more neurologically based inhibitory processes. This interpretation implies that once differentiation takes place, active inhibition is no longer necessary. The older normals would have differentiation or motor skill well established by the age of sixteen.

The present data suggest that changes in SP activity and late components of EPs (e.g. P300) may be related to the development of cognitive and information processing in the child and should be extensively studied.

SUMMARY

Positive activity after the WS was found to be developmentally related. Young children (9–12 yr) had larger PWSP than older children (13–18 yr). The positivity was increased and the RT decreased by a compound WS. The CNV was not found to differ in amplitude between the two groups. While a specific evoked potential to vibration may contribute to the positivity, the evidence favours the interpretation that the positivity reflects the voluntary or reflex inhibition of irrelevant motor activity in the young child.

ACKNOWLEDGEMENTS

Ted Holden, Betty Karrer, and David Hoats collaborated in this study. We wish to thank Dr Herman Spitz and Dr John Winters for the use of facilities at the Edward R. Johnstone Training and Research Center, Bordentown, N.J. and Anne Campbell for her diligent assistance in the computer analysis.

Clinical and Development Aspects of Slow Potential Changes

Co-chairman's Opening Remarks

M. D. LOW

University of British Columbia, Canada

We have come a long way in a good way; most of us having relinquished proprietary rights to certain SP phenomena and most of us having stopped trying to establish a single correlate of the CNV. As a community of scholars we are engaged in a joint enterprise, with divisions of responsibilities and labours. There still remain some problems, however, particularly in making clinical electrophysiology more valuable a contributor to patient diagnosis and management. This leaves us in a quandary when we examine some of the grant experimental designs which have been set up to study the CNV in splendid isolation, and some of us may be locked into that level of effort. Clinicians may themselves at times be creating difficulties by fostering different problems. One cannot forget that replicability is a fundamental tenet of science and that generality of application is essential if any true and useful clinical technique is to be developed. A test that can be verified only in the laboratory that developed it can hardly be a significant contribution to diagnostic medicine.

What has been done in the clinical area of CNV? Apparently only a small dent has been made so far, which may be due to the fact that we have not recruited very many clinicians to this group. Looking at developmental aspects we know a few things but not very many. We know to a degree that the CNV appears in the 8th or 9th year. Slow potentials of some types may be demonstrated in younger children although there is a question as to whether these are the same things we see in the S_1–S_2 interval in a CNV paradigm. However, in contrast to our knowledge of EEG changes in geriatric populations, thus far nothing is known about the appearance of CNVs in old people, which would seem a ripe field for pursuit.

On the clinical side, the data have clustered around 4 areas. In the field of psychiatry, the CNV and other SPs may have been more productive than conventional EEG, although there are questions here which require further resolution. Part of this depends upon diagnostic criteria, upon which agreement must obtain. Studies on behavioural and learning difficulties in children also present us with unresolved problems, since our Vancouver studies appear to yield results different from others, at least within the so-called minimal cerebral dysfunction group. Again, however, definitions may differ, and methodology of research design may differ. The problem of organic lesions is another area of concentration. McCallum's work was probably the first to suggest that CNVs might produce localizing information. Our own, somewhat prolonged, work has involved serial studies on particular patients. In about 75 patients our tentative conclusions are that vertex CNV is destroyed by thalamic or high brain stem lesions, but that hemispheric lesions without diencephalic involvement may or may not affect the distribution of the CNV, unless a great amount of cortex is involved. Studies have also centred around speech, and several reports from different centres tend to be confirmatory, so that little controversy exists here. The problems as they do exist for the most part have room for further work, and there are of course many clinical problems yet to be identified and investigated by SP techniques.

At this point I would like to turn the meeting over to Dr Jerome Cohen for some reminiscences of early days at the Burden Neurological Institute.

JEROME COHEN

Northwestern University, Chicago, U.S.A.

It is appropriate that we are meeting in Bristol on about the tenth anniversary of work leading to the discovery of the CNV, which was at the time of my first sabbatical leave to be spent here.

Many people have wondered why this particular type of discovery could be made here at the Burden Neurological Institute rather than elsewhere. In the first place this is England and the English climate is restrictive and keeps one indoors and at work. Secondly, the work took place in pastoral surroundings away from the middle of the city, and away from the extra-scientific distractions of a big university. The principal point, however, was that the Burden Institute was of the right size to provide a very happy marriage of scientists and clinicians who did not ignore each other's problems.

This interaction, with the clinicians bringing their problems to the research group, which understood and reacted constructively with the clinicians, resulted in better ways of diagnosing patients and monitoring treatment.

At our conference in Vancouver two years ago, Dr McCallum showed a slide of the first CNV that was recognized and recorded in the laboratory. Actually, there is published in the *Annals of the New York Academy of Science* on a conference on evoked potentials in 1963, by Grey Walter, a figure of slow potential shifts occurring between pairs of conditional signals, which were called 'secondary negative surges'. This was the same paradigm that was used to demonstrate the CNV about a year later, when a type TC EEG instrument made recording in the DC mode, or with extended time constants, possible. The CNV then emerged in its true ramp shape and was 'discovered', and then used in a clinical context.

Some Problems and Tentative Solutions to Questions Raised by Slow Potential Changes in Psychiatry

M. TIMSIT-BERTHIER, J. DELAUNOY and J. C. ROUSSEAU
Service de Psychologie Médicale, Université de Liège, Belgium

Since 1967 our research, centred on the study of CNV in psychiatry, has permitted us to replicate some variations in CNV amplitude, described previously by other laboratories (Abraham and McCallum, 1973). Approximately 25 per cent of psychotic subjects were found to have CNVs with amplitudes less than −5 μV. Anomalies were also found relating to CNV duration. Accordingly, in our statistics, approximately 75 per cent of 'prolonged' CNVs were found in psychotic subjects presenting a CNV with an amplitude greater than −5 μV.

Concentrating our research on CNV duration has increased our understanding of 'prolonged' CNVs and has enabled us to differentiate types of CNV according to their mode of resolution. Thus, we distinguish:

Type I. Normal duration: baseline return occurs immediately after the imperative stimulus.

Type II. Long duration: baseline return is delayed and the drop begins immediately after the imperative stimulus.

Type III. Baseline return is delayed, but there is no drop after the imperative stimulus. The curve has a 'plateau' aspect.

Type IV. Baseline return delayed; also the negative deflection increases in amplitude after the imperative stimulus, producing a characteristic 'dome' appearance.

Methodology and detailed results of this research are described elsewhere (Timsit-Berthier, 1973a, b). Our primary goal is to examine three types of problems raised by the discovery of 'prolonged' CNVs.

1. The problem of artefacts.
2. Problems of the stability and variability of these phenomena.
3. The question of significance.

ARTEFACTS

a. *GSR:* The cerebral origin of the prolonged CNV has been challenged by several authors (Picton and Low, 1971; Beaumanoir *et al.*, 1972).

In defence of a cerebral origin the following arguments are offered. The wave forms of curves contaminated by GSRs are different from the four types of curves described above. These contaminated curves were discarded in our research. Taking into account the work of Picton and Hillyard (1972) we have systematically injected atropine (0·1 mg) under each electrode (reference and active). We obtained the same results with and without the drug (*Fig.* 1).

b. *EOG:* As shown by Hillyard and Galambos (1970), the CNV may be contaminated by involuntary ocular movement, very common in mentally ill patients (Wallach and Wallach, 1964).

In order to avoid this complication we have recorded and averaged the EEG and EOG (recorded from right infra- and supra-orbital electrodes) with the same time constant (10 sec) and have eliminated trials in which slow potentials were evident in the EOG. Using this method, we have obtained long duration CNVs in conditions in which the EOG showed no variations. The fact that the EOG can on occasions show parallel wave forms does not mean that it is necessarily the EOG which is the source of

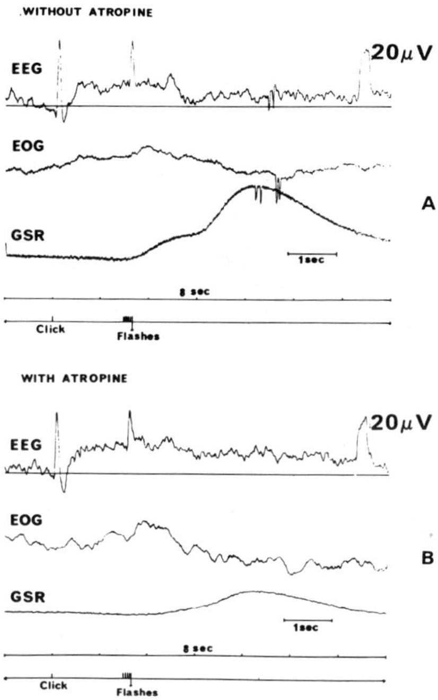

Fig. 1. CNV of a psychotic subject (20 trials; negativity upward). A, Normal recording. B, Recording after subcutaneous injection of atropine under each EEG and EOG electrode. Calibration signal at the end of EEG traces is $-20\ \mu V$.

the slow potential changes seen in the EEG. Indeed, in some cases the opposite may appear. This research is being continued in collaboration with Dr F. Bostem with a view to producing spatio-temporal maps.

Preliminary results indicate that the maximum of prolonged CNV negativity both in the pre-imperative segment and in the post-imperative segment is at its highest at the vertex (*Fig.* 2) and is consequently not ocular in origin.

c. *Medication:* Although it has not been possible to withdraw all medication from the patients examined, there are indications that psychoactive agents are probably not the factor leading to the appearance of post-imperative negativity in psychotic patients, since some patients, free of medication, have CNVs similar to those of patients under medication, while some patients have changed medication without modification in CNVs. Conversely, some CNVs curves have changed without change in medication. Nevertheless, specific research is required on this subject.

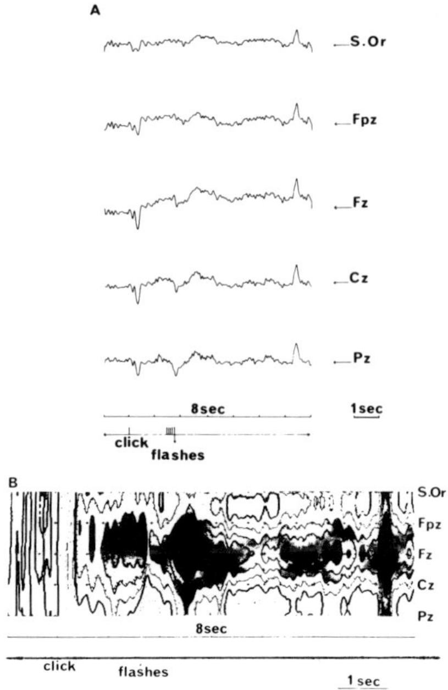

Fig. 2. A, Set of 5 CNVs with common reference derivations (left mastoid) in a rectilinear longitudinal configuration on the scalp (4 cm separates each electrode). B, Spatio-temporal representation by isopotential curves. Shaded area: Negative gradients (Pale grey 0–10 μV; Dark grey 10–20 μV; Black 20–30 μV). White area: positive gradients. Each successive isogradient line corresponds to a difference in slope of 5 μV/cm. (*Based on research carried out in collaboration with Dr F. Bostem.*)

STABILITY AND VARIABILITY

Once technical problems are solved, the question remains does the duration of the CNV and the form of its termination show stability over time and within a given individual? We have attempted to answer this question in two ways:

1. By prolonging the recording, beyond the 20 trials usually employed (short-term stability).
2. By having recourse to longitudinal studies which allow for repeated recordings in the same *S* at intervals of several days.

We have compared six successive CNVs (each the average of 12 trials) during two separate experimental sessions in 19 psychotic patients and 11 control *Ss*. *Table* 1 shows the results obtained. Control *Ss* generally presented stable curves. However, the three subjects who presented at the outset Type II curves evolved towards Type I curves. Psychotic patients had much less stable curves. Only seven of the 19 presented identical CNVs in the six successive sessions. Furthermore, in this group, Type II curves evolved either towards Type I, or towards Type III or IV. Although curve stability or variability is on the whole independent of the type of curve, Type II curves seem to be the least stable.

We have also compared the results of two sessions, separated by 8–15 days, in 30 psychotic patients and 12 normal *Ss*. *Table* 2 shows the results obtained.

Control *Ss* had perfectly stable curves.

Psychotic subjects had considerably less stable curves; half of them presented a modification in the CNV termination mode at the second recording.

Table 1. Short-term variability study. For each subject 6 CNVs were obtained (12 trials). The results of each subject in the 6 sessions are indicated by a cross.

	CNV Type III–IV	*CNV<5 μV*	*CNV Type II*	*CNV Type I*
CAL. Schizophrenia (paranoid type)				× × × ×
LAS. Unspecified neurosis	×			× × × × ×
MAS. Personality disorders (border-line)				× × × × × ×
DEM. Acute paranoid reaction				× × × × × ×
GAL. Paranoid state	× ×			× × × ×
VIN. Acute schizophrenic episode		×	×	× × × ×
LEV. Paranoia			× × ×	× × ×
TAE. Schizophrenia (simple type)	×	× × ×	×	×
DEN. Schizophrenia (paranoid type)	× × ×	× ×	×	
OUL. Acute paranoid reaction		× × × ×		× ×
GUEU. Affective psychose (depressed type)	×		× ×	× × ×
CAN. Schizophrenia (simple type)	× ×	×		× × ×
BR. Puerperal psychosis		× × × × × ×		
JEH. Schizophrenia (simple type)	× × ×	×		× ×
COEN. Schizophrenia (simple type)	× × × × × ×			
DAL. Paranoia	× × × × ×		×	
AC. Paranoia	× × × × × ×			
SOL. Schizophrenia (simple type)	× × × × ×		×	
LEC. Schizophrenia (simple type)	× × × × × ×			
Control subject Number 1				× × × × × ×
Control subject Number 2				× × × × × ×
Control subject Number 3				× × × × × ×
Control subject Number 4				× × × × × ×
Control subject Number 5				× × × × × ×
Control subject Number 6				× × × × × ×
Control subject Number 7			× × ×	× × ×
Control subject Number 8			× ×	× × × ×
Control subject Number 9			× × ×	× × ×
Control subject Number 10	× × × × × ×			
Control subject Number 11	× × × × × ×			

Table 2. Long-term variability study. First row=first experimental session. Four successive rows=second experimental session. The figures indicate the total number of cases presenting each type of CNV. The framed figures represent the cases in which the aspect of the CNV remained stable in both sessions. A=Control subjects; B=Psychotic subjects. Note the greater variability in the psychotic group.

A *Type III & IV CNV*	*CNV* $<5\ \mu V$	*Type II CNV*	*Type I CNV*		B *Type III & IV CNV*	*CNV* $<5\ \mu V$	*Type II CNV*	*Type I CNV*
				1st experimental session				
2	0	3	7		17	5	0	8
				2nd experimental session				
[2]	0	0	0	Type III & IV CNV	[7]	2	0	2
0	[0]	0	0	CNV $<5\ \mu$V	2	[2]	0	1
0	0	[3]	0	Type II CNV	4	0	[0]	1
0	0	0	[7]	Type I CNV	4	1	0	[4]

Stability or variability properties were independent of the types of curve.

One may conclude that, as reported by Callaway (1969) to be the case for evoked potentials in mental pathology, the CNV shows a high degree of variability in this condition. Psychotic patients tend to separate into two groups: those whose 'pathological' curves are permanent and those in whom the 'pathological' curves seem transitory. Although these patients may all be grouped within one general clinical category, it is relevant to ask whether this distinction is related to the reaction to the therapy or the prognosis of the disorder; for example, are the patients presenting permanent, prolonged CNVs the ones who are the most resistant to treatment?

SIGNIFICANCE

CNV research in psychiatry is of increasing practical interest to the clinician. Nevertheless, the detection of an 'abnormal' wave form does not provide a formal diagnosis, since a non-specific anomaly is found in most clinical categories.

The results of all research in our laboratory (Timsit-Berthier *et al.*, 1972; Koninckx, 1972) reveal the following percentages of Type III and Type IV CNVs. Of 92 control *Ss* with a CNV greater than $-5\ \mu$V, 10 per cent had Type III or Type IV CNVs; of 70 psychosomatic patients, 11 per cent; of 178 neurotic patients, 22 per cent; of 63 patients with personality disorders, 30 per cent; of 25 mentally retarded patients, 48 per cent; of 27 patients with affective psychoses, 56 per cent; and of 76 schizophrenics and paranoics 78 per cent had Type III or Type IV CNVs.

On the other hand, the finding of a normal wave form in a patient suspected of advancing psychotic processes excludes almost absolutely that diagnosis.

Of 21 subjects recorded from before treatment, at the height of a critical episode, 3 had CNVs less than $-5\ \mu$V, none had Type I, 1 had Type II, 8 had Type III and 9 had Type IV CNVs. It would seem that a psychotic state, in its initial stage, is accompanied by two particular CNV characteristics; zero amplitude CNV or prolonged duration CNV. In the light of these findings, CNV research in psychiatry could constitute a form of warning signal which would alert the clinician to a pathological state. However, in no instance, as shown from the stability studies, is the type of wave form obtained pathognomonic. The CNV data must always be interpreted in conjunction with clinical findings. Indeed, the significance of similar wave forms could be quite different when recorded in different pathological states (Netchine, 1969).

CNV anomalies must be interpreted in a different way depending on whether they occur in isolation or whether they are accompanied by other anomalies in the event related potentials.

When the CNV is 'abnormal', while MPs and sensory EPs are 'normal', one might reasonably suspect that the attitude of the subject towards the experimental session may be resulting in unusually protracted expectancy or attention. If this is the case, it should be possible to prolong artificially Type I CNVs in normal control *Ss*, for example by giving contradictory instructions and by making the experimental situation less easily resolved.

When the 'abnormal' CNV accompanies abnormal MPs and EPs, one might expect the disturbance to be at a more general level of reactivity in the brain. In such a case, CNV studies can afford only complimentary information, the anomalies being equally apparent from more elementary tests.

The CNV and its Relation to Specific Psychiatric Syndromes

P. ABRAHAM

Royal Victoria Hospital, Netley, Southampton, U.K.

W. C. McCALLUM

Burden Neurological Institute, Bristol, U.K.

J. GOURLAY

The Maudsley Institute of Psychiatry, London, U.K.

INTRODUCTION

Low CNV amplitude and prolonged post-imperative stimulus negativity were reported in psychiatric patients by McCallum and Walter (1968) and Walter (1966). More recently the simultaneous presence of both these features has been found in groups of schizophrenic patients by Small and Small (1971), Timsit-Berthier (1973) and Abraham *et al.* (1974). However, three laboratories have reported one or other of the features to be present to a significant degree in a wide variety of other psychiatric disorders and there is undoubtedly a substantial overlap of values between schizophrenic and normal control groups. This might suggest that mental illness is a multidimensional continuum with schizophrenia at one extreme, but it does not resolve the question of whether the continuum is one of inherited and/or environmentally induced predisposition to a given disorder, or whether it is one of degree of disturbance of function from a normal baseline—or both, rather like epilepsy. Perhaps the more important question is: can a physical phenomenon such as the CNV be used to measure susceptibility to, or current severity of, an illness? If so can it then provide an objective predictor, diagnostic aid and guide to treatment and management in the less floridly ill patients and those from whom it is difficult to obtain a reliable account of their symptoms?

If CNV measures relate to diagnostic features of mental disorder, then there is at least a possibility of contributing to the prognosis and clinical assessment. If, however, the CNV anomalies relate to non-specific features of mental disorder, such as simple depression, which are shared by several conditions, or to some general undifferentiated disturbance of mental function, then the potential contribution to psychiatry of the CNV is limited.

In order to attempt to answer these questions it has been necessary in the present study to pay attention not only to the clinical quality of CNV recording but to use more stringent and standardized procedures for assessing and evaluating the psychiatric disorders concerned.

METHOD

Successful CNV recordings were made from 20 control *Ss* and 42 clinically diagnosed schizophrenic patients according to the method described by McCallum and Abraham (1973). *Ss* were all servicemen and the groups were approximately matched for age, social and marital status, intelligence and ethnic origin. The Present State Examination (PSE), noted for its reliability in measuring and classifying psychiatric state (Wing *et al.*, 1967), was carried out on the first occasion, before treatment had been given, and on the last occasion when most patients had both received treatment and had largely recovered from their acute illnesses. No control *Ss* were receiving treatment. The PSE on the second occasion disregarded all circumstances prior to one month preceding that assessment. It therefore happened that with most patients both the severity of the individual syndromes derived by the CATEGO programme from the PSE data gathered, and the classification derived by the same programme from those syndromes, differed quite markedly from the results of the first assessment. For example, if auditory hallucinations had ceased for a month, that syndrome rated zero; and with good recovery some schizophrenics emerged with neurotic classifications. CNV data only from the two recordings when the PSE was carried out will be presented, in order to relate the physical and clinical findings as closely as possible.

Four features of CNV, recorded from F_Z, C_Z and P_Z locations, were studied. These were:

1. Acquisition Amplitude (AA). The mean amplitude of the CNV over a period of 200 msec prior to S_2.
2. Distraction Amplitude (DA). Similarly calculated from the second average of 8 trials during the distraction sequence.
3. Prolongation. A mean microvolt amplitude measure of the PINV from S_2+350 msec to S_2+3500 msec.
4. Parietal Spread. The difference in microvolts between the amplitude of the CNV measured at P_Z from that at C_Z. (A negative value indicates a higher CNV at P_Z). The significance of this feature has been commented on by Abraham *et al.* (1974).

The baseline in all cases was obtained from the mean EEG activity during 1 sec prior to S_1.

RESULTS

I. Classification

The patients were classified by their present state, i.e. without regard to their family history, previous psychiatric history or premorbid personality, into one of 50 groups. About 50 per cent, who would have been described by most clinicians as florid schizophrenics, fell within the 'Group One'. Allocation to this Group depended largely on the presence of the Nuclear Syndrome (NS), which in turn depended substantially on the unequivocal presence of Schneider's 'First Rank' symptoms. A further 50 per cent of the patients (overlapping the Group One patients) were classified at some stage in one of 14 groups having as a common feature manic or depressive symptoms prominent relative to other symptoms. These have been collectively designated as the 'Manic-Depressive Group'.

1. Acquisition amplitude was significantly lower in Group One than in both control and other patient groups. The variability from *S* to *S* was less in every group on the second occasion. Each group showed consistency in its mean acquisition amplitude on retesting (*see Tables* 1a and b).

Table 1a. Vertex amplitude values on first test and on retest

		Controls	*All patients*	*Group one patients*	*Other patients*
Test I	Acquisition	−18 μV (SD=7·37)	−14 μV (SD=9·14)	−11 μV (SD=7·86)	−17 μV (SD=9·45)
	Distraction	−12 μV (SD=7·16)	−8 μV (SD=8·48)	−6 μV (SD=6·67)	−10 μV (SD=9·73)
Test II	Acquisition	−15·5 μV (SD=6·27)	−14 μV (SD=8·34)	−10 μV (SD=6·12)	−18 μV (SD=8·76)
	Distraction	−12·5 μV (SD=4·03)	−8·5 μV (SD=6·89)	−6·5 μV (SD=5·67)	−10·5 μV (SD=5·67)

Table 1b. Significance of differences (*t* test) between group mean amplitudes

		Controls/ all patients	*Controls/ group one patients*	*Controls/ all others*	*Group one patients/ all others*
Test I	Acquisition. *P*=	0·05	0·005	NS	0·025
	Distraction. *P*=	0·05	0·01	NS	NS
Test II	Acquisition. *P*=	NS	0·005	NS	0·0025
	Distraction. *P*=	0·025	0·0025	NS	0·05

2. The distraction amplitude of the entire patient group was significantly lower than that of the control group, the difference again being most marked in 'Group One' patients, though this distinction was not as sharp as that produced by acquisition amplitude. All groups showed an increased amplitude on retesting and the variability from subject to subject was this time considerably less in every group on the second occasion. The groups were, however, more consistent in their mean distraction amplitude, and in the relationship of this value to that of the other groups, as shown in *Tables* 1a and b, than was the case with acquisition amplitude. The correlation coefficient between AA and DA was 0·7.
3. The mean Prolongation amplitude of the patient group was −3 μV while that of the control group was +2·4 μV. These differences were significant beyond the $P=0{\cdot}0005$ level although it diminished somewhat with retesting. Further analysis showed that Group One patients retained 75 per cent of their prolongations on retesting while the negative shift returned from a mean of −3·9 μV virtually to baseline in the remainder. A similar consistency of prolongation was seen in the 'Manic-Depressive Group'.
4. Mean 'Parietal Spread' values were: Controls=6·75 μV, Patients (other than Group One)=6·22 μV, Group One patients=3·72 μV. The differences between Group One and the other two groups were significant ($P<0{\cdot}05$). The mean values of the patient groups were consistent on both test occasions. The difference between Group One and other patients increased in significance ($P<0{\cdot}025$) on retesting.

II. Syndromes

All Subjects (*Patients and Control*). Thirty-six syndromes were rated on each occasion for each subject, a score of 3, 2, 1 or 0 being allocated according to whether the syndrome was present floridly, definitely, uncertainly, or not at all. Each of the 36 syndromes rated on the first occasion was cross correlated with each of the last three CNV features studied.

One syndrome, Hysteria (HT), did not occur in any subject. Ninety-five out of the remaining 105 tables resulted in correlation coefficients with negative values. That is, increasing syndrome severity was associated with decreasing distraction amplitude, increasing post imperative stimulus negativity and decreasing C_Z–P_Z values (increasing 'parietal spread'). Thirteen of these correlations were significant ($P<0{\cdot}05$). Thus those features of the CNV which serve to distinguish psychotic groups can also be representative of a wide range of disturbance of mental function.

Nine patients had obsessional symptoms at the time of the first test. Obsessional Neurosis (ON) was the only syndrome which correlated positively with all three features of the CNV. It was associated with rising values of 'Distraction Amplitude' and increasing C_Z–P_Z values. These two features also correlated significantly with CNV Acquisition amplitude ($r=0{\cdot}73$ and $0{\cdot}66$ respectively; $P<0{\cdot}01$). The amplitude of the basic CNV in 13 patients who had obsessional symptoms on one occasion or another rose with the severity of these symptoms:

Obsessional Neurosis Syndrome Severity	Doubtful	Definite	Severe
Mean amplitude of CNV in microvolts	8	13	22

Previous studies have indicated that obsessional symptoms on their own are linked with high CNVs.

Patients Only. The marked differences in the CNV variables between different groups of patients defined by the PSE suggested that it could be worth while to investigate the different incidence of syndromes in groups of patients defined by their CNV variables.

A comparison was made of the syndrome profiles of 2 groups of patients, the grouping being determined by the extent of their Parietal Spread. The 15 patients with the lowest C_z–P_z values, i.e. greatest parietal spread, were compared with the 15 having the highest C_z–P_z values, i.e. least parietal spread.

Syndrome scores were calculated for each group on the basis $\sum_{i=0}^{3} in_i$, where i was the level of severity and n_i the number of patients displaying the syndrome at severity i. The group scores were then compared, syndrome by syndrome. In 24 syndromes the scores were higher in the group with the greatest parietal spread, suggesting that this group was more ill in a diffuse way. This difference was significant on a two-tailed test ($P<0{\cdot}01$). The same method applied to the second stage results yielded substantially smaller syndrome scores. Most syndromes, particularly the schizophrenic syndromes, continued to predominate in the parietal spread group.

Fifteen syndromes have been selected for illustration (*Fig*. 1). They include the 11 syndromes showing group differences of 9 or more, the higher score invariably being in the parietal spread group. Syndromes showing differences of 3 or more in the reverse direction (viz. SD, SL, LE, IT, HY) are also illustrated.

Half of the 10 syndromes preponderant in the parietal spread group (NS, AH, PE, RE, SF) have clear schizophrenic connotations.

The results suggest that even within a single group of schizophrenic patients the syndrome loading, particularly of schizophrenic syndromes, increases with parietal spread.

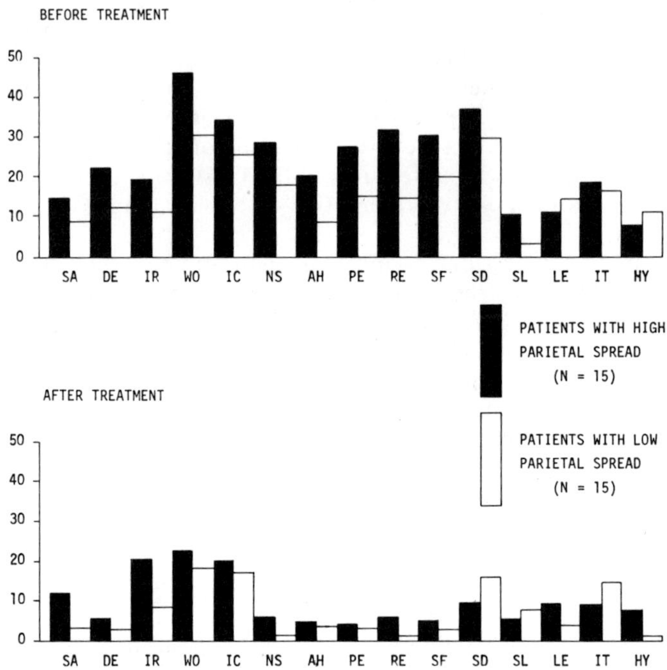

Fig. 1. Histograms showing the relative severity of syndromes detected by the PSE in two groups of patients, one selected on the basis of high parietal CNV spread, the other on the basis of low parietal spread. SA, Situational anxiety; DE, Depersonalization; IR, Ideas of reference; WO, Worry; IC, Loss of interest and concentration; NS, Nuclear syndrome; AH, Auditory hallucinations; PE, Delusions of persecution; RE, Delusions of reference; SF, Sexual and fantastic delusions; SD, Simple depression; SL, Slowness; LE, Lack of energy; IT, Irritability; HY, Hypochondriasis.

DISCUSSION

The syndrome results show that the CNV features studied are related both to a general disturbance of mental function and to a variety of specific, particularly 'schizophrenic', syndromes.

The consistency of the acquisition amplitude and distraction amplitude features regardless of treatment and clinical state suggests that they are related more to the patient's constitution than his current illness. Prolongation on the other hand, tended to disappear with treatment and recovery, suggesting some link with the current state, particularly in those groups of subjects whose illnesses were relatively lacking in psychotic, i.e. schizophrenic or manic-depressive, features. The consistency in groups of subjects with such features provides some support for the PINV findings of Timsit-Berthier (1973), although the PINVs in the present study tended to be morphologically different from those described by the Liège group.

The finding of consistent prolongations in those patients with manic-depressive features parallels the finding of prolongation in manic-depressive patients (Timsit-Berthier, 1973). Small and Small (1971, 1973) on the other hand did not find such prolongations in their manic or their depressive patients. The presence of prolongations in the manic-depressive group of the present study cannot be accounted for merely by the simultaneous presence of schizophrenic symptoms, because in those cases where manic symptoms were overwhelming (most clinicians would have regarded them simply as manic patients at this stage) the prolongation effect was not only present but showed a higher mean amplitude (—5·8 μV) than any other group. Also the syndrome 'Hypomania' (HM) has a higher correlation with Prolongation than with the other two CNV variables, and ranks eighth among the correlations between 'Prolongation' and the 36 syndromes.

Wider use of a standardized and reliable method of psychiatric assessment, such as that described, would greatly facilitate comparison of neurophysiological data from different centres.

Parietal Spread also seems to be linked with the Group One patients, both when they are acutely ill and untreated, and when they have recovered with treatment, suggesting, as in the case of acquisition amplitude and distraction amplitude, a relationship more with underlying constitutional factors than with fluctuations in the illness. There is also a consistent inverse correlation of Parietal Spread with age as shown by the following correlation coefficients:

	1st Recording	Final Recording
PATIENTS	—0·31	—0·45 $P<0{\cdot}01$
CONTROLS	—0·32	—0·20
BOTH	—0·33 $P<0{\cdot}01$	—0·36 $P<0{\cdot}01$

Limitations on number of channels available has regrettably restricted the Parietal Spread measures to differences between only one pair of electrodes. Nevertheless the strength of the relationships revealed strongly suggest the need for further study of this effect, using a more extensive electrode montage to determine the extent of this spread and its significance, both functionally and clinically.

SUMMARY

Clinical evidence has been produced to show that 4 features of the CNV, low acquisition amplitude, low distraction amplitude, prolonged post imperative stimulus negativity and small differences between CNV amplitude at C_z and P_z, are linked with reliably defined psychiatric syndromes and diagnostic groupings. These features are neither caused nor obliterated by treatment and on the whole seem more closely related to the underlying constitution than to phases of illness. It is suggested that the CNV has a definite place in the diagnosis of psychiatric disorder.

ACKNOWLEDGEMENTS

The assistance of members of the staff of the Royal Victoria Hospital, Netley, the Burden Neurological Institute and the U.K./U.S. diagnostic Project at the Maudsley Institute is gratefully acknowledged.

Evaluation of Event-related Slow Potentials in Selected Groups of Psychiatric Patients

B. DUBROVSKY and M. DONGIER

Allan Memorial Institute, McGill University, Montreal, Quebec, Canada

The significance of event-related slow potentials (ERSP) in psychiatry, after some 10 years since the original work of Walter *et al.* (1964), is still very controversial.

The clinical significance of these tests in relation to psychiatry has recently been reviewed in the light of the experience of the group in Liège (Belgium) (Dongier, 1973; Dongier *et al.*, 1973). Most of our observations obtained in Montreal, in a different environment, with different sources of psychiatric patients and with a slightly different methodological approach, are in general agreement with the results obtained in Liège. In our view, they confirm the need for this kind of functional exploration as a tool in psychiatric evaluation from the diagnostic, prognostic and research perspectives. We have therefore extended our studies to a wide range of psychiatric disorders.

METHOD

Two hundred and fifty patients with various psychiatric diagnoses were studied. A detailed description of the methods used has been published elsewhere (Dubrovsky *et al.*, 1973a). However, the following aspects require additional explanation.

Contamination of the CNV by potentials emanating from other sources, e.g. electrical fields of the eyes and the skin, present methodological problems which increase when psychiatric populations are studied. Since the level of ambient light is an important factor in determining the amplitude of the electro-oculogram (Pasik *et al.*, 1965), which is larger in high levels of illumination, recordings were made with the patients in a comfortable darkened environment. To minimize further the involvement of eye potential contamination, only auditory warning and imperative stimuli were used. Davis *et al.* (1972) have described the effects on EP of sequential stimulation of the same sensory modality but we have found no deleterious effects on the CNV.

Eye potentials were monitored by placing two electrodes diagonally across the eye. Using a tangent screen and perimeter, eye movements were calibrated periodically for size and direction and the appropriate sensitivities set. During the course of the experiment, trials showing eye movements greater than 5° were discarded, as proposed by Papakostopoulos *et al.* (1973). A second method of evaluating the effects of eye potentials on the slow brain potential was subsequently introduced. This record from a supraorbital electrode referred to the common ear lobe electrodes, as advocated by Regan (1972), to which the vertex electrode was also referred. The same amplification factor and time constant (9·5 sec) was used for both channels.

The higher incidence of skin potential changes in certain psychiatric populations, such as schizophrenics (Venables, 1960), presents a special problem. This problem was minimized by the use of a vertex-ear lobes derivation, shown by Picton and Hillyard (1972) to be almost devoid of phasic skin potentials. Nevertheless, skin potential changes were recorded between mastoids. Utilizing this information and that derived from the EOG, between 45 and 50 per cent of our records were rejected.

RESULTS

Timsit-Berthier *et al.* (1972) have shown that in most normal *Ss* there is a rapid resolution of the CNV and return to the baseline after the *S* has responded. In a small proportion of a normal group (<15 per cent), return to the baseline is delayed but in these cases the negativity usually begins its decline immediately after S_2.

In the majority of psychotic patients, the CNV is prolonged in the form of a post-imperative negative variation (PINV) (*Fig.* 1). We have arbitrarily defined an 'abnormal' PINV as one extending for more than 2 sec after S_2. Post-imperative negativity may be quite independent of the CNV and obtained without a warning stimulus (Dongier *et al.*, 1973).

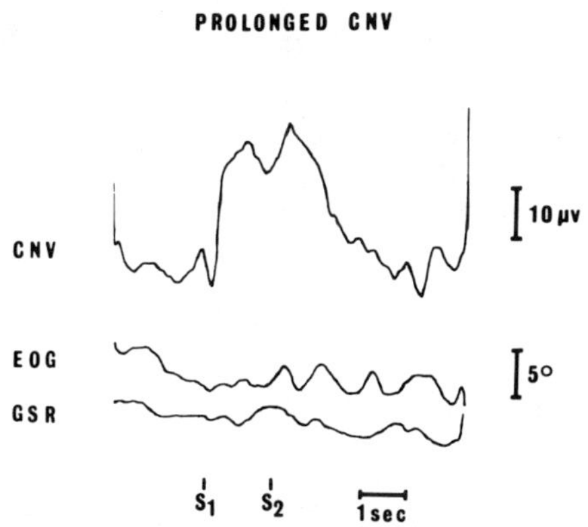

Fig. 1. Average of vertex CNV and PINV with EOG, derived from supra-orbital electrode referred to ear lobes, and electro-dermal controls.

A comparison between the results of the first 113 patients studied in Montreal and the comparable groups studied in Liège is given in *Fig.* 2. In the larger psychiatric groups subsequently studied it was found that 109 out of 146 (75 per cent) schizophrenics showed a prolonged PINV, as compared with 37 per cent of a group of neurotic patients and 15 per cent of normal control *Ss*. If the schizophrenic group is broken down into early schizophrenics (operationally defined by the presence of hallucinations and/or delusions for more than a month and less than 6 months), and chronic schizophrenics, a marked difference is apparent. An abnormally prolonged PINV is present in 65 out of 69 early schizophrenics (94 per cent). This finding replicates the previous reports of Timsit *et al.* (1968, 1970) in Liège. It is of practical value in that in any case where the diagnosis of early schizophrenia is doubtful, the finding of a normal CNV would seem to be a fairly strong argument against it.

An element which was not apparent from the previous studies in Belgium, but which is evident from the present studies, is the prognostic significance of the PINV in neuroses. Of 24 subjects clinically diagnosed as neurotic by psychiatrists from the Allan Memorial Institute, 10 patients categorized at the end of their hospitalization period as having a bad prognosis presented PINVs of abnormal duration. In the remaining 14 cases, considered to have a favourable prognosis, only two abnormal PINVs were found. These findings depend upon an ability to distinguish between 'true neurosis' (phobic, obsessional, conversion manifestation) and chronic personality disorders ('character neurosis'). In the latter cases even when they were severe and seriously affected the patients' social behaviour, slow potential abnormalities were found to be present in only 10 cases out of 43.

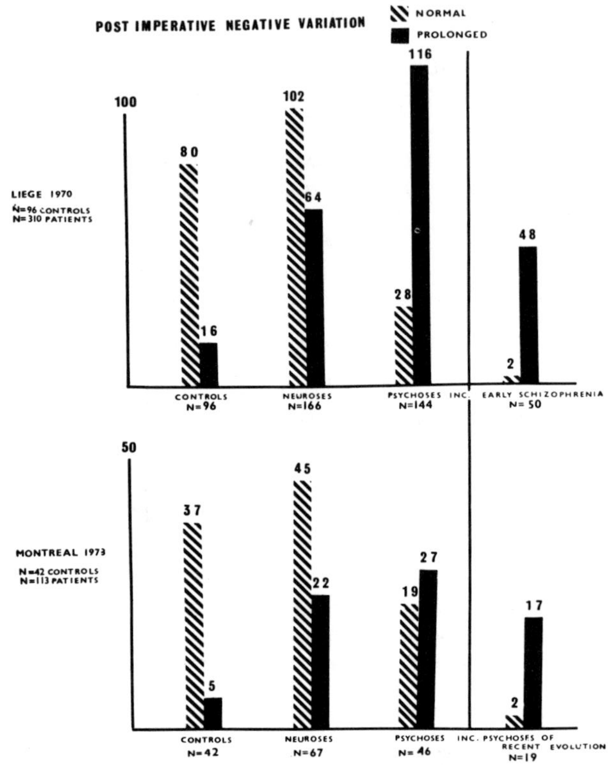

Fig. 2. A comparison of PINV results obtained in psychiatric patients and normal controls in Montreal and Liège.

In a previous comparison (Dongier *et al.*, 1973) of PINV and NPRP* in different groups of patients, the NPRP was found to be a significant feature in psychotics in evolution or remission; it seemed to constitute a permanent feature of the psychotic potential, whereas the PINV tended to follow the evolution of the clinical symptoms and to normalize during the remission period (Delaunoy, 1971). This was not so in the present series. Of 19 psychotics in the acute phase of the disorder, the NPRP was abnormal in 12 and the PINV in 17 cases. Of 18 chronic psychotics the NPRP appeared in only one case,

* NPRP (negativity persisting after readiness potential); operationally defined as negativity lasting more than 500 msec after the initiation of a voluntary motor response.

the remainder showing a normal RP. The PINV showed abnormal prolongation in 7 of the 18 cases.

More precise clinical data is required before an adequate evaluation of the relative diagnostic value of the NPRP and PINV can be made.

DISCUSSION

The possible functional significance of the abnormally prolonged negativity in a significant number of mentally ill patients can only be surmised at present.

Available evidence indicates that the neurophysiological state of the brain may be different during the CNV and the PINV. Evoked potentials elicited during the PINV phase are of lower amplitude and longer latency than those during the CNV. This has been interpreted (Timsit-Berthier *et al.*, 1971) as an index of decreased cortical reactivity during this stage.

More recently we have shown (Dubrovsky *et al.*, 1973b) that hyperventilation selectively affects the post-imperative negativity, significantly decreasing its duration without seriously affecting the amplitude and duration of the CNV in control subjects. Although the PINV is a valuable index of mental pathology, it is not strictly specific for any particular nosological entity as classically described. We are, therefore, exploring possible correlations with specific features of certain syndromes. We agree with Abraham *et al.* (p. 144) that clinical correlations are likely to be more effectively established with well defined symptoms or syndromes (such as auditory hallucinations or delusions of reference) than with broad diagnostic entities such as schizophrenic or manic depressive psychoses. We do not, however, expect to find a simple linear relationship between these potentials and any single psychological dimension.

Another context in which psychiatric concomitants of ERSP are of interest is that resulting from the work of Hubbard *et al.* (1971), who have revealed a powerful influence of beta-adrenergic catecholamines in the generation of steady potential shifts.

These findings suggest the need for a combined, biochemical-neurophysiological approach to explore the possible link between catecholamines, slow potentials and some aspects of mental disease.

Event-related Slow Potentials in Mental Retardates

RATHE KARRER and JUDY IVINS

Illinois State Pediatric Institute, Chicago, Illinois, U.S.A.

Mental retardation refers to a heterogeneous array of individuals whose main common feature is their reduced cognitive and social performance. This reduced performance is often attributed to arousal (Lindsley, 1957; Clausen, 1966), orienting (Luria, 1963), motivation (Zigler, 1967), attention (Zeaman and House, 1963), and information processing (Spitz, 1966). It is probable that some or all of these processes are deficient in some retardates.

The CNV, as an electrophysiological concomitant of an attentional-motivational state in which the organism is set to receive and process information, may be expected to reflect perceptual, cognitive, and performance deficits attributed to the mentally retarded. If it is true that negative changes in steady potential (SP) level accompany increased cortical excitability and attending behaviour, the study of SP activity in the mentally retarded might provide an insight into the bases of their impaired behaviour.

Two experiments were designed to obtain basic information on SP characteristics in the retardate. The studies were also designed to contribute information to understanding the relation of SP changes to behaviour. A simple developmental paradigm was used to compare retardates equated for mental age (MA) and normals equated for chronological age (CA). A reaction time (RT) task and a dark interval threshold (DIT) task allowed recording during the traditional expectancy paradigm of the CNV and during a more complex perceptual process. Each task required different aspects of temporal processing and preparation to respond. These tasks were chosen for their known power to discriminate retardates from normals (retardates slower than normal) and their demonstration of developmental trends (young children slower than older children).

METHOD

The details of the DIT methodology are described by Thor and Thor (1970) with the exception that 20 msec flash durations were used. A hand held push button allowed *S* to initiate the stimulus sequence at his optimal readiness. Each trial consisted of the experimenter's verbal 'ready' (orienting) signal followed by *S*'s fixation of a small red disc and voluntary button press activating the tachistoscope 1 sec later. *S* reported (decision) whether he had perceived one or two flashes. Steady potentials were recorded from F_Z and C_Z sites referred to left mastoid using Grass (Model 5) DC amplifiers and an Ampex (FR1300) FM tape recorder. Vertical eye movement recordings allowed us to discard trials contaminated by eye movement. EEG epochs beginning 200 msec prior to and ending 7·2 sec after the ready signal were averaged in two ways: (a) timelocked to the ready signal, and (b) timelocked to *S*'s button press. From these averages four measures of SP activity were determined: (1) The area under the curve for the initial 0·6 sec after the ready signal, designated the *Orienting-Ready SP*

(ORSP); (2) The area 0·5 sec prior to button press, designated the *Pre-Motor SP* (MSP); (3) The CNV area measured over the 1·2 sec between button press and the longest 2-flash interval; (4) The area under the curve during the 400 msec following the second flash, defined as the *Decision SP* (DSP). The MSP, CNV and DSP were all determined from the button press average.

The RT task procedure is described elsewhere in this volume (p. 132). It should be noted that identical CNV measures were derived in both tasks. Fifteen retarded *Ss* were drawn from the population of the Edward R. Johnstone Training and Research Center. Their ages were 13–18 ($\bar{x}$=16·1) with IQ 48 to 73 ($\bar{x}$ IQ=63; $\bar{x}$ MA=10·2). None had neurological impairment as judged from medical histories. Two groups of 13 normal subjects, one matched to the retardates age (CA) and the other matched to the retardates mental age (MA) have been described previously (p. 132).

RESULTS

Retardates were significantly slower in RT, had higher DIT thresholds, and took longer to make a decision after perceiving the flashes. This confirms the poorer performance previously found by others on these tasks (Thor and Thor, 1970).

It appears (*Fig.* 1) that the retardates have smaller mean CNVs in the DIT task and smaller mean PWSPs in the RT task (*see* p. 133, *Fig.* 1), but these were not statistically significant on either task. Neither were amplitudes of the ORSP, MSP or DSP different in retardates.

Computation of the difference between vertex and frontal amplitudes and latencies indicated that the retardates had a different topography of SP activity in the RT task. Retardates had a significantly smaller C_z–F_z difference of the amplitude and peak time of the PWSP. This was also true for the RT/CNV slope and amplitude although significant only for the former. The frontal CNV slope (rise time) was about 100 msec longer than vertex CNV slope for both normal groups but simultaneous in the retardates, i.e. activity seems to build up in frontal areas at the same time as at the vertex for the retarded.

The topography of the retardate's SP activity was also different in the DIT task. The difference in topography was often such that the retardates and young MA matched children were similar in vertex activity. For example, the slow decision time retardates and MA children had less vertex-frontal difference in MSP than slow decision time CA children. This was due to reduced frontal negativity in the MA normals and retardates. As an additional example, fast decision time older children had larger vertex-frontal differences for CNV and DSP than slow decision time *Ss*. This was due to smaller vertex negativity of the older children. Neither the retardates nor young children reflected these differences that accompanied good performance.

The relationship between SP activity and performance on the DIT task was different for retardates and normals. There was less ORSP positivity for low threshold (better performing) normals than for high threshold normals at the vertex. Conversely, there was more ORSP positivity for low threshold retardates than for high threshold retardates. Low threshold *Ss* had larger vertex MSP negativity than high threshold *Ss* in all groups combined. There was an interaction: High threshold normals had vertex positive MSPs while low threshold normals had vertex negative MSPs. High threshold retardates had less vertex MSP positivity than low threshold retardates. Combining groups, fast decision time *Ss* had larger CNVs and DSPs than slow decision time *Ss* in both frontal and vertex. Slow decision time retardates again had smaller frontal CNVs than age matched slow normals.

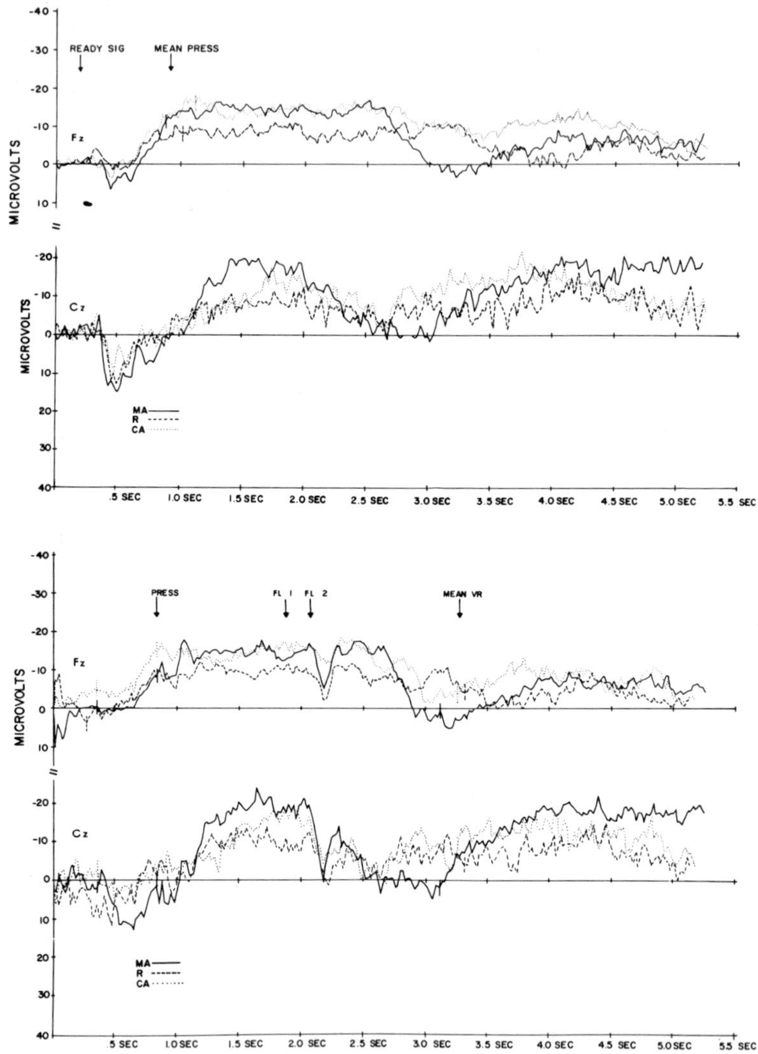

Fig. 1. Group averages of activity time-locked to the ready signal initiating a trial (*upper*) and to *S*'s button press initiating the presentation of the two flashes (*lower*).

DISCUSSION

Contrary to expectations, CNV (and DSP) amplitude did not discriminate groups or high/low threshold *Ss*. However, fast decision time *Ss* had larger CNVs and DSPs than slow *Ss*. Therefore, the CNV and subsequent DSP negativity late in the trial were more related to cognition than to perception.

Perception was related to MSP and ORSP, which occurred earlier in the trial. Faster processing (low thresholds) was preceded by larger negative MSP and smaller positive ORSP. In fact, slow perceptual processing was accompanied by positive MSP.

Although amplitudes of the SP activity did not differentiate the groups, the retardates did not always reflect the same relationships between performance and SP activity found in age matched normals (e.g. the low threshold retardates had larger ORSPs and positive motor readiness potentials). This may imply that retardates use different underlying processes in performing the tasks.

The timing and topography of the SP activity was also different. The CA normals had larger topographical differences and the best performance indicating better differentiation. Fast decision time CA normals had a greater topographical difference because of predominant frontal CNVs. Slow retardates substantiated this trend by having the smallest frontal CNVs. This clearly indicates that faster decision processes are accompanied by greater frontal negativity—a characteristic lacking in retardates.

The pervasive topographical difference between retardates and normals implies a lack of differentiation of processes underlying task performance. This may be due to improper timing of impulses reaching different brain areas or to different underlying processes and/or cognitive strategies. The synchrony, phase, and timing of brain activity is crucial to adequate performance. The topography of frontal and vertex midline sites points to a potentially lucrative payoff from studies using a full montage during motor, and cognitive tasks.

SUMMARY

Retardates, MA-matched normals and CA-matched normals were studied on RT and 2-flash perceptual tasks. The results indicated that retardates differed from CA normals in topography and timing of SP activity. The relation of SP activity to performance was also different in retardates.

ACKNOWLEDGEMENTS

Don Thor, Betty Karrer, and Dave Hoats collaborated in the DIT study. Thanks are due to Herman Spitz and John Winters for facilitating the studies and Anne Campbell for diligent computer analysis.

CNV in Patients with Frontal-lobe Lesions and Mental Disturbances

R. ZAPPOLI, M. PAPINI, S. BRIANI, P. BENVENUTI and A. PASQUINELLI

Clinica delle Malattie Nervose e Mentali, Policlinico Universitario, Careggi, Florence, Italy

Following the work of McCallum *et al.* (1970) who examined the CNV in patients with various brain lesions, study is being made on slow potential shifts in *Ss* with known frontal-lobe lesions and different mental disturbances.

The experiments reported here were carried out in order to obtain information regarding (*a*) possible relationship between site and extension of frontal-lobe lesions, features of the CNV patterns and of some correlates and neuro-psychic symptoms; (*b*) possible utility of the CNV as a diagnostic-prognostic aid in *Ss* with frontal-lobe lesions.

METHOD

Eighteen selected patients (12 males, 6 females) aged 31–68 (mean 49·1) and a control group of 6 volunteer *Ss* (4 males, 2 females) aged 24–36 (mean 28·3) were studied.

The 18 patients included 7 unilateral prefrontal tumours and 11 bilateral frontal-lobe lesions. Of the 7 patients with surgically verified unilateral tumours, 5 were affected by glioblastoma multiforme (3 on right, 2 on left), 1 by a large right cystic glioma and 1 by a right meningioma. In these cases the destructive lesions certainly involved Brodmann's area 9, sometimes extending to areas 10 and 46, and/or Foerster's $6\alpha\beta$ and $8\alpha\beta\delta$ areas. All patients suffered from more or less serious mental disturbances typical of the 'frontal-lobe syndrome', progressive dementia, especially with impairment of vigilance, attentional and cognitive functions, of association processes and temporo-spatial relations, slowness in motor performance and in formulation of ideas, poor memory for recent events and behavioural disorders.

Of the 11 patients with bilateral frontal-lobe lesions, 3 had a large meningioma of the anterior third of the falx which by compressing produced destructive lesions prevalently of the medial surface of Brodmann's areas 9 and 10 and sometimes on the anterior part of Foerster's $6\alpha\beta$ area on both sides. Two had malignant tumours (right glioblastoma multiforme and left angioblastoma) extending from the depth of the prefrontal region of one frontal lobe through the corpus callosum, with extensive infiltrations into the opposite prefrontal areas. Two others had large aneurysms of the anterior communicating artery, involving also the system of both anterior cerebral arteries, with extensive damage in both prefrontal regions. All 7 patients had more or less severe frontal-lobe syndromes including psychic symptoms. Two other patients had marked atrophy of both frontal lobes with progressive dementia, slighter in one case, typical of Pick's disease. Two cases had undergone—one 15, the other 18 years previously—a Freeman and Watts bilateral prefrontal lobotomy for chronic obsessional neurosis, one with high anxiety. This psycho-surgical procedure interrupts especially the interconnections between prefrontal cortex and dorsomedial nuclei of the thalamus.

The CNV paradigm in all cases was the standard S_1 (click)–1 sec–S_2 (moderate intensity repetitive tones) R (button press terminating S_2). 80–100 S_1–S_2 trials were presented at variable i.t.i.s ranging from 10 to 60 sec.

All the patients were able to understand the instruction to stop S_2 as quickly as possible, and during the test procedure produced fairly consistent operant responses, sometimes with short RT.

EEG recording was obtained from C_Z-linked mastoids, F_4–T_4 and F_3–T_3 (10–20 system). Vertical EOG, finger photo-plethysmogram, EMG of acting muscles, stimuli and RT to S_2 were also recorded. EEG and EOG_V were amplified using a Grass 7PIA chopper-stabilized low-level DC pre-amplifier, modified to introduce a time constant of 6 sec, mounted on a Grass 10-channel AC-DC polygraph.

On-line averaging of EEG and EOG_V trials was obtained by a Hewlett-Packard signal analyser with a 5487A 4-channel input. For monitoring purposes this signal analyser was modified to provide on-line averaging on the primary paper records. Additionally, polygraph data were stored on a Hewlett-Packard mod. 3955D FM tape recorder for further off-line analysis.

Average evoked responses to sets of 8 stimuli were taken during trials which appeared relatively free of eyeblinks and other movement artefacts. Amplitude of the CNV was expressed as a mean amplitude during the 150 msec prior to the onset of S_2, the baseline for this measurement being taken as the mean amplitude of the EEG during 800 msec immediately prior to S_1. The experimental procedures have been described in more detail elsewhere (Zappoli *et al.*, 1971).

All 14 patients with frontal lobe tumours or aneurysms were examined before surgery. In 8 the examination was repeated 30–60 days after surgery.

RESULTS

In all control *Ss* normal CNVs were observed at the vertex (Mean —25·6 μV; range −19·5 to −36 μV) in association with short RT (Mean 153·2 msec; range 88–210 msec). Smaller symmetrical CNVs (Mean −13·1 μV; range −6 to −17·5 μV) were recorded also from F_4–T_4 and F_3–T_3. These sometimes showed a different form with a slower return to baseline.

In 10 patients no CNV activity in the usual sense was observed. In a few cases small and inconsistent negative shifts during the i.s.i. were recorded on some trials only. These negative shifts were almost always present only at the vertex. Such shifts appeared in 8–13 per cent of the total trials given during examinations and were generally associated with shorter RT provided that F–T derivations showed no constant asymmetries. These patients (2 with aneurysms, 2 with malignant tumours involving both prefrontal regions, 2 with meningiomas of the falx, 2 with unilateral glioblastomas, 1 with unilateral meningioma and 1 with Pick's disease) all had a rather severe frontal-lobe syndrome with prevalent impairment of vigilance and attentional function. As a group these *Ss* showed the longest RT (Mean 511·6 msec). In 5 patients the examination was repeated after surgery. In spite of improvement in the clinical picture, the CNV still did not appear in 2 of them (1 aneurysm, 1 falx meningioma). In 3, however, a clear CNV appeared after surgery and RTs were generally much shorter. In 2 (1 unilateral, 1 bilateral glioblastoma) the CNV appeared only at C_Z (*Fig.* 1A, B), with amplitudes between −14·5 and −25·7 μV, often with slow return to baseline after S_2. The larger CNVs coincided with the shorter RTs, and the negative shifts tended to become variable and fairly rapidly to disappear after the first 3/5 series of trials.

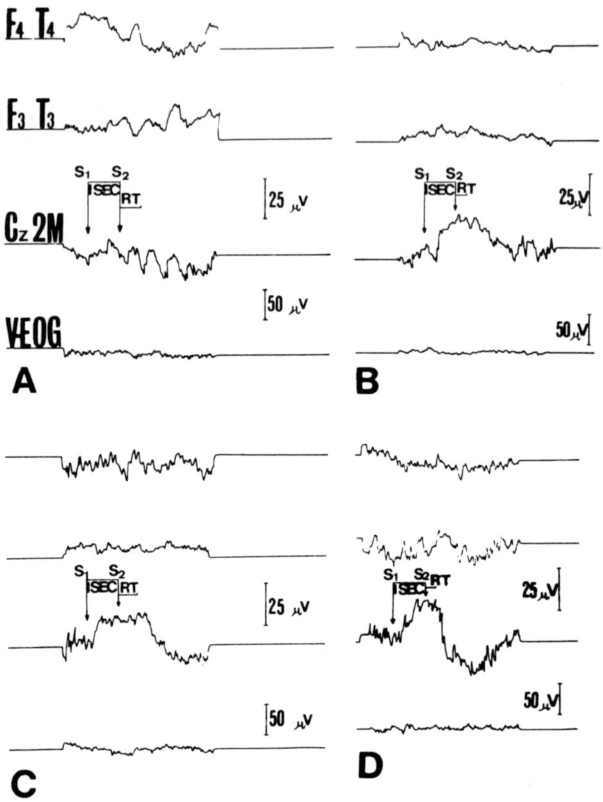

Fig. 1. A, B, Male, aged 31. Glioblastoma multiforme involving both frontal lobes at the level of the prefrontal regions. A before and B after surgery. C, Male, aged 33. Right prefrontal glioblastoma multiforme before surgery. D, Male, aged 53. Left prefrontal glioblastoma multiforme before surgery.

Mean RT: A. 670 msec; B. 370 msec; C. 603 msec; D. 330 msec.

In a second group of 6 patients a fairly clear CNV was observed at the first examination. It was almost always of variable amplitude, however, and tended to disappear after the first 4/6 series of trials. The larger amplitudes appeared almost always in association with the shorter RT. All these *Ss* had a frontal lobe syndrome, less severe on the whole than that of patients in the previous group, particularly with regard to vigilance level, attentional and associative processes and motor performance (Mean RT=348·9 msec). In 4 of these cases (2 unilateral gliomas, 1 meningioma of the falx and 1 Pick's disease with moderate dementia) the CNV appeared both at the vertex (Mean amplitude −18·8 μV; range −12 to 28·5 μV) and in F_4–T_4 and F_3–T_3 (Mean amplitude −10·3 μV; range −6 to −15·5 μV). In 3 of these *Ss* the CNV appeared in the F–T derivations without particular asymmetries (*Fig*. 2C), often with a slower return to baseline in comparison with the vertex. In the 4th case (the only left prefrontal glioblastoma in the group), the i.s.i. negative shifts could be clearly observed only in the non-affected hemisphere (*Fig.* 2A, B). This phenomenon persisted even after surgery, the only effects of which were to increase the amplitude and consistency of the CNV recorded from the non-affected hemisphere and to decrease RT. Similar increases of CNV amplitude and constancy at the vertex, F_4–T_4 and F_3–T_3, and decreased RT were observed after surgery in the other 3 *Ss*, evidently connected with the improvement in mental condition (*Fig.* 2C, D). Increase in amplitude of the CNV averaged 44·7 per cent at the vertex, 54·2 per cent in the F–T derivations; the average RT reduction was 35·1 per cent.

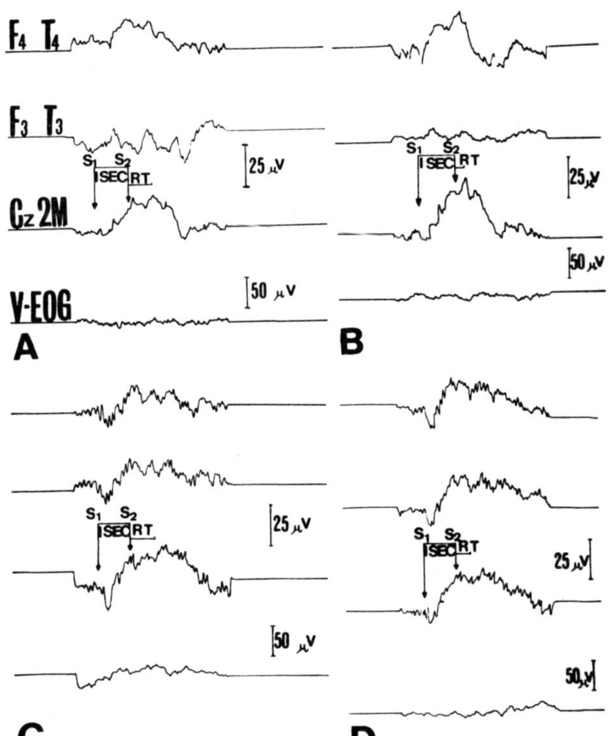

Fig. 2. A, B, Male, aged 42. Left prefrontal glioblastoma multiforme. A before and B after surgery. C, D, Female, aged 40. Large meningioma of the anterior third of the falx involving both prefrontal regions. CNV amplitude in i.s.i.; as in *Fig.* 1 averaged responses to 8 trials. A in F_4T_4—10 μV; in C_z-linked mastoids—12·5 μV. B in F_6T_6—18·7 μV; in C_z2M—25 μV. C in F_4T_4 and F_3T_3—10 μV; in C_z2M—14 μV. D in F_4T_4 and F_3T_3—14 μV; in C_z2M—17·5 μV. Mean RT: A. 740 msec; B. 270 msec; C. 800 msec; D. 470 msec.

In the other two patients of the group (one right and one left prefrontal glioblastoma) the CNVs were elicited only at the vertex (*Fig.* 1C, D), and in one case the negative shifts tended to persist at length after S_2. In these two patients post-surgical polygraphic controls were not carried out.

Both the cases who had been treated with Freeman-Watts bilateral prefrontal lobotomy showed i.s.i. negativity with normal features both in F–T derivations and at the vertex, but of low amplitude.

DISCUSSION

The results lead to various and complex considerations. We would like to draw attention to some of the more important findings.

Absence, presence and features of CNV activity in the great majority of our patients appeared to be related essentially more to mental disturbances than to site and extension of destructive lesions of the frontal lobes that had caused them. The better the level of vigilance, attentional functions and motor performances, the more constantly were the CNVs elicited. When patients with prefrontal lobe tumours improved after surgery there was generally a corresponding improvement in CNV activity and RT.

In many cases there was a difference in behaviour and features of the slow potentials at the F–T derivations and at the vertex. Often the negative shifts were present at the vertex only. Sometimes this occurred in cases where CNV activity was completely absent before surgery. The highly destructive lesions of the prefrontal regions, even in unilateral, seem therefore often able to cause suppression of CNV activity, especially in F–T derivations.

It may be presumed from these findings that meso-disencephalic structures are essentially involved in the formation of i.s.i. negativity recordable from C_z-linked mastoids, and this probably reflects non-specific arousal or attentional processes.

We were unable to observe in our patients clear asymmetries in the form and amplitude of the CNV, resulting in a marked reduction in amplitude on the side of the brain lesions, as reported by McCallum *et al.* (1970); McCallum and Cummins (1973). We observed only in one case with left prefrontal glioblastoma that before and after surgery there was complete absence of the CNVs in F_3–T_3, while in the other derivations they appeared more or less constantly.

Both cases with bilateral interruption of the inter-connections between prefrontal cortex and dorsomedial nuclei of the thalamus showed slow potential shifts both in the F–T derivations and at the vertex, with very reduced amplitude in one *S* with high anxiety. This fact seems to confirm the results of McCallum and Walter (1968) of frequent reduced amplitude of the CNV in chronic high anxiety neurotics.

In conclusion, no consistent relationship was observed between CNV appearance and features on the one hand, and type, side or extension of the frontal lobe lesions on the other. It was noted, however, that there existed an inverse correlation between slow potentials and psychic disturbances. These findings do not as yet allow any definitive conclusions as regards the possible use of CNV data obtainable by the standard S_1–S_2–R paradigm as a diagnostic aid in brain diseases of this kind. CNV studies can, however, serve as a useful clinical tool, especially for subjects with various brain damages causing impairment of attention and vigilance-arousal mechanisms.

SUMMARY

Eighteen patients with varied frontal-lobe lesions and mental disturbances, and a small normal control group were studied.

In 10 patients no i.s.i. negative slow potential shift was observed on first examination. In 3 of these after surgery a clear CNV appeared, although in 2 cases only at the vertex.

In a second group of 6 patients i.s.i. negativity was observed in 4, both in the F–T derivations and at the vertex, while in 2 the CNVs were elicited only at the vertex. In one patient in this group with left prefrontal glioblastoma, before and after surgery no i.s.i. negativity was observed in F_3–T_3, but appeared at the other derivations.

In post-surgical controls (8 cases) it was observed that concomitant with an improvement in vigilance, attentional function and motor performance, CNV (activity was more evident and constant and RT shorter.

Two chronic obsessional neurotics treated by Freeman-Watts bilateral prefrontal lobotomy both showed CNVs at F–T derivations and at the vertex.

Results are discussed in the context of the possible clinical use of CNV studies in neuro-psychiatry.

ACKNOWLEDGEMENT

Apparatus used in these investigations was acquired in part by a grant (Cont. n. 70.0186.04/115.1067) from the CNR.

Stuttering and CNV

G. ZIMMERMANN and J. R. KNOTT

University of Iowa, Iowa City, Iowa 52242, U.S.A.

At one time stuttering was regarded as a result of lack of lateral (hemispheric) cerebral dominance, but due to inadequate experimental procedures for direct testing of the hypothesis, this fell by the wayside. Since CNV and motor potential techniques have been applied to speech processing, a re-evaluation of the dominance theory seemed possible.

The paradigm involved measuring vertex and left and right inferior frontal CNVs, generated in a 1500-msec interval (i.s.i.) between visual presentation of a single word, and a signal to speak the word. Eight stutterers and 6 normal speakers were studied. DC amplifiers were used and artefact-free trials averaged.

Non-stutterers showed CNV at vertex, in the i.s.i., and a larger negative shift at the left inferior frontal electrode than the right.

Stutterers, when speaking 'normally', showed a CNV at vertex, but *no shift* at either left or right inferior frontal electrode.

Stutterers, when stuttering occurred, showed *no* CNV at vertex in the i.s.i., an initially large negative shift, followed by positivity, in the left hemisphere, and a positive shift in the right hemisphere.

The application of SP techniques yield data which re-open the dominance theory of stuttering for consideration, and demonstrate speech-related lateral SP differences in normal speakers.

Editors' note: For a fuller report of this study *see:* Zimmermann, G. and Knott, J. R. (1974). Slow potentials of the brain related to speech processing in normal speakers and stutterers. *Electroenceph. Clin. Neurophysiol.* **37,** 599–607.

Electroencephalographic Localization of Conative Aspects of Language Production in the Human Brain

MORTON D. LOW, JUHN A. WADA and MORGAN FOX

University of British Columbia, Vancouver, Canada

The language system in the brain of man is presumed to reside primarily in the left hemisphere, and Broca's area is presumed to function as a coding unit for articulation. These presumptions are based on considerable clinical lesion data and much indirect evidence.

The question of the specific relationship of handedness to speech lateralization remains open, however, and it is unclear, for example, what proportion of functionally 'pure' left-handers also have left hemispheric speech lateralization, and whether functional mixed handedness is ever or always an indication of mixed speech dominance. One consequence of these uncertainties is that prior to brain surgery for lesions bordering upon Broca's area, some determination must be made of the patient's speech dominance. Such determination has heretofore depended upon both clinical history and complex procedures such as carotid amytal injection, the latter having slight but definite risk of harm to the patient.

The report of McAdam and Whittaker (1971) suggested that motor speech function may be localized by the detection of asymmetries in the Bereitschafts-potential (BP) or motor potential preceding the spontaneous articulation of polysyllabic words. Their observation suggested to us that the other cerebral slow potentials such as the CNV might serve for the localization of language production in the human brain. Because of the unique combination of previous research experience and interest and the clinical necessity for evaluation of certain epileptic patients in our laboratory using cartoid amytal speech localization, we were able to design an electrographic test for cerebral speech dominance using carotid amytal testing as a final control.

METHOD

Ss included 22 healthy volunteers, selected as 11 'pure left-handers' and eleven 'pure right-handers' according to the criteria of Annett (1967), and 11 epileptic patients admitted to the Vancouver General Hospital for evaluation for surgical treatment of medically intractable seizures. Of the 11 patients, 8 were right-handed and 3 left-handed. All *Ss* were between 17 and 30 years of age.

Each *S* was run in at least two recording sessions. If no clear indication of lateralization was obtained in two sessions a third recording was done. Only four *Ss* were run three times.

During each session, scalp EEG activity was recorded from International 10–20 System electrode positions F_7, F_8, T_1, T_2, T_3, T_4 and C_z using a contralateral ear reference. The amplifier system (Beckman) time-constant was 4 sec. During the test series *S* sat in a comfortable chair in a sound deadened room facing a translucent screen with his eyes open.

All averaging was done off-line from tape-recorded EEG using a Fabritek 4-input signal averager. Averages were written out with an X–Y Plotter for comparison and measurement.

Each test session lasted about thirty minutes and included 6 paradigms. The paradigm of interest was the standard cued reaction-time $S_1 \rightarrow S_2 - - R$, where S_1 was a single stroboscopic light flash, S_2 was a 1-sec tone burst at 1500 Hz, the $S_1 \rightarrow S_2$ interval was 1·4 sec and R was an instructed word.

Paradigm five consisted of 33 trials of $S_1 \rightarrow S_2$ with the *S* responding to the tone by saying the word ‘hi’. Paradigm six was the same with *S* responding to S_2 by saying his given name. These trial sets were preceded by a series of sixty flashes alone, sixty tones alone, thirty-three presentations of flash followed by tone and thirty-three $S_1 \rightarrow S_2$ pairs with *S* responding to S_2 with a button-press.

Averaging was done in blocks of 32 trials with a 5-sec epoch beginning 0·5 sec prior to the onset of S_1. X–Y plots of averaged traces from T_1, F_7 and T_3 electrode positions were compared to traces from T_2, F_8 and T_4 positions, using area measurements as the units for comparison (Low, 1969).

Following EEG testing the epileptic patients eventually were subjected to bilateral carotid amytal tests for language dominance. Determination of dominance with carotid amytal tests was made without the knowledge of the predictions from EEG evaluation.

RESULTS

Epileptic Patients

Two *a priori* assumptions were made in assessing these data. The first was that the intracarotid amytal test as performed by one of us (J.A.W.) would provide conclusive evidence regarding cerebral dominance for language production. The second assumption was that the averaged traces would be asymmetrical, showing greater negative shifts on the dominant side. An example of

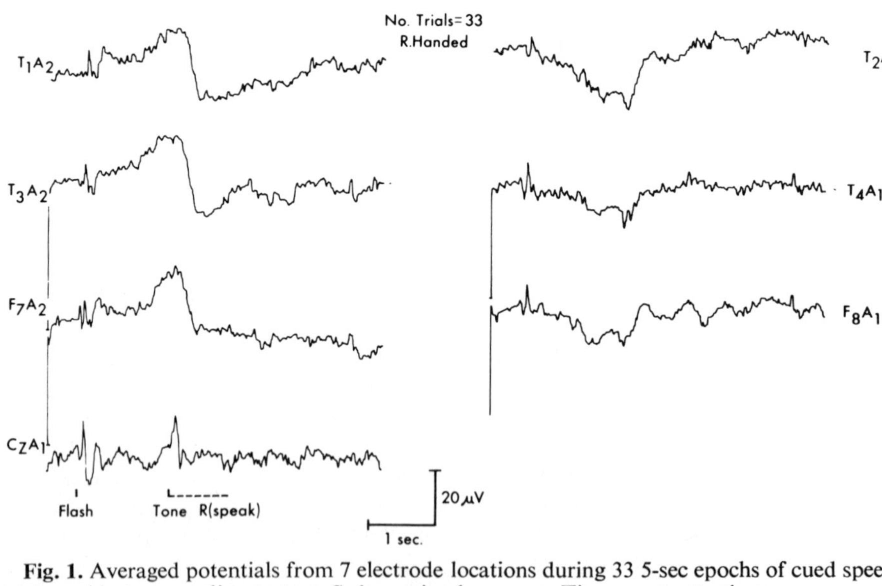

Fig. 1. Averaged potentials from 7 electrode locations during 33 5-sec epochs of cued speech trials; subject responding to tone S_2 by saying her name. Time constant=4 sec.

Table 1. Comparison of CNV-Speech Test Results with Carotid Amytal Test for Language Dominance in 11 *Epileptic Patients*

CNV Test Prediction		*Carotid Amytal Test Result*
R. handed / L. dominant	5	5+1* confirmed
L. handed / L. dominant	2	2
R. handed / R. dominant	3*	2
L. handed / R. dominant	1	1

* CNV test error.

the raw data obtained is illustrated in *Fig.* 1. EEG testing correctly predicted the conclusion of carotid amytal tests in 10 out of the 11 patients as in *Table* 1. The one CNV test error was the prediction of right dominance in a right-handed patient shown ultimately to be left dominant.

Volunteer Subjects

The areas of averaged negative shifts occurring between S_1 and S_2 were measured for electrode positions T_1 and T_2. The results (overall mean values including measures from both runs for all *Ss*) were as shown in *Table* 2. The T_1 and T_2 means differed significantly at the $P=0{\cdot}05$ level.

While there was a significant difference in grouped temporal CNV area measurements which was in the expected direction, i.e. greater negativity on the side opposite *S's* handedness, not all individual tests showed this relationship. In fact, assuming that greater negativity indicated hemispheric dominance for language production, the 11 right-handed *Ss* included 8 who were left hemisphere dominant, 1 who was right dominant and 2 with no lateralization. The 11 left-handed *Ss* included 6 who were right hemisphere dominant, 3 who were left dominant and 2 with no lateralization (*Table* 3).

Table 2. Comparison of mean speech-CNV areas recorded from T_1 and T_2 electrode positions in 11 right-handed and 11 left-handed normal subjects. Mean values include measures from two recording sessions for each subject. Units of area are result of integration of negative amplitude as a function of time over 1·2 sec following S_1

Handed	*Electrode Location*	*Mean CNV Area (N=22)*
Right	T_1	12·3* units
	T_2	7·5
Left	T_1	7·0
	T_2	15·3

* T_1 *v*. T_2 means significant difference at P 0·05.

Table 3. Relationship of handedness to cerebral dominance (as indicated by speech-CNV asymmetry) in 22 normal adult subjects

Handedness	*CNV Assessment*		
11 Right-handed	8 L. dominant	1 R. dominant	2?
11 Left-handed	3 L. dominant	6 R. dominant	2?

DISCUSSION

Our work was specifically not an attempt to replicate the experiments of McAdam and Whittaker since we could not see any easy way to avoid the technical difficulties inherent in precisely defining *the point* of beginning articulation for backward averaging.

There is some support for our findings in the work of Zimmermann and Knott (p. 164) in that they have demonstrated slow potential asymmetries recorded from inferior frontal regions preceding speech in normal subjects and they have related certain variations in these asymmetries to pathological speech.

Because of the potential value of such testing in patients being evaluated for epilepsy surgery, we are extending our studies to include data from normal children (including left- and right-handed twin pairs) and from children with focal epileptic lesions.

We have preliminary data from normal children between 6 and 12 years of age which suggests that the asymmetry of temporal slow potentials preceding speech is present before the expected age of appearance of the vertex CNV.

The findings raise interesting questions regarding the presumed specific relationship of handedness to cerebral speech dominance. They suggest that tests like carotid amytal speech localization may be replaced by completely innocuous EEG type recording and appear to provide a means for the electrophysiological exploration of the mechanisms of language production, both normal and abnormal in intact man.

SUMMARY

To explore the possibility that CNV-like slow potentials might be recorded prior to speech, and that such potentials might show hemispheric asymmetries systematically related to hemispheric dominance for language production, experiments were carried out with 33 *Ss*. Twenty-two normal *Ss* were selected according to traditional criteria as 11 'pure' left-handers and 11 'pure' right-handers. Eleven were patients awaiting bilateral carotid amytal speech localization tests as a preliminary to epilepsy surgery. Scalp recordings were made of cerebral slow potentials from the inferior-posterior frontal areas during the 1·4 sec following a cue and prior to articulation of an instructed word. Computer analysis of these potentials showed an asymmetry which allowed accurate prediction of the results of carotid amytal testing for speech dominance in 10 of the 11 patients (2 correctly determined as right-hemisphere dominant) and which was significantly correlated with handedness in the non-patient volunteers.

ACKNOWLEDGEMENTS

The authors are indebted to Siew Young Th'ng, R.E.T. for her invaluable technical assistance, and Ms Frances Burton for her assistance in manuscript preparation. This work was supported by Grant MA-3313 from the Medical Research Council of Canada.

Bereitschaftspotential in Parkinsonian Patients

L. DEECKE, H. H. KORNHUBER and G. SCHMITT

Abteilung Neurologie, Universität Ulm, German Federal Republic

The human brain potentials preceding voluntary movement—Bereitschaftspotential (BP), pre-motion positivity (PMP), and motor potential (MP)—(Deecke *et al.*, 1969) have not been studied so far in patients with diseases of the central motor system. Since these potentials have been attributed to certain brain structures (Kornhuber, 1971; Deecke *et al.*, 1973), it appears tempting to study pre-movement brain potentials in patients with deficiencies in these regions.

A first attempt was begun in patients with Parkinson's disease, and especially in cases of hemi-Parkinsonism, giving the opportunity to let each patient serve as his own control. At this early state of the study, only findings concerning the Bereitschaftspotential are conclusive.

METHOD

Experiments were carried out in 22 Parkinsonian patients showing more akinesia than tremor. Ten of them were hemi-Parkinson cases. Patients lay comfortably in an EEG chair performing quick voluntary volar flexions of the index finger by pulling the trigger of a pistol, at the very first displacement of which an electric pulse was generated. Self-paced movements were performed at irregular intervals, mean intervals between movements being 3–4 sec. Right and left index finger flexions were investigated alternately in blocks of 100 trials in the same experiment, starting conditions being randomized. Totals of 400–500 movements of either side were averaged. The electromyogram (EMG) of the right and left flexor indicis muscle was recorded, rectified and averaged in order to determine the exact onset of movement. During experiment, the patient fixed his gaze on a given point and the electro-oculogram was recorded. All samples containing eye movement and other artefacts were marked on line by an electric signal. When reverse-time analysing the tape-stored data (Kornhuber and Deecke, 1964) contaminated samples were automatically rejected.

Brain potentials were recorded using Beckman electrodes attached with collodion to left-, mid-, and right precentral and to left-, mid- and right parietal regions with reference to linked ears. In addition, bipolar recordings between left and right—and between left and mid-precentral electrodes were made. Overall time constant 1·2 sec.

RESULTS AND DISCUSSION

1. Hemi-Parkinson Patients

The healthy hemisphere of the patients with hemi-Parkinsonism generated an overall BP 150 (amplitude measurement taken 150 msec prior to onset of EMG activity) averaging $-1{\cdot}32\ \mu V$ (SE=0·54), whereas the pathological hemisphere generated only $-0{\cdot}78\ \mu V$ (SE=0·54). The difference was significant on a two-tailed test ($P<0{\cdot}025$). A similar difference was found with P_0 (amplitude at the onset of EMG activity). It averaged $-2{\cdot}44\ \mu V$ (SE=0·7) over the healthy as compared with $-1{\cdot}66\ \mu V$ (SE=0·62) over the pathological hemisphere. This difference was also significant on a two-tailed test ($P<0{\cdot}05$).

The BP 150 of the healthy hemisphere when it was contralateral to the movement averaged $-1{\cdot}83\ \mu V$ (SE=0·61) as opposed to $-0{\cdot}55\ \mu V$ (SE=0·57) on the pathological hemisphere. The difference was significant ($P<0{\cdot}05$). The P_0 of the healthy hemisphere when it was contralateral to the movement averaged $-2{\cdot}8\ \mu V$ (SE=0·77), that of the pathological hemisphere was only $-1{\cdot}8\ \mu V$ (SE=0·77). The difference was significant on the basis of the one-tailed hypothesis only ($P<0{\cdot}05$).

Thus, hemi-Parkinson patients showed reduction of the BP in leads contralateral to the akinetic side of the body. This is in contrast to normal *Ss*. In *Fig*. 1, brain potentials in the normal *S* preceding right index finger movements are compared with those preceding left finger movements. The negativity is always larger over the precentral region contralateral to the movement. In bipolar recordings (left precentral–right precentral) the responses preceding left and right finger movements appear as virtual mirror images of one another.

The situation is different in hemi-Parkinsonism. A typical example is given in *Fig*. 2. When the healthy (right) index finger was moved, a clear preponderance of the left precentral BP is seen, yielding a marked deflection of the bipolar left *v*. right precentral derivation. When the akinetic (left) finger was moved negativity was not larger over the contralateral (right) precentral region. Both showed the same minimal BP, and, consequently, an absence of response in the bipolar derivation.

Thus, in hemi-Parkinsonism the side of the larger potential depends on the side of the pathological hemisphere and not on the side of the movement.

2. Bilateral Parkinson Cases

In bilateral Parkinsonism there was a gross bilateral reduction of the BP.

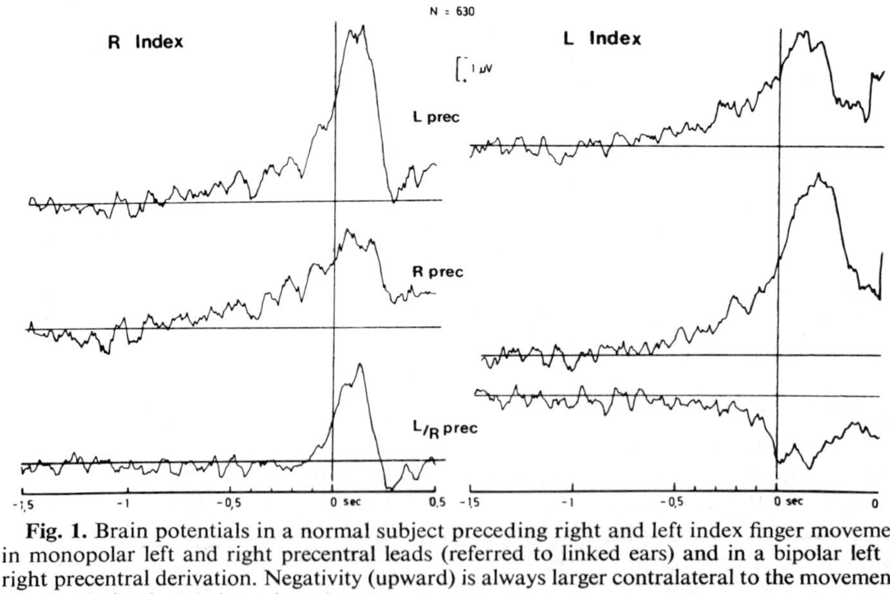

Fig. 1. Brain potentials in a normal subject preceding right and left index finger movement in monopolar left and right precentral leads (referred to linked ears) and in a bipolar left *v*. right precentral derivation. Negativity (upward) is always larger contralateral to the movement, bipolar derivations being mirror images.

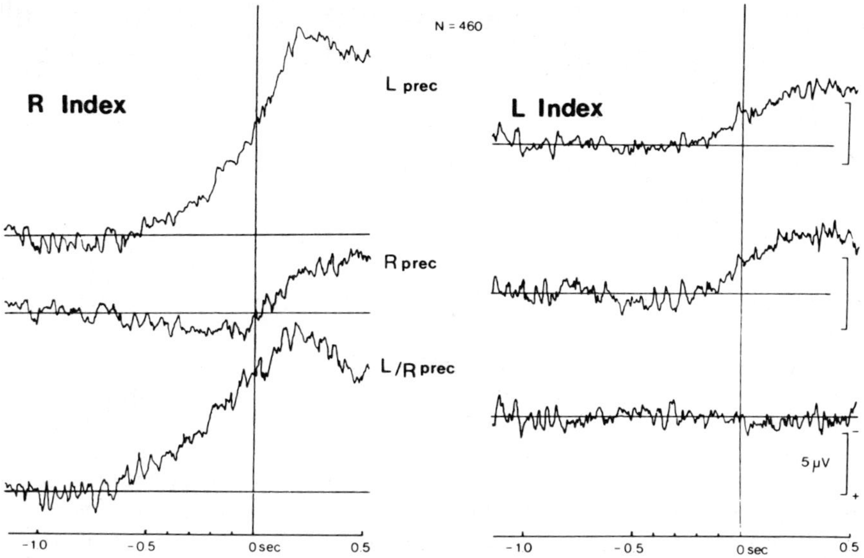

Fig. 2. Brain potentials in a patient with hemi-Parkinsonism (left side of the body) preceding right and left index finger movement. Larger negativity over the healthy hemisphere (L prec). In contrast with *Fig*. 1, movement of the left (akinetic) finger results in no greater BP negativity at the contralateral (R prec) electrode. Consequently the bipolar (L prec—R prec) trace shows no downward deflection in this condition.

For instance P_0 precentral contralateral averaged only $-0{\cdot}11$ μV (SE=0·5). If this figure is compared with the $-5{\cdot}3$ μV (SE=0·5) which we have obtained from normal *Ss* in other experiments, the difference is, despite different age distributions, highly significant ($P<0{\cdot}005$, two-tailed). (Experiments concerning Bereitschaftspotential in different age groups are currently being undertaken.)

The results show that patients with Parkinsonism, which is a disease of subcortical origin, display significant changes of cortical phenomena preceding movement. These changes are most obvious in hemi-Parkinson cases showing a lateralization of pre-movement negativity different from normals, whereas the clinical EEG of these cases does not show systematic side differences of the intrinsic cortical waves. Further studies including evaluation of the other pre-movement components, PMP and MP, are necessary, and experiments are presently being undertaken to compare the Bereitschaftspotential in Parkinson patients before and after L-dopa treatment.

SUMMARY

The Bereitschaftspotential (BP) preceding voluntary finger movement was studied in unilateral and bilateral cases of Parkinson's disease, showing more akinesia than tremor. Right and left index finger movements were compared in the same experiment. Hemi-Parkinson patients showed significant reduction of BP amplitudes in leads contralateral to the akinetic side of the body.

This lateralization of negativity, always to the healthy hemisphere, is different from normals, in whom lateralization is found towards the side contralateral to the movement. In bilateral Parkinsonism, a bilateral absence or diminution of BP was found.

Event-related Slow Potentials in Aphasic, Dyslexic and Normal Children During Pictorial and Letter-matching

DAVID A. OTTO,* KAREN HOUCK, HERBERT FINGER and SANDRA HART

Man-Machine Integration Branch, NASA-Ames Research Center, 239–3, Moffett Field, California 94305, U.S.A.

The need for research on developmental aspects of event-related potentials was stressed at the 2nd International CNV Congress held in Vancouver in 1971. In addition to the lack of normative maturational data, the utility of ERPs in the clinical assessment of developmental disorders needs to be determined. Low and Stoilen (1973) reported pessimistically that 'CNV did not reflect minimal cerebral dysfunction, 'school failure' or 'specific learning disorder'. Cohen (1973b) found a lack of CNV in children with learning disabilities. Small *et al.* (1971), on the other hand, were able to differentiate negative EEG patterns in normal and autistic children. Other event-related potentials such as the P300 and RP apparently have not been studied in normal or disordered children.

The present study was undertaken to examine ERPs in normal, aphasic and dyslexic children during complex pictorial and letter-matching tasks. A three-stimulus paradigm adopted from Posner and Boies (1971) was used to differentiate general and selective attentional components related to preparatory and encodement functions. This design was chosen to test whether children with aphasic or dyslexic language disabilities show general or selective attentional deficits.

METHOD

Twenty children aged 5·7–12·7 years were studied at the Scottish Rite Institute for Childhood Aphasia including an experimental (E) group of 9 children enrolled in therapy and one diagnostic referral (mean age=9·3 years) and a control (C) group of 10 normal children (mean age=9·5 years). Seven aphasic children, varying widely in the extent of linguistic impairment, and 3 dyslexic children were included in the E group.

Ss were seated in an electrically-shielded, sound-dampened chamber about four feet in front of a screen on which slides were presented by means of a Kodak Carousel Projector. Subjects viewed 3 series of slides, each containing 40 pairs. A small display device mounted centrally in the screen provided a continuous fixation target, warning light (S_1), and feedback following response. The first slide (S_2) appeared after a 500 or 1000 msec warning interval and the comparison slide (S_3) after a 1300 msec encodement interval. Each stimulus was displayed for 200 msec. *Ss* pressed one of 2 buttons after S_3 to indicate same or different.

* Present address: Environmental Protection Agency, Mason Farm Road, University of North Carolina, Chapel Hill, North Carolina 27514, U.S.A.

This study was conducted at the Scottish Rite Institute for Childhood Aphasia; Stanford University School of Medicine; Palo Alto, California, U.S.A.

Series I required a physical match of coloured pictures (familiar objects such as cars and houses). Series II entailed physical matching of lower case letters, while series III entailed name matching of mixed upper and lower case letter pairs.

Monopolar EEG recordings were obtained from C_Z and P_Z referred to linked mastoids. Vertical eye movements were recorded sub- and supra-orbitally from the left eye. Analog recordings of EEG, EOG and stimulus/response events were made with an Ampex 1300 tape recorder and Grass P511EF pre-amplifiers set at a bandpass of 0·1 to 30 cps half amplitude.

Signal averages digitized at 20 msec/point with eye movement compensated by weighted linear summation with vertical EOG were constructed for each *S* and condition. Mean amplitudes of eight 320-msec epochs were measured relative to pretrial baseline as illustrated in *Fig*. 2. A selective analysis of single trial data segregated by same/different and correct/incorrect categories is currently in progress and will be reported later.

RESULTS

Behavioural data for 16 *Ss* are summarized in *Table* 1. The 2 youngest aphasic children were unable to complete series III and were not included. To maintain balanced ages and sample size, the 2 youngest control *Ss* were also excluded. Es made more errors than controls on both same and different trials of all series

Table 1. Behavioural data. Percent errors and reaction times for same and different trials during Pictorial matching (series I), Letter matching-physical (II), and Letter matching-name (III)

		Control Group (N=8)		
		Same	*Different*	*Mean*
Percent errors	I	5·75	13·25	9·50
	II	3·25	4·00	3·63
	III	5·00	0·75	2·88
	Mean	4·67	6·00	5·34
Reaction time (msec)	I	1307	1456	1382
	II	1369	1496	1432
	III	1711	1628	1670
	Mean	1462	1527	1495
		Experimental Group (N=8)		
		Same	*Different*	*Mean*
Percent errors	I	10·25	19·5	14·88
	II	9·50	16·25	12·88
	III	23·25	23·25	23·25
	Mean	14·33	19·67	17·00
Reaction time (msec)	I	1788	1967	1878
	II	1440	1649	1545
	III	2011	1966	1989
	Mean	1746	1861	1804

(ANOVA: $F=7{\cdot}22$ (1, 14), $P<0{\cdot}025$). Control *Ss* made most errors on pictorial matching while experimental *Ss* made most errors on letter-name matching. RTs of experimental *Ss* were also considerably longer than those of control *Ss* on all tasks ($F=4{\cdot}59$ (1, 14) $P<0{\cdot}05$). RTs of both groups were longest on series III. Control *Ss* responded quickest on series I while experimental *Ss* were quickest on series II and performed best on physical letter matching in terms of both errors and RT.

ERPs from both groups were labile, due in part to the inability of children under 9 to control eye movements. EOG compensation reduced eye movement artefact substantially, but not completely. Data are being reprocessed to eliminate all trials with EOG artefact. As many as 75 per cent of the trials must be discarded for children aged 5–9. The following measurements are based on EOG-compensated data.

Fig. 1 illustrates EOG, C_Z and P_Z averages of 6 *Ss* during letter name-matching. Subject GH was aphasic; RM and JO were dyslexic. RM had responded well to reading training and was ready to terminate therapy. His records could not

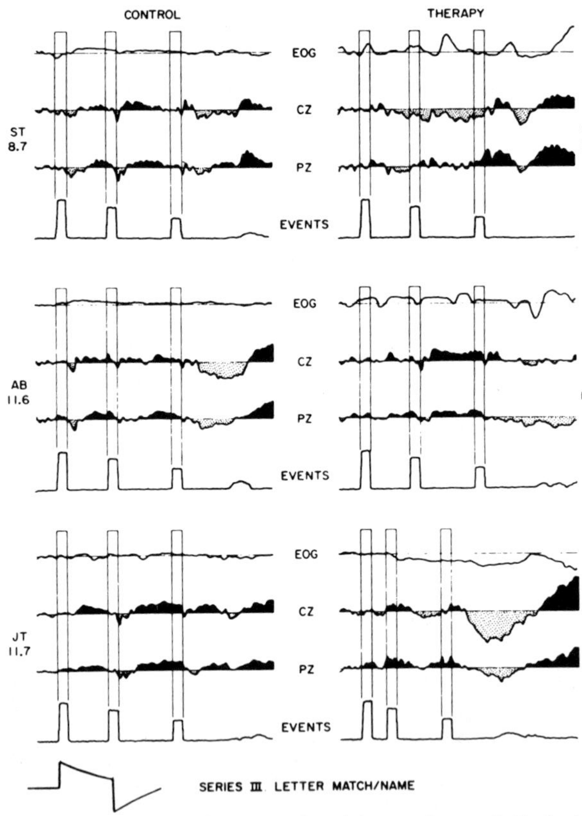

Fig. 1. Representative averages of 3 control and 3 experimental *Ss* during letter matching. Ages indicated in left and right margins. C_Z and P_Z records were EOG-compensated to minimize eye-movement artefact. Solid areas negative, stippled areas positive with respect to baseline. Sequence and duration of warning (S_1), standard (S_2) and comparison (S_3) stimuli indicated on event channel. Averages of 16 trials digitized at 20 msec/pt. Calibration: 1-sec, 50 μV, negative up.

be distinguished from controls. GH and JO showed positive shifts at C_Z during the encodement interval, characteristic of the experimental group.

ERP measurements averaged for each group across tasks are plotted in *Fig.* 2. ERPs of controls shifted in a negative direction during the warning interval at C_Z and P_Z. Both groups shifted negatively during the encodement interval at P_Z, but positively at C_Z. Sustained positive shifts predominated during the response interval. Experimental *Ss* showed larger positive shifts than controls during the encodement interval (Mann-Whitney test at C_Z (B3): $U=10$, $P(8, 7)=0{\cdot}04$). Controls tended to shift more negatively than experimental *Ss* during the warning interval, although this trend reached significance during letter physical-matching only (C_Z (A2): $U=8$, $P(8, 7)=0{\cdot}02$).

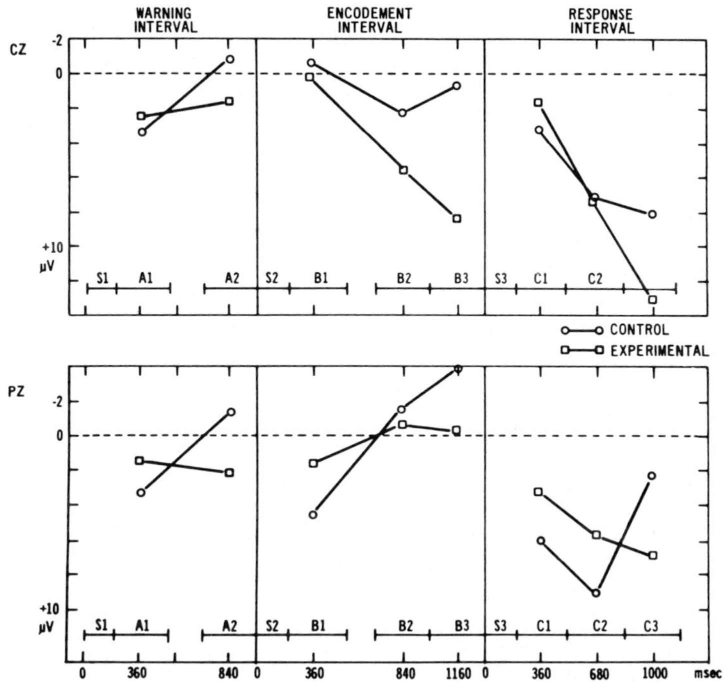

Fig. 2. Mean ERP amplitude of control and experimental groups during 8 320 msec measurement epochs designated A1–B3. S_1, S_2 and S_3 indicate relative position and duration of stimuli. Group means differed significantly at C_Z (B3) during the encodement interval. Measurements pooled across series I, II and III.

In order to assess the effects of age on ERPs linear correlations were computed for each measurement epoch. Results appear in *Table* 2. Substantial negative correlations were found in the control group for A2, (C_Z, P_Z), B1 (P_Z), B2 (C_Z), and B3 (C_Z). No significant experimental group correlations were observed. Group differences during the warning and encodement intervals cannot be accounted for simply by age.

DISCUSSION

A 3-stimulus paradigm was used to differentiate the effects of general arousal from selective attention or cognitive processing. S_1 signalled subjects to get ready, but provided no information concerning the subsequent same-different match. The S_1–S_2 period thus constituted a preparatory interval in which to mobilize attention to the task. This segment corresponds closely to the typical CNV paradigm. If the experimental group were deficient in general attention, less negativity should have been observed during the warning interval in this group than in controls. A trend in this direction was found, particularly in the letter physical-match task, lending some support to the hypothesis that aphasic children have a general attentional deficit.

S_2 delivered the first piece of information necessary for responding. Subjects were required to form and store an internal representation of S_2 for comparison with S_3. If shifts in general arousal were contained within the warning interval, differences observed during the S_2–S_3 segment could be attributed to selective attentional or 'encodement' functions. Sustained positive shifts were observed at the vertex in both groups during the encodement interval although the shape and time course of positivity differed significantly. A progressive increase in positivity throughout the interval was found in experimental *Ss*, while a lower amplitude shift which initially increased, then declined, was found in control *Ss*. The sharpest difference between groups was observed at the end of the encodement interval, suggesting that it might reflect a decrement in short-term memory among aphasics.

Positive shifts observed during the encodement interval may be protracted P300 waves. Sutton *et al.* (1965, 1967) have shown that P300 reflects the significance or information content of stimuli. They also found that the latency and duration of P300 varied widely relative to the resolution of uncertainty. It seems likely that sustained positive shifts observed in the present study also reflected stimulus-relevant information processing. Error scores and RTs indicated that controls were much more efficient than experimental *Ss* in encoding stimulus information. Greater amplitude and duration of positivity in experimental *Ss* may reflect the increased difficulty or uncertainty of this group in the encodement process.

Inverse correlations between age and ERP amplitude were found in the control group. Young children tended to show positive shifts during the warning and encodement intervals, while older children showed negative shifts similar to CNV. Cohen (1973) reported that CNV matures about age 20 in normal *Ss*. In view of the slow maturation of this waveform, the CNV *per se* may not be a reliable index of childhood disabilities. Other waveforms, such as the sustained positive pattern described here, may prove more useful in the study of developmental disorders.

SUMMARY

ERP patterns were studied in 10 normal, 6 aphasic, and 3 dyslexic children during pictorial and letter-matching. A 3-stimulus design permitted differentiation of general and selective attentional components during warning and encodement intervals. Experimental *Ss* made more errors and responded more slowly than controls. ERP amplitudes in controls correlated inversely with age. Sustained positive shifts were observed in both groups during encodement. Experimental *Ss* showed greater positivity than normals during warning and encodement intervals, indicating impairments in both general and selective attention. It was suggested that sustained positive shifts were extensions of the

P300 wave reflecting the uncertainty of aphasics and dyslexics in processing stimulus information.

Table 2. Pearson product-moment correlations of age *v.* ERP amplitude during warning (A1, A2), encodement (B1–B3), and response (C1–C3) intervals. Correlations are presented for control (C) and experimental (E) groups

		Measurement Epochs							
		Warning Interval		*Encodement Interval*			*Response Interval*		
		A1	*A2*	*B1*	*B2*	*B3*	*C1*	*C2*	*C3*
C_z	C	−0·30	−0·82†	−0·52	−0·67*	−0·88‡	−0·62	−0·09	−0·13
	E	0·44	0·33	0·01	−0·46	−0·64	0·50	0·48	0·49
P_z	C	0·08	−0·75*	−0·82‡	−0·11	0·20	−0·29	0·09	0·01
	E	0·45	0·07	0·35	−0·35	−0·34	0·48	0·09	0·31

* $P<0{\cdot}05$.
† $P<0{\cdot}02$.
‡ $P<0{\cdot}01$.

ACKNOWLEDGEMENTS

The authors are very grateful to Drs Norman and Jane Mackworth for encouragement and assistance in planning this study and to Dr Jon Eisenson, Director of the Aphasia Institute, and his staff for their generous support. Special thanks are also due to Drs James Anliker and Mark Patton, NASA-Ames Research Center, for the loan of equipment. Dr Otto held a National Research Council Associateship at NASA-Ames Research Center. This study was supported in part by NINDS Grant 5P 15 NS 07514 NSPB to Dr Eisenson.

CNVs in Hyperactive Children: Effects of Chemotherapy

N. J. C. ANDREASEN, JON F. PETERS and JOHN R. KNOTT

Department of Psychiatry, Division of EEG and Neurophysiology, University of Iowa, Iowa City, Iowa 52242, U.S.A.

Since Walter (1966) first described the CNV in children, subsequent investigations have indicated that changes occur with age. Initially, it was suggested that the CNV is poorly developed in children under 8, assuming an adult configuration at about 12 years. Gullickson (1973b) has elicited CNV-like slow potentials in 2- and 3-year-old children. Low and Stoilen (1973) compared CNVs in normal children with a group considered to have 'minimal cerebral dysfunction (MCD)'. There were no significant differences. All children 10 and over had CNVs, those 7 and younger had either no CNV or small, inconsistent ones, and the 8- and 9-year-olds had generally small CNVs. Fifty-six per cent of the MCD children and 29 per cent of the controls were reported to have abnormal resting EEGs. Cohen (1973) reported CNVs in children with learning disabilities and stated that it was diminished or absent in 60 per cent of his sample. The average age of children with no apparent CNV was $8\frac{1}{2}$ and that of children with a CNV was 10. Small *et al.* (1971) have reported that autistic and normal children have different CNV behaviours.

This investigation was undertaken to compare CNVs in hyperactive children with learning disabilities while on and off medication, and compares the CNVs of the hyperactive children with age-matched 'normal' children.

METHOD

The experimental *Ss* were seven consecutive male admissions to the Child Psychiatry Inpatient Service at the University of Iowa, ranging in age from 8 years, 5 months to 12 years, 4 months. They were diagnosed as hyperkinetic on the basis of having all the following symptoms since early childhood: decreased attentiveness, restlessness, distractibility, excitability, emotional lability, and poor impulse control. All had normal IQs but were considered to have learning disabilities based on performance of at least two years below expected grade level. 'Soft' neurological signs were present in 6 cases and include: impaired graphesthesia, synkinesia or choreiform movements at rest. EEGs obtained on 5 of the *Ss* were interpreted as being within broad normal limits for their age, while 1 (recorded elsewhere) was 'questionably abnormal with excessive high voltage slowing and some bilateral posterior and temporal spike and wave activity.' Clinically, he was seizure free. The background EEG during CNV recording was normal. *Table* 1 summarizes the clinical data.

Initial CNVs were obtained on the experimental *Ss* while on cerebral stimulants and again after they had been discontinued for two days. These drugs are rapidly metabolized and cleared, and ordinarily clinical symptoms markedly increased within twelve hours after the last dose. Seven normal males from the neighbourhood of one of the investigators served as age-matched controls.

The experimental and control *Ss* had about the same degree of familiarity with the investigators.

Table 1. Clinical profiles of patients studied

Age	*Grade*	*I.Q.* *V*	*P*	*FS*	*Learning Disability*	*'Soft signs'*	*Bender*	*Medication*
8 yr 5 mth	1			88	Reading	many	+	Ritalin 15 mg, bid
8 yr 10 mth	2	99	93	96	Reading, spelling	some	+	Ritalin 10 mg, bid
9 yr 7 mth	3	94	97	95	Reading	none	+	Ritalin 20 mg, bid
11 yr 2 mth	5	103	99	101	Reading, spelling	some	+	Ritalin 20 mg, bid
11 yr 8 mth	6	101	107	104	Reading, math	some	+	Ritalin 20–25 mg, bid
11 yr 9 mth	6	120	97	101	Math, writing	many	+	Dexadrine 10 mg, bid
12 yr 4 mth	7	119	125	124	Math, writing	some	+	Dexadrine 5 mg, bid

The experimental paradigm consisted of an auditory warning stimulus (S_1) at either 1500 or 800 Hz followed 1500 msec later by S_2 (Grass PS2 flash, intensity '1'). *Ss* were instructed to respond to S_2 by pressing a telegraph key only when it followed the 1500 Hz tone. All *Ss* indicated that they could tell the tones apart with no difficulty. The high and low tones were randomly intermixed in blocks of 40 trials for a total of 80 in each session. Each *S* was given a brief rest between trials 40 and 41.

CNVs were recorded from the vertex, referred to linked mastoids, with a Grass 7P1A DC pre-amplifier, using silver–silver chloride electrodes attached with collodion. Eye movements were recorded from electrodes placed above and below the eye. All trials were recorded on FM tape as well as directly on the Model 7 polygraph. Trials free of eye movement and other artefacts were averaged off line with a PDP–12, utilizing 15–20 trials for each *S*, although for one (11 : 09, *Fig.* 1) only one artefact-free trial could be obtained while off medication.

RESULTS

Fig. 1 shows the averaged CNVs recorded from each experimental *S* on and off medication and from the age-matched controls. CNVs were independently scored by each author as absent (—), questionably present (?+), present but small (+), and present and large (++). There was a high degree of agreement between the authors as to the presence or absence of a CNV, with minor divergence as to the amount. *Table* 2 summarizes the final scoring, as agreed on by all authors.

As shown in *Table* 2, CNVs differ in the hyperactive children while on and off medications. Off medications, 2 children had CNVs during the response trials, and 1 had a questionable CNV in the non-response trials. When on medication, 6 children had clear negative shifts on response trials and 3 of these also showed CNVs on non-response trials. One child on medication had no CNV during

Table 2. Comparison of CNV in hyperactive children with learning disabilities with and without cerebral stimulants, and normal controls

	Patients						Normal Controls			
		OFF Medication		ON Medication						
Age	*Clinical* severity*	*Response trials*	*Non-response trials*	*Response trials*	*Non-response trials*	*Ability to perform task*	*Age*	*Response trials*	*Non-response trials*	*Ability to perform task*
8 yr 5 mth	4	–	–	+	–	Poor	8 yr 10 mth	?+	?+	Poor
8 yr 10 mth	3	–	–	?+	?+	Good	8 yr 11 mth	++	++	Good
9 yr 7 mth	2	+	–	+	+	Good	9 yr 8 mth	?+	–	Good
11 yr 2 mth	3	–	–	+	–	Good	11 yr 0 mth	?+	+	Good
11 yr 8 mth	4	–	–	–	+	Poor	11 yr 10 mth	–	+	Good
11 yr 9 mth	4	–	–	+	–	Poor	11 yr 11 mth	–	+	Good
12 yr 4 mth	2	+	?+	++	+	Good	12 yr 7 mth	++	++	Good
Some CNV		2	1	6	4		Some CNV	5	6	
No CNV		5	6	1	3		No CNV	2	1	

* 4=very severe; 3=moderately severe; 2=moderate.
See text for details of CNV scoring.

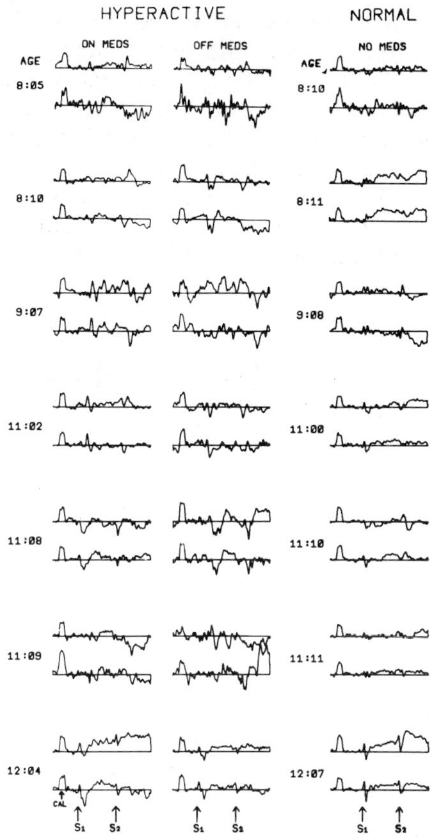

Fig. 1. Averaged CNVs recorded from each of the experimental *Ss* while on and off medication and from each normal control. The top tracing in each pair is from the response trials. Negativity is up. Averages are for 4·096 sec (256 points). Abbreviations: S_1, S_2=Stimulus 1 and Stimulus 2; cal=calibration pulse of $-25\ \mu V$.

the response trials, but showed a shift during the non-response trials. Thus, all the children developed CNVs during one of the conditions while on medication. On the other hand, 5 of the normal children had some CNV during response trials, and 6 showed negative shifts during non-response trials.

DISCUSSION

Cerebral stimulants appear to 'normalize' the CNV in hyperactive children with learning disabilities. Among the hyperactive children, those who were more impaired clinically were more likely to have questionable or absent CNVs. The one experimental *S* with a superior intelligence quotient, who was also the oldest, had the most prominent CNV. There was no demonstrable relationship between clinical EEG evaluation and CNV magnitude.

No distinct age-related patterns were observable, although the oldest child in each group showed the largest CNV. Many of the older hyperactive children did no better than the younger ones, either on or off medication.

It is of additional interest that, in this differential response type of design in which adults routinely show larger CNVs in the response condition (Irwin *et al.*, 1966), such differentiation was not observed in either group of children. It is suggested that this may be due to intrinsic differences in the ability to process stimuli, with generalization occurring at younger ages, and in children who are hyperactive. This generalization may occur at the input level, requiring a later inhibitory process to prevent the overt manual response.

The data also suggest further interesting paths, as well as some caveats, for future CNV studies. Drugs may indeed have a potent effect on the CNV in addition to clinical course. This is a critical variable to examine in future studies on the psychiatric correlates of CNV.

ACKNOWLEDGEMENT

Appreciation is expressed to Dr Maurice I. Mendel, Department of Otolaryngology and Maxillofacial Surgery for use of the PDP–12.

INTRACEREBRAL RECORDING OF SLOW POTENTIAL CHANGES IN MAN AND ANIMALS

Animal Studies

Chairman's Opening Remarks and Review

CHARLES S. REBERT

Stanford Research Institute, Menlo Park, California 94025, U.S.A.

CNV research using scalp-recording techniques has produced a voluminous literature detailing its characteristics and its responsiveness to experimental manipulations. However, human scalp recording has its limits and some of its cortical implications require verification by intracerebral recording. For example, a good deal of evidence implies that the generation of CNV is some function of activity in the diencephalic and brainstem nonspecific projection system. Scalp recording suggests the existence of diverse CNVs with differential sources of genesis and a differential genesis under some circumstances of CNV, RP and P300.

Intracerebral recording allows the identification of subcortical regions involved in the cued RT task and their functional interrelationships with each other and the cortex. It is also the only means of determining physiological mechanisms of SP genesis and the relationships between SPs and other electrical events (e.g. a single or multiple unit activity). Recognizing the usefulness of direct recording from the brain, several investigators undertook CNV studies from electrodes implanted intracerebrally for therapeutic purposes in patients. Because such patients tend to be severely disturbed, and because there are limits to the experimental manipulations possible, interest developed in using primate models of the cortical CNV. Later, subcortical recordings of SPs were undertaken in both humans and subhuman primates. Those developments are discussed in greater detail below, but first it is desirable briefly to review intracerebral CNV recording in the wider context of SP research in general.

SP research can be divided into three broad categories: (1) studies on SP relationships to sensory input, overt behaviour, and covert psychological processes; (2) studies of SP consequences of indogenous or induced tissue abnormality; and (3) studies on mechanisms of SP generation and relationships to other electrocerebral events. This review will consider primarily the first category.

A variety of behaviourally relevant, SP-evoking circumstances are recognized, including: (1) SPs evoked by sensory stimulation; (2) SPs following reinforcing or punishing events; (3) SPs occurring 'spontaneously' in relation to arousal, attention, and the sleep-wakefulness spectrum; and (4) SPs conditioned during specific training procedures like classical conditioning or the cued reaction time (RT) task.

Sensory stimulation—including visual, auditory, tactile, and gustatory—produces SP shifts in the cortex and in subcortical regions. The forms of such responses are variable. In the rabbit, visual stimulation produces a complex change in the cortex, consisting of a multiphasic, positive–negative shift, with a positive after-effect that is emphasized by barbiturate anaesthesia. Positive and negative SPs appear to be independent in that they can be differentially produced by appropriate stimulus parameters (Goldring and O'Leary, 1951). In the cat's auditory cortex the relatively simple, negative SPs to sound relate specifically to auditory functioning but are neither a massive response of auditory cortex nor a widespread cortical response to arousal (Gumnit, 1961). Lickey and Fox (1966) reported that, although several sensory regions of the cat's cortex responded to stimulation of either of three sensory modalities with positive and negative SPs, the response of the cortical area representative of a particular modality was electrically negative with respect to responses in other areas. A more diffuse and longer-lasting SP was isolated by algebraic analysis and was attributed to the activity of non-specific arousal mechanisms.

Vastola (1955) and Rebert (1973a) showed that visual stimulation produced negative SPs throughout most of the extent of the lateral geniculate nucleus, with positive potentials occurring in a small region above or in upper regions of the geniculate.

Reinforcement effects can be seen in relation to gustatory input which produces positive or negative SPs in the cortex. Rowland (1968) reported that food-deprived cats exhibited negative or positive SPs in several cortical regions during consummatory responses, and that the amplitudes of such shifts depended on the quality of reward, degree of deprivation, and the presence of oral sensitivity. Despite the variety of response characteristics (latency, duration, polarity), each cerebral location reacted similarly to experimental manipulations (e.g. food incentive value).

An oscillatory response of 6–10 Hz, referred to as 'postreinforcement synchronization' (PRS), occurs in the striate cortex of cats on receipt of reinforcing stimuli. PRS is accompanied by a positive SP shift, reward contingent positive variation (RCPV). RCPV was found to depend on the quality of reinforcement, being reduced when water was substituted for milk. PRS and RCPV were also shown to depend on the presence of visual input and to be mediated by cholinergic mechanisms (Marczynski, 1972).

Whereas RCPV occurs *during* the consumption of reward and is thought to be limited to parietal-occipital cortex, a large negative shift related to reinforcement was observed by Steinmetz and Rebert (1972) in the frontal region of the monkey's brain. The shift occurred on *cessation* of consumption. The animals bar-pressed to obtain 1 cc of orange juice during the presence of a cue lamp and when several consecutive reinforcement trials were presented, the premotor cortex exhibited a stepwise increase in negativity during those trials. Amplitudes reached -300 μV after 5 trials and subsided to baseline when non-reinforcement commenced. The amplitude of the shift varied with reward quality, being lower with diluted orange juice. No such shifts were observed in subcortical regions, indicating that the phenomenon does not reflect changes in whole brain metabolism. But the shift might reflect a non-neural event related specifically to extradural recording.

SP shifts associated with consumption or intracranial reward were also observed by Wurtz (1966a) and Kawamura *et al.* (1967) and in subcortical structures such as the lateral hypothalamus, medial amygdala, and midbrain reticular formation (MRF) by Irwin and Rebert (1970).

Spontaneous alteration of arousal level and the sleep-wakefulness spectrum have frequently been found to be related to SPs. Caspers (1961), for example, reported that the cortical surface shifts in a positive direction with onset of sleep and more positively with transition from deep sleep to REM sleep. The cortex becomes negative with wakefulness. Wurtz (1965b) also found a positive shift with the transition from deep sleep to REM sleep when using a subcortical reference, but with respect to several extracerebral references, both the cortical surface and many subcortical regions shifted negatively on occurrence of REM or wakefulness (1965a). Kawamura and Sawyer (1964) and Kawamura and Pompeiano (1969) also showed surface negativity with REM sleep and, like Wurtz, emphasized the diffuse nature of cerebral and spinal shifts with changes in sleep-wakefulness. Wurtz (1966b) interpreted such results as indicating a whole-brain metabolic source of the sleep-wakefulness SPs, but Kawamura and Pompeiano (1969) found that sectioning the spinal cord, which exhibited polarities opposite to those of the cerebrum, abolished the positive cord SP. They interpreted such results as favouring a neuronal–glial interpretation of the shifts, with more depolarized cells in the brain and more hyperpolarized cells in the cord.

In the foregoing studies, phasic increases in surface negativity are consistently reported to accompany orienting and other indexes of increased alertness.

Finally, SP changes are ubiquitous accompaniments or consequents of learning. They have been studied primarily in the *classic conditioning paradigm.* Almost all such studies have employed shock as the US and either tone or light as the CS. Recent studies demonstrated the occurrence of SP changes in the cortices of rabbits, rats, cats and dogs, and in pigeon striatum. Chiorini (1969) observed negative SPs in the cat's somatosensory and motor cortices in response to a tone CS, which decreased in amplitude during habituation. In one group of animals that learned a CR, cortical SP amplitudes increased in negativity during training, whereas no changes occurred in a non-learning group. In animals that learned, SPs were largest in cortices contralateral to UCS application during training but not during habituation or extinction. SP amplitudes declined during overtraining, increased on initiation of extinction, but declined again during extinction. In all cortical areas, SP amplitude was larger on trials preceding a CR than on trials where a CR was absent. Symmetrical changes during habituation and extinction were attributed to activity in the MRF, whereas the cortical asymmetry developed during training was attributed to the specific involvement of ventral thalamus in the learning process.

Pinto-Hamuy *et al.* (1969) found *positive* SPs developed during early training in both sensory and motor cortices of the rabbit and declined when high performance levels were reached. SPs were largest in the sensory projection area of the CS, in contrast to other sensory projections. Eleven of 13 *Ss* exhibited significant but low level correlations between SP and the occurrence of CRs. During late training, CRs began to occur in the absence of SPs, whereas SPs usually occurred only in association with CRs. Such dissociations suggest that SPs are not necessary to the operant response. Whether they are necessary for the establishment of associative bonds remained unanswered.

Negative SPs were observed in rat cortex during tone-shock pairing by Pirch and Barns (1972), but no tendency was observed for potentials to decline in late training. However, only 4–5 training sessions of 20 trials each were given.

Pigeon striatum exhibited predominantly negative SPs during tone-shock pairing in one study (Durkovic and Cohen, 1968), but the acquisition function was peculiar in that negative shifts evoked by the CS, on the average, were smaller during training than during pretraining. However, larger negative SPs were evoked in experimental birds than in pseudo-conditioning and stimulus-control birds. The occurrence of SPs in the nonlaminated structure of the pigeon striatum indicated that a radial orientation of neurons is not a necessary condition for the occurrence of slow shifts, a conclusion supported by subcortical recordings in mammals (Rebert and Irwin, 1969).

Negative SPs were also recorded in the cat's anterior sigmoid (AS), coronal (C), and anterior lateral (AL) cortices with tone-shock pairing by Skinner (1971). With overtraining, the SPs were diminished or of longer latency. Cryogenic blockage of the inferior thalamic peduncle (ITP) abolished the SP in AS cortex but had only a slight effect on the other regions. The disappearance of responses in C and AL regions during overtraining, their continuation in AS, and the differential effect of ITP block on the three cortical regions indicated that in the cat, as in the monkey (Borda, 1970), at least two separate SP sources exist in frontal cortex, and they appear to be activated by different subcortical mechanisms.

Rebert and Irwin (1969) did not observe SPs in the motor cortex of cats during classic appetitive or aversive conditioning, but later results (Irwin and Rebert, 1970) suggested that this was due to low motivation of the animals or small rewards; higher level rewards lead to SPs in the motor cortex. Deep cortical and subcortical SPs were also investigated. The splenial gyrus exhibited positive conditioned SPs only in aversively motivated animals, whereas positive SPs in the medial geniculate developed only in appetitively motivated cats. Negative shifts occurred in both conditioning groups in the ventralis postero-medialis and increased in amplitude through days 5 and 6, showing a sharp decline thereafter. The point of decline appeared to be associated with a levelling off of the behavioural conditioning curve. Systematic increases of positive SPs occurred in the MRF for both conditioning and pseudoconditioning groups, suggesting some sensitization effect there, but little relationship to conditioning *per se*. Neither study showed effects of conditioning on SPs in the ventral hypothalamus or amygdala. However, there were differential effects of the tone CS on the ventromedial nucleus (negative shifts) and surrounding tissue (positive shifts) and an evocation of positive SPs by the CS alone in the amygdala. The motor cortex, lateral hypothalamus, medial amygdala, and reticular formation all exhibited larger SPs to high than to low reward. The lateral hypothalamus in both studies exhibited during conditioning increasing positive SPs which were not influenced by either the magnitude or hedonic character of the reward. Associative factors appear to be crucial in producing lateral hypothalamic SPs; in other regions motivational factors prevail.

CNVs are dependent, at least in part, upon *S*'s knowledge that an association exists between warning (WS) and imperative (IS) stimuli. Usually, with human *Ss*, that knowledge is verbally mediated, and there is little evidence of a learning effect on CNV amplitude. When verbal mediation is omitted, however, the CNV can be seen to develop (rather rapidly) as the *S* perceives an association between the WS and IS (Hillyard and Galambos, 1967). When classic conditioning procedures are employed, the CNV develops rapidly but disappears during later stages of conditioning, as noted above in many animal studies (Walter *et al.*, 1964). A possible explanation of that effect is that the WS is transformed into an IS in the classic conditioning paradigm (Rebert and Knott, 1970).

Rebert (1972c) showed that in the monkey the earliest indication of learning of the simple association of WS and IS occurred in non-specific ascending reticular structures. Slow potentials developed in 2–3 days in the non-specific system but occurred only after 4–6 days in the cortex. Some caution needs to be exercised in interpreting that difference because the trends were not developed with respect to asymptotic amplitudes, and the non-specific nuclei produced larger SPs. The same differential manifestation of learning was evident for the *discrimination* of WS and discriminative stimulus (DS). The result is similar to that of Halas *et al.* (1970) who studied multiple units during classic conditioning.

The fact that the subcortical conditioned SPs occur earlier in training than cortical SPs raises an interesting question concerning the inter-relationships of those brain regions. Donchin has questioned whether learning in different nuclei could be considered to be autonomous events, or whether the total brain learns, and the learning is manifest in different nuclei at different times. Several investigations show that learning can occur in isolated spinal cords (Patterson *et al.*, 1973) or in functionally decorticate rats (Bures and Buresova, 1960) and that the cortices, or cerebral structures in general are not necessary for simple associative learning to occur. It is possible that, in the cued RT task, subcortical structures 'learn' independently of the cortex. At any rate, the differential responsiveness of the non-specific system and the cortex during training indicates that the cortical potential is not a simple consequence of ascending non-specific influences. That conclusion is also supported by the dissociation of cortical and reticular SPs reported elsewhere (Rebert, p. 191).

Hablitz (1973) also presented learning curves of CNVs in monkeys over blocks of 2-week training periods. His 'central' placement exhibited larger shifts in the first block than the 'frontal', but at 6 weeks amplitudes were reversed.

Intracerebral recording has also been employed to investigate how the CNV is distributed in the cortex and what brain regions other than the cortex are involved in mediating the preparatory process.

At the scalp the CNV is largest at the vertex, smaller parietally, and smallest at frontal leads (Weinberg and Papakostopoulos, p. 34). Many early reports of CNV involving both scalp and intracerebral recording emphasized the anterior distribution of CNVs (Walter, 1964; Walter *et al.*, 1964; Walter 1965a, b; Walter *et al.*, 1967). It was also suggested that some frontal regions exhibited CNVs, whereas others did not. In the first study of CNVs in monkeys, Low (1969) showed that short-latency, large CNVs occurred in the frontal cortex. Smaller and longer latency waves were recorded from somatosensory cortex, and yet smaller waves from more posterior regions. Borda (1970) showed that in monkeys CNVs occurred in frontal gyrus and anterior motor strip (being more consistent frontally) but were absent in the inferior parietal gyrus. A similar result in monkeys was reported by Donchin *et al.* (1971) when a standard CNV paradigm was used. McSherry and Borda (1973) also indicated that CNVs recorded from frontal sites occurred with greater amplitude and consistency than those recorded from the motor cortex.

Rebert (1972c) recorded from the region of cortex dorsomedial to the arcuate region (premotor cortex—Brodmann's area 6) potentials that were three times as large as those obtained from the motor cortex, suggesting that the premotor area may contribute dominantly to the scalp-recorded CNV. Hablitz (1973) undertook the most thorough analysis of CNV distribution in frontal cortex to date, although postcentral recordings were not made. Recordings were made from (1) anterior frontal gyrus, anterior to the arcuate region ('anterior frontal'); (2) a region near the upper extent of sulcus arcuatus—apparently in Brodmann's area 6 ('frontal'); (3) anterior motor cortex ('precentral'); and (4) motor cortex ('central'). Although CNVs were recorded at all electrode sites (in the food-reinforced task), two major generators were delineated at 'frontal' and 'central' sites. Potentials at 'anterior frontal' and 'precentral' sites could not be distinguished from volume-conducted activity from the major generators. Rebert (1972c) reported a tendency statistically not significant for the 'frontal' potentials to be larger than 'central' ones. 'Frontal' potentials recorded during a shock avoidance task were not different from those in the food-reinforced task, but a large 'precentral' CNV developed during aversive conditioning that was very small during appetitive conditioning, suggesting yet a third region of CNV genesis in that task.

Recent intracerebral recordings in man indicate that a substantial CNV is generated in the postcentral gyrus during a standard cued RT task. In agreement with data from monkeys, it was also found that prefrontal areas generate very small CNVs (Papakostopoulos and Crow, p. 201). It appears that CNVs evoked in a standard cued RT task are generated primarily in a region extending from near the arcuate sulcus to the postcentral gyrus with little activity more posteriorally and with two dominant foci in Brodmann areas 4 and 6. However, cortical mapping endeavours have been scant, and most of the data from both humans and monkeys only include placements near the midline.

The form and distribution of SP responses in the cortex depend on particular task requirements. For example, in a study by Donchin *et al.* (1971), there was little evidence of postcentral CNVs in the monkey during a standard cued RT task, but a large positive-negative complex developed there when the monkey was required to make a motor response at the onset of the WS. At the same time, frontal CNVs were greatly reduced. Similar findings are reported in man when the *S* initiates a trial or is required to respond to the WS (McAdam and Rubin, 1971; Otto and Leifer, 1972b; Hamilton *et al.*, 1973). The effect occurs, however, only with *initiation* of movement. *Tonic* muscle tension (e.g. key-pressing) has no effect on CNV (McCallum and Papakostopoulos, 1972) unless the task involves a relatively high degree of effort, in which case CNV amplitude is *increased* (Rebert *et al.*, p. 126). Perhaps the imposition of a response requirement at trial onset changes the task sufficiently to question the appropriateness of referring to the evoked potentials as CNVs.

Intracerebral recording in humans has also revealed the presence, in one patient at least, of beta bursts from several cortical regions that occur more frequently during certain portions of the CNV paradigm than under more normal circumstances. During initial sessions, the bursts occurred most often following the *S*'s motor response, but decreased in incidence in subsequent sessions. They were restored, however, by instituting the equivocation procedure. In general, the bursts seemed to be related to the possibility of checking the outcome of predictions, but their relationship to CNV was not clear (Storm van Leeuwen and Kamp, 1973).

McSherry (1971) attempted to define the intracortical CNV generator by recording from several cortical layers. He observed three types of shifts, including a surface negative variation with no shift near the soma, surface negativity and deep positivity, and no surface shift with deep positivity. Polarity inversions occurred in the outer millimetres of cortex, suggesting a dendritic–somatic dipole generator. Using a model suggested by Creutzfeldt *et al.* (1966), McSherry concluded that the patterns of SPs noted above related to dendritic depolarization, dendritic depolarization with somatic IPSPs, and somatic IPSPs, respectively. Thus, it might be concluded that a general decrease in cellular activity occurs in the foreperiod. No studies of cortical cellular activity have been undertaken specifically in the CNV paradigm, but Fuster (1973) reported that in a similar situation (delayed responding) the most common response of frontal cells during delay was *increased* activity. And, frontal cells decreased firing in the presence of distracting stimuli, correlating with well-known effects on human CNV (Tecce and Scheff, 1969). Fuster also noted a lack of increased cellular firing during delay in the parietal area and a strong relationship of parietal cell activity to movement.

As indicated by Caspers and Speckmann (p. 195), during cortical paroxysmal activity the surface of the cortex becomes negative as a function of the mean depolarization of pyramidal tract cells. However, the form of the extracellular SPs observed in deep cortical layers is more similar to intracellular potentials than is the surface recording, indicating that the surface potential is also some function of nonsomatic membranes. Their data are compatible with the notion of increased cortical excitability during surface negative shifts, as suggested above.

Some preliminary investigations of unit activity associated with the 'readiness' potential also suggest a predominance of increased activity during preparation (Gantchev, personal communication), but Rowland and Dines (1974) found cortical SPs to be dissociated from cortical multiple unit activity during 'expectation'. Other findings concerning the relationship of cortical surface negative shifts to neuronal activity are also mixed (*see* discussion in Rebert, 1972b; Somjen, 1973), and it is not possible to draw any conclusions about neuronal activity during the CNV at this time.

Subcortically generated SPs may have a more direct relationship to neuronal activity than SPs recorded at the cortical surface. Simultaneous measurements of SP and neuronal population responses in the cat's lateral geniculate nucleus indicate that those two measures correspond closely with respect to initial phases of evoked discharges. However, continuation of SP responses after the cessation of cellular discharge suggests a differential genesis of the two responses (Rebert, p. 191). Such data are compatible with the notion of SP production by glia through potassium efflux from active neurons (Somjen, 1973).

The cortical EP complex preceding CNVs in monkeys appears as a consistent phenomenon in several reports. It consists of a simple positive–negative–positive series, the second positive wave often being larger and of longer duration than the first. The complex continues to be evoked by a discriminative stimulus (DS) to which CNVs have been extinguished (Rebert, 1972c) and appears in response to miscellaneous laboratory noises. It may represent an evaluative response to potentially significant stimuli. The initial EP complex and CNV are not necessarily coexistent and appear to be independent events, as suggested earlier by Rebert and Knott (1970).

Few subcortical recordings of SP changes have been made during behavioural performance in animals or man. Haider *et al.* (1968) recorded from the thalamic ventrolateral nucleus of human patients and showed that a negative wave there accompanied or preceded onset of the cortical CNV. A second negative wave occurred before or following the IS. It was conjectured that the thalamus initiates and terminates the CNV. Such a conclusion is compatible with a similar notion developed by Chiorini (1969) and with Skinner's (1971) finding that a cortical SP in cats is abolished with cryogenic blockade of the inferior thalamic peduncle.

Rebert (1972c) showed that negative SPs occurred in several regions of the ascending reticular system (midbrain reticulum, posterior hypothalamus, midline thalamus, thalamic reticulum) during the cued RT task; whereas positive potentials occurred in the caudate nucleus, medialis dorsalis, cingulate gyrus, amygdala, and preoptic area. Some nuclei, e.g. the pulvinar, were unresponsive.

Recently, SP recordings have been made from several human subcortical nuclei in the cued RT task (McCallum *et al.*, 1973) with results very similar to those obtained in the monkey. McCallum *et al.* observed positive potentials in or near the caudate nucleus and negative potentials in the lateral mesencephalic nucleus. As in the monkey, the cortical and caudate SPs had a relatively slow onset in contrast to rapidly rising brainstem SPs. In the monkey, both the cortex and caudate exhibited complex EPs before development of the SP, whereas only a brief and usually simple EP preceded SP onset in the non-specific projection nuclei. The similarity of results from humans and monkeys in terms of SP polarity, latency, and general wave forms encourage further extrapolation from monkey studies to man.

It is evident that the CNV is some complex function of inputs from many subcortical regions. Yet, there are wide gaps in our knowledge of cerebral mediation of the cued RT task. Data concerning cortical and subcortical distributions of SPs are sketchy, and there is a dearth of information about neuronal activity during the CNV. The significance of positive and negative SPs in terms of cellular activity is only barely suggested by recent findings. A variety of physical events can give rise to SPs in the brain, but given the possibility that some SPs represent glial responses, and that the multiple unit measure reflects neuronal population responses, it would seem desirable to record those events simultaneously to determine whether or not SP and multiple units exhibit parallel responsiveness to experimental manipulations. Other questions deserving attention are the role of synaptic mediator substances in CNV genesis and the extent to which there are causal relationships among nuclei found to be reactive in the cued RT task.

Slow Potential Changes in the Monkey's Brain during Reaction Time Foreperiod

CHARLES S. REBERT

Stanford Research Institute, Menlo Park, California 94025, U.S.A.

It was reported previously (Rebert, 1972c) that negative SP shifts were observed in the premotor cortex (PMC) and in several non-specific projection nuclei (midbrain reticular formation—MRF, nucleus reuniens—NR, posterior hypothalamus) of the monkey's brain during anticipation, while positive SPs were observed in regions such as the preoptic area (PO), caudate nucleus (CN), cingulate gyrus (CNG), and nucleus medialis dorsalis (MD) of the thalamus. Those observations were made in 6 monkeys, but electrode placements were not uniform, and some regions were represented by only one or two placements.

Several nuclei (CN, MD, MRF, PMC and NR) were studied again in four naïve monkeys to determine whether SPs like those observed before would recur.

One electrode was also used to study the amygdala (AMG), a structure previously shown by Irwin and Rebert (1970) to be unreactive during training phases of classic conditioning in the cat, but which, on the basis of a hypothesis developed from subcortical recordings in monkeys, was expected to exhibit positive potentials during a reaction-time anticipatory task. More specifically, in previous data there appeared to be a general consistency of response polarity and wave form from several components of the ascending activating system, all showing rather rapidly rising negative SPs. In contrast, several nuclei (*vida supra*) emitted positive SPs in the foreperiod. Although all cerebral regions may ultimately interconnect, and the differentiation of systems may be somewhat arbitrary, it nevertheless appeared that the positive shifting nuclei were closely related to the rhinencephalon. For example, the CN and MD are related to the frontal lobe, and MD is related to the PO region. The CN receives projections from the CNG, which also projects to the MD and so on. It can be hypothesized that the non-specific arousal system and the rhinencephalic system are reciprocally responsive in the cued RT task. Given that hypothesis, it was predicted that positive SPs would be recorded from yet other rhinencephalic regions. The AMG was studied in the present animals to test that suggestion.*

Trial-to-trial variations of SPs in different nuclei were observed to assess the extent to which different regions might be functionally related.

Supported by Grant NS 08248 from the U.S. Public Health Service.

* There are, of course, only three options: no SP shift, positive or negative shifts, so there is a high chance probability that the hypothesis would be confirmed. Many regions need to be sampled adequately to test the hypothesis. In addition, negatively shifting nuclei might produce positive shifts in related nuclei, and it is questionable to conclude that all positive shifting nuclei belong to one system, whereas negative shifting nuclei belong to another. Nevertheless, the specific hypothesis that nuclei clearly identified as rhinencephalic will produce positive shifts (interpreted as decreased neuronal activity—Rebert, 1973c) is reasonable to propose.

Presumably, negative and positive evoked SPs (in relatively homogeneous nuclei) reflect increased or decreased cellular activity, but the exact relationship of SP polarity to direction of cellular response is unknown. Studies of SP correlates of neuronal population responses in the lateral geniculate nucleus of the cat (Rebert, 1973c) suggest that subcortically generated negative SPs may reflect neuronal excitation, and vice versa. Despite the inappropriate tip size (≈250 μ) of the DC-recording electrode used in this study, an attempt was made to record multiple unit (MU) responses from nuclei in one animal that showed exceptionally robust SPs.

METHOD

Techniques employed were as reported previously (Rebert, 1972c, 1973b). Briefly, 4 female adult stumptailed macaque monkeys (*Macaca arctoides*) were studied in a discriminative reaction time foreperiod task using 'Tang' as reinforcement (1 cc/bar press). Delivery of reinforcement following a bar press was made contingent on the presence of a 10-sec duration cue lamp which served as

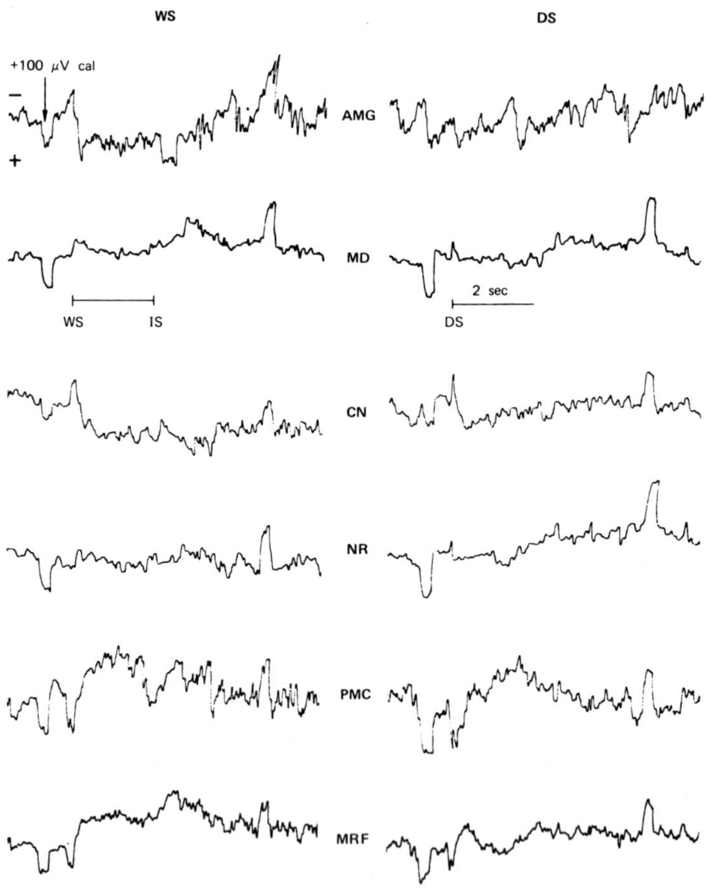

Fig. 1. Examples of SPs recorded on a single trial from electrodes aimed at the amygdala (AMG), n. medialis dorsalis (MD), caudate nucleus (CN), n. reuniens (NR), premotor cortex (PMC) and midbrain reticular formation (MRF).

the imperative stimulus (IS). This was subsequently (during training) preceded by a 2-sec warning stimulus (WS). Another tone, the discriminative stimulus (DS) was presented alone for half the trials. Tone frequencies were 1000 and 600 Hz, and the frequency used as WS or DS was varied among the animals. Fifteen trials each of WS and DS were given every day in pseudo-random sequence. Bar presses made during the WS or DS terminated the trial. Potentials recorded from the structures enumerated above with silver–silver chloride electrodes, like those described by Rebert and Irwin (1973), were amplified by Grass Model 5P–1 DC preamplifiers, recorded on a Precision Instruments 6208 recorder, and analysed with a Linc-8 computer. Reaction times in msec were obtained with a Computer Measurements Corporation 726C counter.

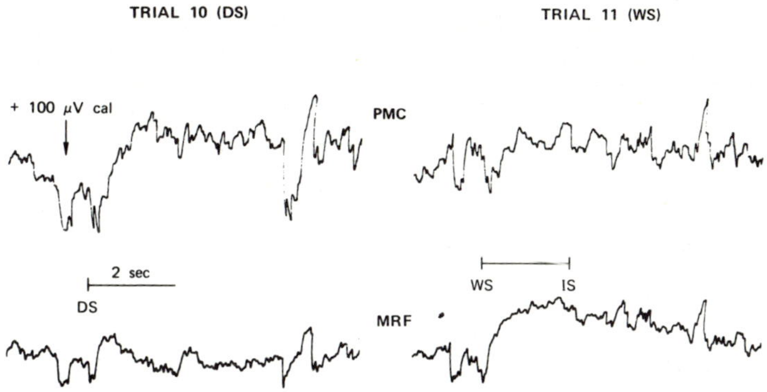

Fig. 2. Examples of SPs recorded from the premotor cortex (PMC) and midbrain reticular formation (MRF) on two consecutive trials during training day 24, showing dissociation of the two responses. Such differences occurred frequently.

RESULTS

Histological verification has not yet been undertaken, but electrodes were apparently placed successfully in the AMG, CN, MRF, and PMC, but not in NR. No conclusion can be made about the electrode aimed at MD. The inference is made that the former electrodes were correctly placed, since potentials like those observed previously from CN, MRF, and PMC were again obtained. Given correct placements there, it is extremely likely that the AMG was also penetrated since it is such a large structure, and coordinates for it were obtained in the same manner as for the other placements. The NR is relatively small, and previous findings indicated that atypical recordings could be obtained as little as 1 or 2 mm from it. Responses obtained from the electrodes aimed at MD varied from animal to animal, indicating that interanimal placements were probably not exactly the same.

As shown in *Fig.* 1, positive SPs were recorded in the CN and AMG, and negative SPs in the MRF and PMC. The recordings have not yet been averaged, and the general nature of responses from the MD and MT are indeterminable from single trial observations. However, little SP activity was evident in those areas (*Fig.* 1).

Fig. 2 illustrates the trial-to-trial dissociation that often occurs between the MRF and PMC. Any pattern of shifts may occur, i.e. large SPs may occur in both structures, SPs may be absent in both, or SPs may occur in one but not the other.

Results of MU recordings were somewhat equivocal. A clear increase of unit activity was associated with the negative SP observed in the MRF, and CN positivity was associated with decreased unit firing late in the foreperiod. However, the positive SP recorded from the AMG was not clearly associated with unit activity. A brief increase of activity was noted, followed by a very slight decrease.

DISCUSSION

These results suggest that when electrodes are similarly placed, SP polarity is consistent from animal to animal. Therefore, some confidence can be placed in results from unique recording sites, i.e. given the same placement in other animals, the same results can be expected. Much confidence can be placed in the conclusion that the MRF generates negative SPs and the CN positive ones, as has been demonstrated in several animals. Results from the AMG, which exhibited positive SPs, are in accordance with the prediction that such a polarity would be observed there because of the AMG's rhinencephalic classification.

In general, it appears that the relationships between the several recording sites are highly labile. Specifically, the shift in the PMC does not appear to result from ascending projection from the MRF, although the cortical potential might relate more closely to other components of the arousal system. Trial-by-trial correlations and other techniques need to be employed for quantitative evaluation of intracerebral inter-relationships during anticipation.

Although derived from large electrodes and a minimum of trials, results of the MU recording generally confirmed the suggestion that negative SPs observed in the MFR represent cellular excitation, and that CN positivity represents a decrease of cellular activity. Pursuit of such a line of research could greatly clarify the significance of intracerebral SPs, at least subcortical ones, for Sheafor and Rowland (1973) have indicated that in the cortex, an SP develops during early phases of 'expectancy' without evidence of cellular activity. That result is compatible with McSherry's (1971) notion that the CNV represents both dendritic depolarization and soma hyperpolarization, a configuration of membrane events that might minimize changes in cell firing rates, yet produce surface negativity.

SUMMARY

Slow potential (SP) changes were recorded from several subcortical regions of the monkey's brain. Negative SPs developed in the premotor cortex and midbrain reticular formation (MRF) during the foreperiod of reaction time, whereas positive potentials developed in the caudate nucleus (CN) and amygdala. Results confirmed earlier findings, and substantiated the hypothesis that positive potentials would occur in rhinencephalic structures. Dissociations of cortical and reticular changes were noted.

Preliminary recordings of multiple units from the MRF and CN confirmed previous findings in the cat's lateral geniculate that subcortical negative shifts in some regions represent increased neuronal activity and vice versa.

DC Shifts, EEG Waves and Neuronal Membrane Potentials in the Cat Cerebral Cortex during Seizure Activity

H. CASPERS and E.-J. SPECKMANN

Institute of Physiology, University of Munster, German Federal Republic

In extension of previous investigations on the origin of slow potential changes in the cerebral cortex (Caspers, 1959; Speckmann *et al.*, 1972; Somjen, 1973) the present studies deal with the following subjects: (i) relationship between DC shifts and membrane potential (MP) changes of single neurons; (ii) relationship between MP changes and AC potentials; (iii) relationship between the DC component and AC potentials superimposed on DC shifts.

In these studies seizure activity was chosen as a model, because the activity of a single neuron incidentally recorded from can be regarded as sufficiently representative for the behaviour of the whole neuronal population under this condition. Furthermore, seizure discharges are associated with distinct and reproducible DC shifts as well as with characteristic AC potentials.

METHOD

The experiments were carried out on adult cats anaesthetized with pheno-barbital (40–80 mg per kg), relaxed by gallamonium (1–2 mg per kg) and ventilated artificially, the body temperature being kept constant at 37°C.

The cortical DC potential with the EEG waves superimposed was led both from the surface and deeper layers of the sigmoid gyrus against a reference point in the nasal bone. Simultaneously single pyramidal tract (PT) cells in the immediate vicinity of the active DC electrode were recorded intracellularly. For details of the techniques applied *see* Speckmann *et al.* (1972) and Caspers and Speckmann (1972). Generalized seizure activity was elicited by repeated intravenous applications of 50 mg per kg pentylenetetrazol (PTZ).

The tracings were evaluated on a PDP–12 computer by: (i) simultaneous sampling of cortical DC potential and of neuronal membrane potential on two channels; (ii) selecting defined discharge patterns from the DC recording which remained on the display unit as a guide pattern; (iii) searching for corresponding patterns in the DC recording and superposition under visual control; (iv) simultaneous superposition and averaging of the data on the second channel (neuronal membrane potentials). In some experiments steps (iii)–(iv) were performed automatically.

RESULTS

Repeated applications of 50 mg per kg PTZ induced generalized tonic-clonic convulsions which lasted 0·5–3 min and usually recurred at intervals of 1–10 min. Each single fit was accompanied by a surface negative DC shift, often preceded by a short positive deflection (*Fig.* 1*A*). The DC shifts were associated with characteristic activity changes of PT cells. The DC displacements (*Fig.* 1*C*) ran parallel to series of paroxysmal depolarization shifts (PDS). Both tracings demonstrate a significant correlation between the mean neuronal depolarization and the mean negative DC deviation during the seizure attack (*Fig.* 1*B*).

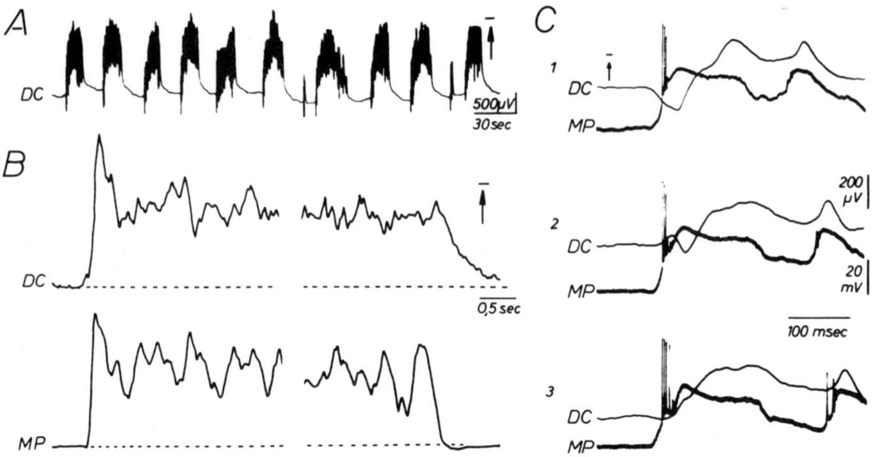

Fig. 1. A, Negative cortical DC shifts preceded by a short lasting positive deflection during successive tonic-clonic convulsions induced by repeated applications of 50 mg per kg pentylenetetrazol (PTZ) in the cat. B, Simultaneous recordings of cortical DC shifts (DC) and of membrane potential changes (MP) of a pyramidal tract (PT) cell immediately below the active DC electrode during convulsive attacks. The average is of 12 consecutive seizure discharges. Triggering was at the commencement (*left*) and at the termination (*right*) of the fit. Gap between tracings 20–40 sec. C, Paroxysmal depolarization shift in a PT neuron (MP) and associated fluctuations of the DC potential in a surface lead at the onset of seizure activity.

Nevertheless, differences exist between those two bio-electric phenomena in both polarity and time course, particularly at the commencement and the termination of seizure activity. In intracellular recordings from PT cells onset of seizure discharges was marked by a PDS. This steep decrease of the membrane potential coincided with one or two rapid deflections in the surface lead which preceded the sustained negative DC shift. In spite of the uniform depolarization of the cell, both polarity and time course of these prepotentials showed considerable variations (*Fig.* 1*C*). With the termination of a seizure discharge intracellular recordings always showed a steep and longer lasting re- and hyperpolarization, whereas the negative DC shift from surface leads characteristically returned slowly to the preictal level (*Fig.* 1*B*). In recordings from deeper cortical layers it increased in slope and turned subsequently to a transient positive shift (*see* Speckmann *et al.*, 1972). This finding corresponds in principle to observations by Ferguson and Jasper (1971). As a whole, extracellular field potentials and MP fluctuations of PT cells were more closely related in deeper cortical layers, especially in layer V, than in surface leads.

The DC shifts which occurred during seizure activity coincided with polymorphic EEG potentials, which were possibly related to single PDS in the unit leads. To determine whether there was such a relationship, five basic patterns which appeared repeatedly in the same animal and in subsequent experiments were selected from the variety of EEG seizure discharges. These patterns consisted of monophasic negative waves, monophasic positive waves and of diphasic mixtures of the two (*Fig.* 2*A*). Computer analysis showed that each of the selected EEG patterns coincided with a neuronal PDS independent of its actual shape and polarity (*Fig. 2A*). The conclusion is that the form of EEG potentials associated with the depolarization of the PT cells must be influenced by other mechanisms. Polarity and time course of evoked cortical potentials such as the direct cortical response (DCR) can be influenced by electrical and chemical polarizations, i.e. by shifting the local DC component to the negative or positive side (Caspers, 1959). A surface negative DCR decreases in amplitude and finally reverses in polarity with increasing cathodal polarization. Therefore it seemed reasonable to test the relationship between the polarity of the EEG patterns and the DC level during seizure activity. The results of a typical experiment are presented in *Fig. 2B*. The averaged curves display a train of successive seizure discharges in the EEG corresponding to pattern 1 in *Fig. 2A*.

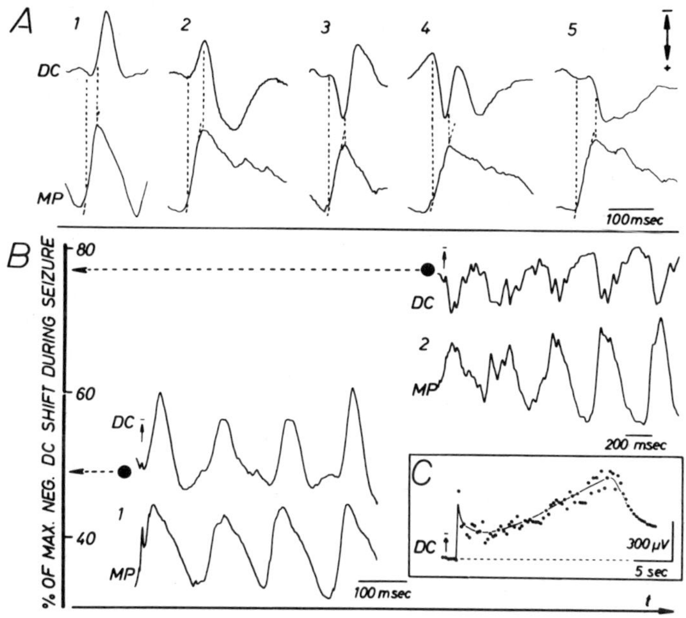

Fig. 2. Relations between seizure patterns of EEG waves in a DC recording, membrane potential changes of PT neurons and DC level during interictal and ictal convulsive discharges elicited by repeated i.v. application of 50 mg per kg PTZ in the cat. A, Selected standard seizure patterns in the EEG (DC 1–5) and their relationship with simultaneously recorded paroxysmal depolarizations of a PT neuron (PDS) in the immediate vicinity of the active DC electrode. Each trace is the average of 20 measurements in the same experiment. B, Trains of neuronal PDS associated with surface-negative EEG spikes at a low DC level (1) and with surface-positive EEG spikes at a higher negative DC shift (2). C, Insert shows time course of cortical DC shifts during consecutive convulsive attacks in same experiment as in A and B.

As found with isolated waves, each EEG deflection is strongly correlated with a neuronal PDS. The sequence of monophasic negative EEG potentials occurred at a cortical DC level equivalent to only 48 per cent of the maximum negative displacement encountered at the beginning of successive fits. In the further course of seizure attacks the amplitude of the surface negative DC displacement tended to increase (*Fig. 2C*). The tracings in *Fig. 2B2* were recorded in such a period. Like the single discharge in *Fig.* 2*A*5, the EEG deflections synchronized with neuronal paroxysmal depolarizations were positive in polarity. In this phase the negative DC shift amounted to 77 per cent of its maximum value. As indicated in *Fig.* 2*B*, selection of EEG patterns according to the EEG level at which they were recorded showed a clear distribution in that surface positive waves occurred with maximum probability at high negative DC shifts, while negative EEG deflections predominated at lower levels.

DISCUSSION

The investigations showed that the negative cortical DC shift during seizure activity is related to the mean depolarization of PT cells, the impaled part of which can be presumed to be the body. The correspondence of the two bioelectric phenomena is especially striking, if the DC component is recorded from the layer of the impaled neurons. These observations suggest that these slow extracellular field potentials are generated mainly by neuronal membranes. Differences can also be observed in DC recordings from the surface of the cerebral cortex, particularly at the beginning and end of seizure activity. Such dissimilarities indicate that the DC shifts recorded from the cortical surface cannot be attributed solely to MP fluctuations of the cell soma in deeper cortical layers. Since the DC shift at the surface and in deeper layers of the cerebral cortex differ mainly in respect of the time course of decay, it may be assumed that non-somatic membranes contribute significantly to the DC shifts in surface leads. There is increasing evidence that a complex generator of dendrites and glial cells plays a role in this context (Creutzfeldt, 1969; Speckmann *et al.*, 1972; Somjen, 1973). Parallel depolarizations of the PT cell body and the superficially located generator by afferent impulses would explain the correspondence with surface negative DC shifts during seizure activity. The variability of EEG patterns associated with a constant neuronal depolarization can be explained by differences in the DC level at which they occur.

The correlation between EEG and DC changes can be interpreted by the following hypothesis. It may be assumed that high voltage negative DC shifts at the cortical surface are initiated by a strong asynchronous excitatory input to the generator structures in superficial cortical layers. In this condition the majority of relay neurons mediating the afferent impulses to the DC generator structures can be presumed to be highly active. An additional synchronized afferent volley to the cortex during a seizure discharge is likely to induce a cathodal blockade of the relay cells resulting in a considerable reduction of the excitatory input to the superficial generator structures. At the same time the strongly synchronized impulses impinging directly on the generator structures, which are already depolarized, become less effective. In this way a surface positive deflection develops reflecting the absence of facilitation. This is supported by greater number of interneurons from the cortex. These show that the pause in unit firing after the beginning of an EEG seizure discharge becomes more pronounced, when the amplitude of the surface positive component of the EEG spikes increases (Jami, 1972; Dichter *et al.*, 1973). When there is a less pronounced asynchronous excitatory input to the cortex, resulting in a lower negative DC shift, the generator structures in the superficial cortical layers can be activated further by a synchronized afferent volley. The EEG spike associated with a paroxysmal neuronal depolarization tends then to be negative.

The proposed interpretation of the findings would explain the 'phase shifts' between unit activity and EEG potentials during convulsive discharges often described in the literature (c.f. Creutzfeldt, 1969). Moreover, the observations suggest that EEG waves and DC shifts can originate from the same generator.

SUMMARY

The relationship between cortical DC shifts, EEG waves and neuronal potentials during generalized seizure activity was studied in cats with the following results:

1. The negative cortical DC shift occurring with a series of seizure discharges is related to the mean depolarization of pyramidal tract cells. The closest correspondence is found if the DC component is recorded from the same layer as the impaled unit. DC tracings obtained in surface leads differ from MP fluctuations of PT cells at the beginning and end of a convulsive attack.
2. Sharp EEG waves of various shapes superimposed on the DC shifts coincide with neuronal paroxysmal depolarizations independent of their actual form and polarity. The variability of EEG potentials can be attributed to differences in the DC level at which the various seizure patterns occur.

Human Studies

Co-chairman's Opening Remarks

W. STORM VAN LEEUWEN

In the following section we will be endeavouring to leave phenomenology and to search for the mechanisms behind the phenomena. It would appear that the slow potentials may reflect a state in a certain population of neurons which permits a particular type of behaviour to develop. On the evidence of our intracerebral studies in both dogs and man this would also seem to be the case in respect of certain forms of rhythmic activity in the brain. Consequently I think that it is important to study the slow potential activity in relation to the rhythmic activity occurring in various parts of the brain. Using paradigms having two or more stimuli, we have been able to demonstrate such relationships in patients with implanted electrodes. We have already reported relationships involving bursts of beta activity, but we have also noted apparent relationships between both alpha and theta activity and slow potentials.

The animal work in this field has been important, but new developments of technique now permit us to carry out these studies in man, and the availability of modern amplifiers with high input impedance even allow the study of intracerebral slow potential changes in humans in relation both to behaviour and ongoing intrinsic activity.

Electrocorticographic Studies of the Contingent Negative Variation and 'P300' in Man

D. PAPAKOSTOPOULOS and H. J. CROW

Burden Neurological Hospital and Institute, Bristol, U.K.

A positive component of some evoked responses has been called P3 or P300 or association cortex potential (Sutton *et al.* 1965; Donchin and Smith, 1970; Ritter *et al.*, 1972). It has attracted the interest of a number of investigators, and its appearance and amplitude have been correlated with a number of psychological variables such as orienting, stimulus certainty or selective attention (Donchin and Cohen, 1967; Sutton *et al.*, 1967; Ritter *et al.*, 1968; *see* Karlin, 1970 for review).

Previously reported studies have made use of scalp electrodes; and only some have referred to the possibility of records being contaminated by eye movement potentials, although eye blinks can produce positive potentials with long latencies at the vertex (Eisengart and Symmes, 1971; Roth, 1973).

We have taken the opportunity afforded by chronically indwelling subdural electrodes, implanted as part of a multifocal, leucocoagulation treatment procedure for intractable mental suffering (Crow *et al.*, 1961; Crow, 1973), to examine the contingent negative variation (CNV) and the P300 in man.

Paradigms used to evoke responses with a P300 component all have something in common with paradigms used to evoke the CNV. We have therefore paid particular attention to relationships between the P300 and the CNV, including their spatial distribution over the cortex. To assist in differentiating the two phenomena an experimental paradigm has been used in which CNVs are terminated either by a motor action or by a decision *not* to perform the motor action. This paradigm has been previously described by Papakostopoulos *et al.* (1970).

METHOD

The data have been collected from 5 psychiatric patients implanted with gold electrodes (4 mm × 150 microns) for treatment purposes. The electrodes and the method of implantation were described by Crow *et al.* (1961). Recordings were obtained from a 16-channel Elema Schönander Electroencephalograph fitted with universal amplifiers, each with an input impedance of 100 megohms so that the effective time constant of the intracerebral electrodes is sufficiently long for recording the slow potential.

The timing of brief auditory, visual or somatosensory stimuli was controlled by a PDP–12 computer which also collected and stored the data for subsequent off-line analysis.

The experimental design used is that described by Papakostopoulos *et al.* (1970) and Papakostopoulos and McCallum (1973). Briefly it consists of the presentation of the various stimuli: (1) without instructions (neutral); (2) with the instruction to press a switch on receiving an imperative stimulus (S_3) which is consistently preceded by a warning stimulus (S_1), the i.s.i. being 2 sec; (3) with the instruction that when a third stimulus (S_2) appears between S_1 and S_3 the subject should 'change his mind' and *not* press the switch on receiving the imperative signal (S_3). The inhibitory signal (S_2) in the present series of experiments was always presented halfway between S_1 and S_3. Forty-eight trials were made with each patient, pseudo-randomly mixing 24 when S_2 was present with 24 when it was not. Various permutations of auditory and visual stimuli were used as S_1, S_2 and S_3 for each *S*.

RESULTS

Evoked potentials to neutral visual, auditory or somatosensory stimuli are very similar when recorded from the same cortical electrode located at the frontal or parietal association cortex. *Fig.* 1, Ia shows superimposed averaged (N=8) evoked responses of the three sensory modalities. They present a characteristic main negative component preceded and followed by a positive deflection. The second positivity is followed by another but smaller negative

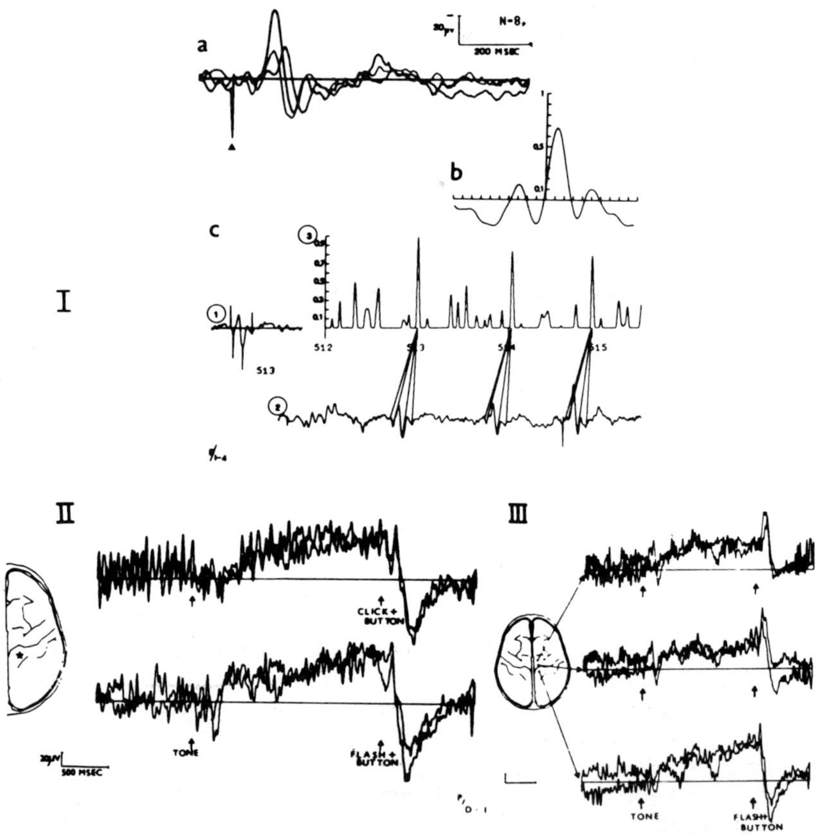

Fig. 1. I, Evoked responses to auditory, visual and somatosensory stimuli; (a) from prefrontal cortex; (b) cross correlogram of the visual and auditory responses; (c) recognition indices (3) of the three responses, using the auditory response (1) as template. II, CNV and P300 from the same cortical electrode (*). In the upper trace the imperative stimulus was auditory, and in the lower trace was visual. III, Distribution of the CNV and P300 from cortical electrodes 0·8 cm apart.

component. The amplitude and delay of these components varies according to modality, but the pattern is similar for all as can be seen by the high cross-correlation value (0·7) between auditory and visual response (*Fig.* 1, Ib) and by the high value of the recognition index (Weinberg and Cooper, 1972a, b) for all responses (*Fig.* 1, Ic). The described pattern of the evoked responses changes under certain circumstances. A consistent change is the appearance of a prominent positive component (P300) peaking about 300 msec after the presentation of the imperative stimulus. The appearance of this wave in all 5 patients under all situations tested coincides with the disappearance of both the second positive and negative component of the neutral evoked response. The appearance of the P300 component is independent of stimulus modality as can be seen in *Fig.* 1, II when it develops after a click (upper trace) or after a flash (lower trace). In both of these situations the CNV precedes its appearance. In *Fig.* 1, III simultaneous recordings were made from 3 subdural electrodes separated by a distance of 0·8 cm in an anteroposterior line. The most posterior electrode clearly recorded both potentials, CNV and P300. The most anterior electrode recorded only the CNV, the P300 being absent. The different spatial distribution of the two potentials is also apparent in *Fig.* 2, I (left part) where the activity at 6 electrodes is simultaneously recorded in another patient. The localized appearance of the P300 in the parietal cortex can further be seen in a third patient, *Fig.* 2, II,

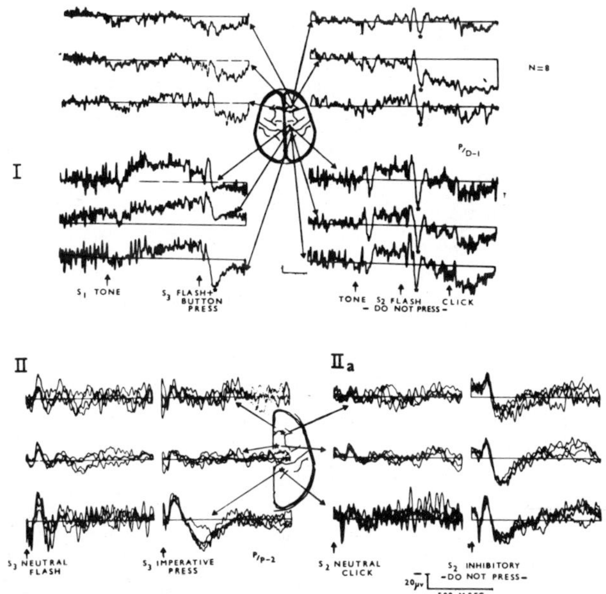

Fig. 2. I, Differential distribution of the CNV and P300: *Left*, when the patient pressed to S_3. *Right*, when the patient, receiving an S_2 before S_3, did not press. II, Distribution of evoked responses: Left, to a neutral flash, and right, to an imperative flash. Both were preceded 2 sec by a tone. IIa, Distribution of evoked responses: Left, to a neutral click; and right, to an inhibitory click. Both were preceded 1 sec by a tone.

where superimposed averages indicate the stability of the phenomena. On the left are the evoked responses to neutral flashes. On the right are the responses of the same electrodes to flash stimuli, presented as S_3 in an S_1–S_3–press paradigm, S_1 being a short tone.

The localized vertex P300 sweeps forward over the frontal and prefrontal cortex in circumstances where the patient changes the decision to press (imposed by S_1). In the condition when an inhibitory stimulus (S_2) is interposed between S_1 and S_3 the CNV cuts off with the decision to inhibit action and a prominent P300 appears following S_2, spreading forward over frontal and prefrontal cortex (*Fig.* 2, IIa, right). Similar behaviour of the CNV and P300 was recorded from all 5 patients tested.

DISCUSSION

The results would seem to leave no doubt that the P300 is a cerebral event, independent of stimulus modality, separate from the CNV and localized in the parietal cortex in conditions which generate a CNV terminated by motor action.

It can also be concluded that the CNV and P300 are manifestations of different neuronal systems and that the P300 is specifically related to a neuronal system with an inhibitory function.

Some authors (Waszak and Obrist, 1969; Karlin *et al.*, 1970) noticed an augmentation of the localized P300 in situations with an inhibitory context. In the light of our data their findings could be explained as being due to an increased spatial distribution of the P300 on the cortex (rather than an increase of amplitude), resulting in an apparent augmentation of amplitude when recorded from scalp (Cooper *et al.*, 1965). The differences in slow potential cut-off in a vigilance situation reported by McCallum (p. 46) would also be consistent with an inhibitory interpretation of the P300 component. The possibility of a relationship between the observed inhibition of spinal monosynaptic reflexes and the P300 invites further experimentation.

The spread of the P300 to prefrontal cortex in situations demanding alterations of a pre-set plan is compatible with views of the frontal lobe functions formulated from observations of behavioural changes in animal and man. Disturbance in alternation and delayed-response performance is the common symptom in rodents (Gross *et al.*, 1965), carnivores (Warren, 1964) and monkeys (Pribram, 1961) after frontal lobe damage. Exactly similar changes have not been described in man with frontal lobe injuries but Teuber summarizing symptoms of a number of people with frontal lobe injuries, revealed the presence of particular perceptual-motor deficits (Teuber, 1966). For example, the task of body orientation (Semmes *et al.*, 1963), which needs rapid and repeated shifts in right-left reference, is impaired in *Ss* with frontal lobe lesions.

SUMMARY

The CNV and the P300 component of the evoked responses recorded from the cortex of 5 patients implanted with gold electrodes for treatment have been studied. Both potentials are cerebral in origin, and are independent of stimulus modality. We conclude that the CNV and P300 are mediated through different neuronal systems and fulfil different functions; and that the P300 is an electrophysiological indicator of an inhibitory brain function.

Distribution of CNV and other Slow Potential Changes in Human Brainstem Structures

W. C. McCALLUM, D. PAPAKOSTOPOULOS

Burden Neurological Institute, Bristol, U.K.

H. B. GRIFFITH

Frenchay Hospital, Bristol, U.K.

The Contingent Negative Variation has generally been regarded as a cortical phenomenon; a view which has been reinforced by numerous reports of CNVs recorded directly from the cerebral cortex in both man and animals. However, there have been suggestions, based largely on animal studies (Rebert and Irwin, 1969; Borda, 1970; Rebert, 1973a; and McSherry, 1973b), that subcortical structures, possibly in the thalamus and/or the brainstem and midbrain reticular system, might be responsible for mediating this, and other, slow potential changes.

Haider and his collaborators (Haider *et al.*, 1968, 1969; Ganglberger and Haider, 1969; and Haider, 1970) reported evoked potential changes from the ventro-oral nucleus accompanying the appearance of the CNV at the cortex. More recently this group has indicated that slow potential changes may be seen in some thalamic nuclei.

McCallum *et al.* (1973) gave a brief report of the presence of CNV-like changes appearing in human midbrain structures and of corresponding positive slow potential changes in caudate nucleus and other structures rostral to the thalamus. The present paper will give a fuller account of these changes and of others related to the making of voluntary motor responses.

METHOD

Two patients with electrodes implanted in brainstem structures for the purpose of carrying out surgical procedures were studied. Electrode placements were not identical in the 2 cases, but their overall distribution was very similar. As the results obtained from both patients were virtually identical those presented will for convenience and continuity be drawn principally from one patient. This patient was a 57-year-old lady who had suffered a cerebral thrombosis two years previously giving rise to a moderate left hemiparesis and the appearance of a thalamic syndrome. There was painful hyperaesthesia with sensory impairment over the entire left half of the body and face.

Thirty-four gold electrodes were available for study. These were 150 micron in diameter, 4 mm in length, with a vertical separation of 2 mm. It was calculated that the most rostral electrodes were located within or adjacent to the caudate nucleus and the pes coronae radiatae, possibly with some extension into the lateral ventricle. The most caudal electrodes were in the lateral mesencephalic nucleus, peripeduncular nucleus, substantia nigra, pes pedunculi, medial and lateral lemniscus. The intermediate electrodes extended through the following thalamic nuclei; reticular, dorso-oralis, zentrolateralis, ventrocaudalis, ventro-intermedius, ventro-oralis and the medial geniculate body. There were also extensions into the internal capsule. All recordings were carried out before any surgical intervention, other than the introduction of the electrodes, had taken place. Throughout the tests both patients were, in spite of pain or discomfort from their disorder, co-operative, attentive and capable of an adequate level of response.

From a preliminary series of tests a common reference composed of a group of seven electrodes was selected on the basis of low levels of intrinsic and evoked activity. Simultaneous scalp recording was carried out from silver–silver chloride electrodes at C_Z and Fp_Z, each referred to linked mastoids. The pre-frontal electrode was used as a control for eye movements.

High gain universal amplifiers with input impedance of 100 megohm and time constants of from 1·2 to 5 sec were used for recording. Averaging was performed by a PDP–12 computer. Stimuli used were an 80 dB auditory click presented through lightweight earphones, single or repetitive flashes delivered by a stroboscope 30 cm in front of the patient's face and electrical stimulation of the median nerve at the wrist.

Testing was carried out in three phases spread over a period of several days. *Phase I* consisted of the recording of evoked potentials to irregular, single stimuli in the three sensory modalities. The interstimulus interval ranged from 5 to 15 sec. Twenty-four trials were given in each modality. Four electrodes, which showed clear flash responses, were selected to investigate the effects of habituation. At this stage 160 flashes were presented at a regular 1·5 sec interval and a further 80 with an intertrial interval of 1·0 sec.

During *Phase II* paired presentations of Click (S_1) followed by repetitive flashes (S_2) were used. For the first 24 trials the patient was not required to make any response to these stimuli. Thereafter she was asked to press a button as quickly as possible with her right hand to terminate S_2. The S_1–S_2 interval was 1 sec. Forty irregularly spaced trials were presented at intervals ranging from 5 to 15 sec.

In *Phase III* the patient was asked to make voluntary button presses as and when she wished. A 3-sec epoch of EEG data was sampled from 6 selected brain stem electrodes and stored by the computer on a single trial basis. The 3-sec epoch was composed of 1·8 sec prior to the press and 1·2 sec after the press. One hundred and twenty such trials were stored to button pressing by the right hand. A further 120 right-hand presses elicited an immediate flash stimulus from the stroboscope. Finally a series of 144 voluntary foot pedal presses were made, 72 with the right foot and 72 with the left. These gave rise to no stimuli.

RESULTS

Evoked Potentials. Clear evoked potentials to single flashes were recorded from electrodes located in the areas of the lateral ventricle, pes coronae radiatae, substantia nigra, nucleus peripeduncularis, nucleus lateralis mesencephali, lateral lemniscus, medial geniculate, pulvinar and the ventrocaudal nucleus of the thalamus.

Responses in the region of the pulvinar and medial geniculate showed an early component at 42–48 msec. The main negative peak had a latency of 92 msec and an amplitude of from -40 to -60 μV. It was preceded by a positive peak at 62 msec. A few millimetres distant, in the area of the pes peripedunculi, this negative component had a latency of 106 msec and the positive 72 msec. Response to click was considerably smaller, more variable in form, and was observed principally in the areas of the lateral ventricle, pes coronae radiatae, caudate nucleus, substantia nigra, some ventral and ventrocaudal nuclei of the thalamus and possibly the medial geniculate. One form of the response, characteristic of the most rostral electrodes, showed two negative peaks, one at 54 msec and one at 92 msec. Some of the most caudal electrodes showed late slow potentials of 200–300 msec duration. Evoked potentials to stimulation of the right median nerve gave rise to evoked potentials in the areas of the lateral ventricle, pes coronae radiatae, nucleus peripeduncularis, substantia nigra, nucleus lateralis mesencephali, medial lemniscus and the ventrocaudal nucleus of the thalamus. Other somatosensory data could not be evaluated because of stimulus artefact.

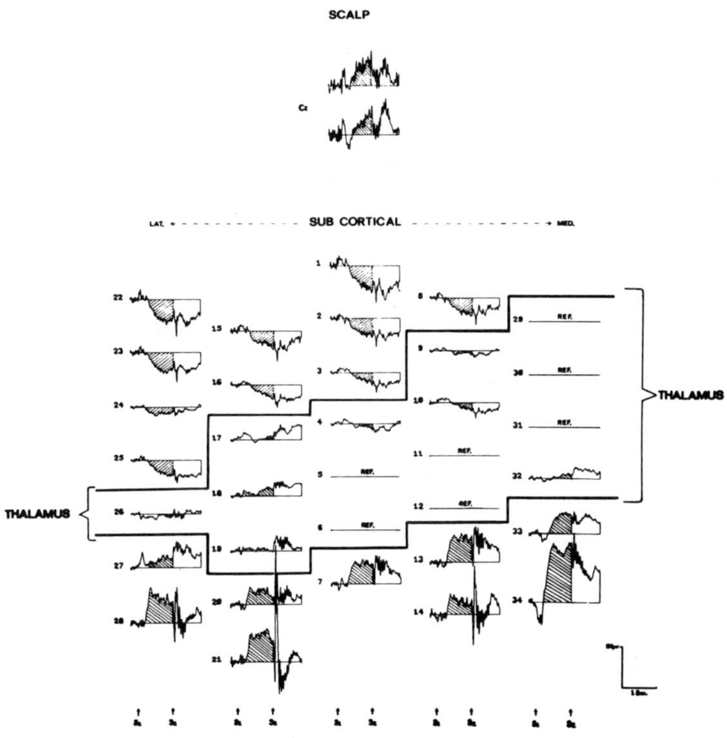

Fig. 1. A diagrammatic layout of CNVs recorded from the scalp and brainstem. Electrode C_Z is referred to linked mastoids. Subcortical electrodes have as a common reference the seven thalamic electrodes indicated. S_1 is a Click; S_2 are repetitive flashes terminated by a button press. Calibration is -20 μV. Note the presence of a prominent CNV at the most caudal (midbrain) electrodes and a corresponding positive variation at the most rostral, extrathalamic electrodes.

Selected electrodes in the areas of pes peripedunculi, pulvinar and ventrocaudal nucleus showed little evidence of habituation with regular flash stimulation at either the 1·5 or 1·0 sec intervals. However, in the area of the lateral ventricle and pes coronae radiatae a flash evoked potential with a negative peak latency of 112 msec was absent on at least half of the averages with regular stimulation at intervals of 1·5 sec and was no longer observable when the intertrial interval was reduced to 1·0 sec.

CNV. Widespread slow potential changes were observed from brainstem electrodes during the S_1–S_2+Response condition, those from the most rostral electrodes being positive with respect to the intrathalamic reference and those from the most caudal (mesencephalic) electrodes negative. The intermediate electrodes in each sheaf, i.e. those in thalamic nuclei, showed very small or no slow shifts (*see Fig.* 1). The form and duration of the slow changes from the caudal electrodes resembled closely the vertex CNV while the positive shift seen at the rostral electrodes constituted a phase reversal of the caudal response components. Amplitudes and onset latencies of the slow potential shifts observed are given in *Table* 1. In view of the short intrinsic time constant of gold electrodes, the amplitudes of slow shifts measured subcortically were probably about 20 per cent less than the true values.

To determine whether the phase inversion seen between rostral and caudal electrodes was essentially due to the choice of an intrathalamic common reference, six subcortical electrodes showing clear evidence of slow potential changes were selected and were referred, on separate channels, both to the intrathalamic reference and to the scalp mastoid reference. Simultaneous recording of CNVs from this dual montage indicated that the waveform and polarity of slow potential change at a given electrode were the same with both references.

Table 1. Amplitudes and latency of SPs in various brain structures

	Slow Potential Change	
Electrode Location	*Amplitude (Mean level for* 200 *msec preceding* S_2)	*Latency to onset*
Scalp (Cz)	−9 μV	470 msec
Subcortical (rostral to thalamus)	+4 μV to +10 μV	340–360 msec
Thalamus	+2 μV to −3 μV	—
Mesencephalon	−5 μV to −25 μV	230–280 msec

Bereitschaftspotential. Six electrodes were selected for study of this voluntary response, 3 rostral to the thalamus in the area of the lateral ventricle and pes coronae radiatae and 3 caudal to the thalamus in the areas of the pes peripedunculi, nucleus peripeduncularis, pulvinar and medial geniculate. Clear slow potential shifts were observed to precede the motor response at all 6 electrodes. At those rostral to the thalamus the shifts were positive going, at those caudal to the thalamus they were negative. Onset of the slow shifts was from 1·0 to 1·25 sec prior to the start of the motor response as measured by the EMG. Peak amplitudes of the positive shifts were from 2·5 to 4 μV and of the negative shifts −3 to −5 μV. When the press elicited a flash the anticipatory negative shift tended to be slightly larger. The evoked potentials to the flash showed a somewhat different distribution from those seen in Phase I. They included a protracted late component which was positive at the rostral electrodes and negative at the caudal ones. The pattern of anticipatory responses to button pressing with a foot pedal was similar to that with the hand, although the negative shifts at the caudal electrodes were less well developed and shorter in duration than those to hand responses. There was, however, at these electrodes an extended and accentuated negativity throughout the period of motor response. For comparison of the 6 electrodes in each condition *see Fig.* 2.

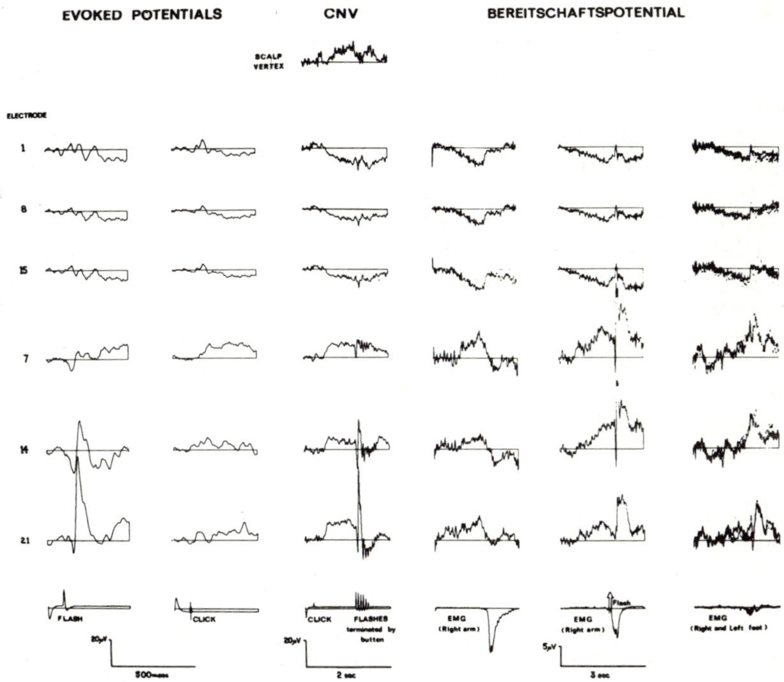

Fig. 2. Comparison of 6 selected brainstem electrodes showing sensory evoked potentials, CNVs and Bereitschaftspotentials. Electrodes 1, 8 and 15 are in the area of the lateral ventricle and pes coronae radiatae. Electrodes 7, 14 and 21 are in the area of the cisternum, pes peripedunculi and nucleus peripeduncularis.

DISCUSSION

The absence of histological confirmation and the size of electrodes used mean that quoted locations of electrodes are 'best estimates'. Inevitably this must impose limitations on the detailed comparisons with animal studies.

The various sensory responses, the CNV and the Bereitschaftspotential appeared to have similar, but not identical, distributions throughout the brainstem. There was little evidence of slow potential changes or evoked potentials in the particular thalamic nuclei in which electrodes were located. Such small changes as were observable reflected in form and polarity the more prominent activity of the adjacent extrathalamic structures. Electrodes located outside of the thalamus tended to show responses in more than one sensory modality and, in at least seven cases, in all three modalities tested. Where responses were seen within thalamic nuclei these tended not only to be smaller but to be specific to one modality (either auditory or visual). Only at one thalamic electrode, believed to be located in the nucleus reticularis thalami, were responses seen to both auditory and visual stimuli. This electrode showed no slow shift in the CNV situation.

The diversity of evoked responses in both form and latency and the varying capacity for habituation suggests a substantial degree of functional differentiation at the subcortical level.

In the case of the Bereitschaftspotential the rostrally distributed positive shifts preceding the movement were similar in form and amplitude irrespective of whether the button press elicited a flash or not, or whether it was made by the foot or the hand. However in certain more caudal, midbrain areas there was accentuation of the shift when the flash was present; the flash response itself changed in form and distribution in these circumstances. At these caudal electrodes the negative shift to foot pressing differed from that to hand pressing; the onset was later and there was a continuing negative shift throughout the period of foot pressure. The pattern of response was complex but consistent and was virtually identical for presses made with either foot.

The CNV findings are consistent with those of Rebert and Irwin in Macaque monkeys performing a discriminative foreperiod RT task. The diversity of responses and the differences in distribution according to the nature and context of the stimuli and tasks involved suggest that the midbrain structures concerned are capable of more than an undifferentiated arousal response. The small positive shifts seen in the area of the lateral ventricle showed less variation with condition and even simple sensory stimulation appeared to give rise to a small, but sustained positive displacement in these areas following the earlier response components. The polarity reversal itself is not easily accounted for, but seems clearly to be a biological phenomenon and not a mere artefact of recording technique.

Attention has tended to focus on cortical levels of information processing and frequently the impression is given that the role of the deeper structures is essentially a crude and non-specific one. The results of the present study should serve to shed doubt on such a view and to demonstrate that many of the electrophysiological phenomena held to be manifestations of higher cortical function are also to be found as deep as the midbrain level, and may in some instances appear there earlier in time than at the cortex.

SUMMARY

Sensory evoked potentials, Contingent Negative (and Positive) Variations and Bereitschaftspotentials (Readiness potentials) have been recorded from areas of the human brainstem and midbrain. The distribution of these phenomena in the deep structures is similar, but not identical. The evidence suggests that not only may a number of complex cortical events be mediated at this level, but that quite well developed functional differentiation exists there.

AN EXPERIMENTAL APPROACH TO BRAIN SLOW POTENTIALS

In organizing the Bristol Slow Potential Congress an attempt was made to depart from the usual pattern of international meeting in which a seemingly endless succession of papers are read with minimal opportunity for discussion. There seemed to be a general feeling that such meetings achieve a level of communication which is generally no better than that attained by the written word, and frequently is worse. Nevertheless, the conviction remained that bringing a diverse and highly experienced group of scientists together for a period of time and allowing maximum opportunity for interaction and discussion should be capable of a more fruitful outcome. Various strategies were employed to bring about this state of affairs, one of which was to provide the opportunity for live, experimentally orientated sessions.

The plan was that there should be five experimental topics or areas, which should be the subject of a laboratory presentation to all participants attending the Congress. In some the aim would be to isolate areas of outstanding controversy, to illustrate these experimentally and to propose experimental solutions, which could be the subject of general discussion. In others the aim would be to present a selection of techniques and recording methods appropriate to the study of evoked and slow potential changes and to highlight some of the methodological problems in a way which would enable the combined expertise of the group to be brought to bear on their solution.

It was stressed that neither prescriptive statements nor rigid dogma was to be offered; the intention was to present one laboratory's solution to some of the methodological problems faced by all SP experimenters.

Following preliminary discussions, well before the Congress, it was agreed that the major issues to be treated in this way should be *CNV–P300 relationships*, which had for some time generated controversy, particularly with regard to the question of whether P300 waves were ever more than the termination of a preceding CNV, and *CNV-Bereitschaftspotential relationships*, where, although the controversies were less strong, doubts remained about which components of the BP/MP complex unequivocally preceded motor movement and which accompanied or followed it. These questions were coupled with outstanding questions about the extent to which CNVs and PBs were linked or independent phenomena. It was further proposed that there should be an experimental presentation on *Slow potentials and psychiatry*, which would aim to replicate some of the recent findings in the clinical field and to evaluate these. At the more technical level demonstration experiments would be run to illustrate *Methodological issues* and *The use of telemetry in evoked and slow potential research*, with a view to pooling the assembled expertise and to achieving some degree of standardization, which would facilitate the evaluation and replication of future experimentation in this field.

Responsibility for each of the experimental presentations rested with an individual or individuals, highly experienced and of international eminence in the areas concerned. These were:

1. Methodological issues: Dr Ray Cooper.
2. The use of telemetry: Prof. Willem Storm van Leeuwen.
3. CNV-Bereitschaftspotential relationships: Prof. Hans Kornhuber and Dr Luder Deecke.
4. Slow potentials and psychiatry: Prof. Maurice Dongier and Dr Bernardo Dubrovsky.
5. CNV–P300 relationships: Dr Emanuel Donchin.

Each presentation was conducted in the laboratories of the Burden Neurological Institute at Bristol under the direction of one of the individuals listed above, assisted by the scientific and technical staff of the Institute and utilizing the recording and computing facilities available. Special computer programmes were prepared by Mr Philip Pocock, Mr W. J. Warren and Dr Ray Cooper to meet the individual requirements of each group.

All of the presentations involved live on-line experimentation with human subjects in a normal laboratory environment with the exception that the proceedings were televised by a battery of cameras and conveyed over a closed-circuit link to monitors viewed by the main body of congress participants in another part of the building. Two-way communication enabled participants to comment and discuss both technical details and major experimental issues during the course of the experiment and to suggest modifications to the design or questions which could be answered experimentally during the session. A commentator/coordinator in the laboratory provided a link between the experimental team and audience.

Methodological Issues and the Use of Telemetry

These two sessions were for convenience combined, the former part being represented by multichannel scalp EEG recordings from a normal female subject and intracerebral EEG recording from a patient, and the latter by telemetred EEG scalp recording from the commentator/coordinator while engaged in describing the recordings involved and in interaction with the audience. On-line averaging and other analysis was carried out on a PDP–12 computer which also sampled and stored the telemetred data for subsequent off-line averaging.

Methodological issues were demonstrated and discussed and the following points made:

Electrodes: Impedance to be less than 5 kohm between pairs, measured at 40 Hz to avoid polarization of electrodes.

Time Constants: For CNV experimentation a rough guide would be for minimum time constants to be three times the inter stimulus interval.

Eye movement artefacts: Usually controlled for by use either of compensation techniques or rejection techniques as described by McCallum and Walter (1968), Papakostopoulos *et al.* (1973) and Winter (1974). Each has advantages and disadvantages. One of the fundamental problems is that eye electrical artefacts do not come from a single source.

Averaging: Features discussed were: (1) *Sample rate.* This should be not longer than half the period of the highest frequency component which requires to be resolved. Usually a considerably shorter interval than this is adopted; for the experiments shown it was based on 512 points per channel. (2) *Baseline normalization.* Achieved by continuous sampling of intertrial data and averaging for a specified period prior to each trial. (3) *Equalization of data.* Achieved by sampling a known calibration signal (20 μV) prior to recording and comparing its digitized average values with a value which was optimum for storage and display purposes, then calculating for each channel a factor which would transform the sampled average to the optimum standard and applying it to all incoming data.

Experimental procedures for continuous involvement tasks (vigilance) and for examining relationships between slow potentials, autonomic changes and spinal reflex activity were demonstrated, together with analysis techniques such as the Recognition Index, described by Weinberg and Cooper (1972b), which constitute an alternative to signal averaging.

Intracerebral electrodes were used to make scalp-cortex comparisons and to demonstrate infra-slow activity at approximately 3 waves/min, similar to that described in animals by Aladjalova (1957, 1964) and in man by Cooper *et al.* (1973). Similarities and differences between adjacent electrodes were noted and the point demonstrated that scalp derivations tend to reflect the average activity from an underlying area of cortex having a diameter in the order of 2–3 in.

THE TELEMETRY EXPERIMENT

The second aspect of the methodology presentation dealt with the recording of evoked and slow potential changes by telemetry. The major problem was to find a 'natural situation', which did not result in *S* being orientated towards a 'laboratory experiment'. The aim was to have *S* engaged on a task which would occupy him fully and which was not contrived, hence the choice of the commentator responsible for coordinating the presentations and acting as link with the remote audience.

Stimuli were provided via the system adopted for relaying questions from the audience to the commentator (*S*). This was done by a special line, the signal that someone wished to ask a question being a 'bleep', similar to those in general use in hospital call systems. A convention was adopted that two tones, 1 sec apart, would be *S*'s signal to take a question. Occasional single tones were introduced which were to have no significance for *S*. On hearing the second of the paired tones the commentator had immediately to acknowledge the signal by raising his hand, thereafter asking for the question to be put. The commentator's *verbal* response might occur almost immediately or be extensively delayed depending on factors in the experimental demonstration.

Initiation of the 'bleep' signal caused an interrupt to occur on the PDP–12 computer, which changed temporarily to a special programme, designed to sample and store *S*'s EEG for a period of 2 sec, then returning to its current, ongoing experimental programmes until the next 'bleep'.

The 8-channel telemetry system, designed by Mr Kamp, was a modified version of that used by Walter *et al.* (1967) and described by Kamp (1963a, b).

Three separate conditions were averaged, each being 16 trials. The first average was of two-tone trials during which *S* had not been actively involved

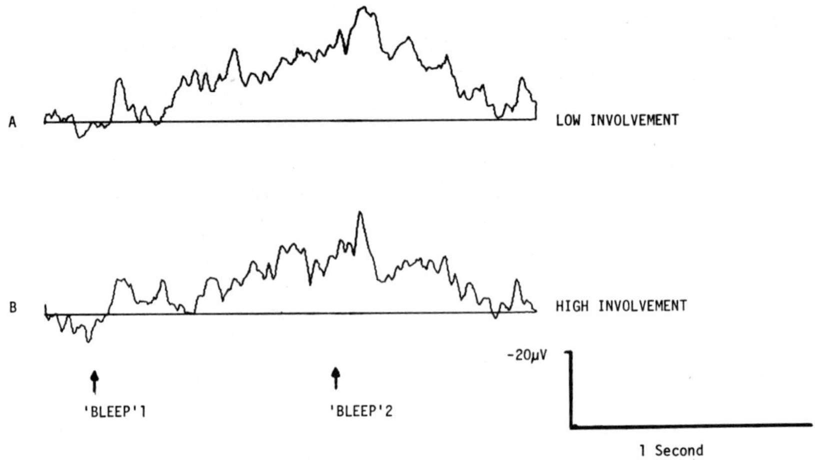

Fig. 1. Average CNVs recorded by telemetry. N=16; Vertex-linked mastoids. *S* was commentator for the methodology demonstration in A a low involvement stage and B a high involvement stage of the commentary when interrupted by the 'bleep' tones. He acknowledged the second bleep by raising his hand.

with the ongoing experimentation at the time they were presented, the second was of similar trials during periods of high involvement and the third was of single tone trials requiring no response (*Fig.* 1). A small slow potential shift occurred at the vertex in all three conditions following the evoked potential to the first tone. The shift was largest in the condition of relatively low involvement. While eye movement occurred between and during trials, it could not account for the negative shifts seen at the vertex.

CNV–Bereitschaftspotential Relationships

The aim of this presentation was to demonstrate some of the potential changes which precede and accompany the initiation of voluntary motor movements and to compare the form and distribution of such potentials with those of the CNV. An experimental paradigm was devised consisting of: (*a*) a condition in which the subject made voluntary self-paced finger movements, and (*b*) a condition in which a similar series of voluntary presses were followed 1·5 sec later by the onset of a tone which had to be terminated as rapidly as possible by the subject pressing the lever a second time. The first condition was designed to permit recording of a simple BP, the second to link the BP and CNV in one situation.

A clear BP was obtained in the voluntary press condition (*Fig.* 2). This consisted of an ascending, negative ramp, starting approximately 500 msec prior to the onset of EMG activity and reaching a level of 6 µV prior to EMG onset at the Vertex (11 µV over Parietal areas). A small positive-going deflection could be seen which peaked at the time of EMG onset, but which in this instance clearly began before the earliest recorded EMG activity. This wave was recognizably that described as premotion positivity (PMP). Following this transient positive deflection the negative ramp continued to rise to a peak 15–20 msec after EMG onset, coinciding approximately with the peak of EMG activity. Thereafter a positive cut-off was observed. In distribution the BP was most prominent at the vertex and over central areas, fairly prominent over parietal areas and almost totally absent over the frontal regions.

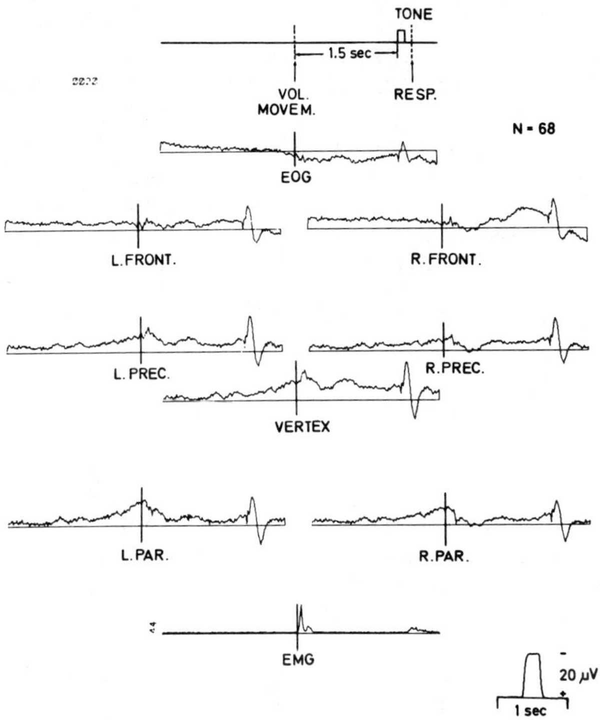

Fig. 2. Average responses in the combined BP–CNV situation. N=68. *S* makes a voluntary button press using his right index finger. This was followed 1·5 sec later by a tone terminated by a second button press with the same finger.

In the combined BP–CNV paradigm the BP was similarly distributed, but the following negativity (CNV) showed a slightly different distribution. Like the BP, the CNV was seen at its largest over the vertex and central areas, amplitudes being in the order of —7 μV. However, the CNV was proportionately less prominent over parietal areas and more prominent over frontal areas. This pattern of the CNV having a more anterior distribution than the BP was consistent with the general body of evidence accumulated by the Ulm Group during their studies. Deecke's view was that nevertheless the BP and CNV were not very different entities. He doubted that it was very helpful to postulate a central generator for the BP and a more frontal generator for the CNV. He considered it more likely that both were dependent upon projections from subcortical generators, the probable candidates being the mesencephalic reticular formation, the non-specific thalamic system, which could account for the selective aspects of the phenomenon, or the limbic system.

Slow Potentials and Psychiatry

The clinical correlate selected for demonstration was the postimperative negative variation (PINV). An attempt was made to replicate its prolonged occurrence in schizophrenia.

Interest first centred on the PINV following the discussion of 'atypical' CNV wave forms at the first CNV Congress held in Liège (Dargent and Dongier, 1969). Dongier (1969) described several instances and Timsit *et al.* (1970) made an extensive study of the PINV in schizophrenia.

In the experimental demonstration one patient with a diagnosis of schizophrenia was used as subject.

In the event a lower than expected level of PINV resulted. This was atypical of the results obtained in Liège and Montreal and was substantially lower than previously shown by the patient himself. Various factors may have contributed to this result, not least being the distracting influences of a demonstration situation and a degree of discomfort experienced by the patient due to heat and strain associated with eye fixation.

As a result of the low PINV values recorded the second phase of the experiment, which was to have demonstrated the effects of hyperventilation on the PINV, was not carried out. Dr Dubrovsky explained that in his experimetnal series hyperventilation had been found to reduce the amplitude of the PINV. Nevertheless, in response to requests from participants two experimental modifications were made to test particular hypotheses.

The first of these was based on the hypothesis that the presence of post imperative negativity might be related to sustained button pressing, as opposed to a sharp, transient press, and that for some reason schizophrenic patients might have a tendency to sustain pressing longer than normals. Dr Dubrovsky explained that this hypothesis had already been tested in Montreal with negative findings. However, *S* was instructed to make a prolonged press to S_2 during a series of such trials. No significant increase in the PINV was found.

The second modification involved the use of an S_2 stimulus close to threshold value which, it was suggested, might enhance the negativity. This modification resulted in a slightly lower amplitude CNV but little difference in the PINV. The evoked potential to the tone became much less prominent.

CNV–P300 Relationships

The following is a narrative summary of the experiment presented to the Congress by Dr Donchin and his colleagues. It was a variant of the CNV–P300 experiment discussed in the correspondence which preceded the Congress. Its design was finalized, by the majority of participants in the correspondence, at a meeting held in Bristol immediately prior to the Congress. An edited version of the pre-Congress correspondence leading to this experiment is included as an Appendix.

The experiment was carried out on one experienced CNV subject (female), who was well rehearsed. Event-related potentials were recorded from F_4, C_3, Cz, C_4, P_3, P_4 and O_2 positions (linked mastoid reference). EOG monitoring was included.

EXPERIMENTAL ARRANGEMENTS

S was supine on a bed with her head resting on a pillow. A Grass photic stimulator lamp was mounted 50 cm over *S*'s head and a television monitor was mounted 3 feet above the foot of the bed, i.e. 6 feet from *S*'s eyes. The TV screen was divided by a horizontal line into a black and a white half. The vertical order of the two halves was reversible. Display 'A' occurred when the upper part of the screen was white; the inverted display was Display 'B'. Tones could be presented to *S* through earphones. A microswitch was held in each hand.

EXPERIMENTAL DESIGN AND EXPERIMENTAL PROCEDURES

The experiment consisted of a series of trials, grouped in runs. On each trial *S* was presented with one of two tones. During Reaction Time (RT) runs *S* had to press the button in her right hand following a high tone and the button in her

Table 1. Experimental design

Condition	*Stimulus Sequence*	*Pre-warning*
Reaction Time	Random	Warn
		No Warn
	Alternate	Warn
		No Warn
Predict	Random	Warn
		No Warn
	Alternate	Warn
		No Warn

left hand following a low tone. During the Predict (P) runs, the *S* was to guess which of the two tones would be presented on the following trial. During RT trials, the tone was repeated if the subject failed to respond within a specified interval. During P trials the tones were always repeated. Within any run some of the tones were preceded, at an interval of 1500 msec, by a warning flash from the photic stimulator. On other trials the tones were presented without warning. *S* was always informed by the television monitor whether she would be forewarned of the arrival of the tone. Display 'A' informed *S* that arrival of the tone would be preceded by the warning flash. Display 'B' indicated absence of the warning flash. One of the two patterns was always on.

Runs could also differ in the sequence of presentation of tones. In some runs tones were presented in a regular (alternating) pattern: low, high, low, high, etc. On other runs the tones were presented in a pseudorandom order, where the probability that a high or low tone would be presented on any trial was equal to 0·5 and was independent of the outcome of the previous trial.

The entire experimental structure can be seen in *Table* 1. There were two main experimental conditions, the RT and the P condition, which differed in the task *S* had to perform following the stimulus. In each condition the imperative stimuli could occur in a Random (RAN) or an Alternating (ALT) series and in each of these *S* might receive a warning tone (W) or no warning tone (NW).

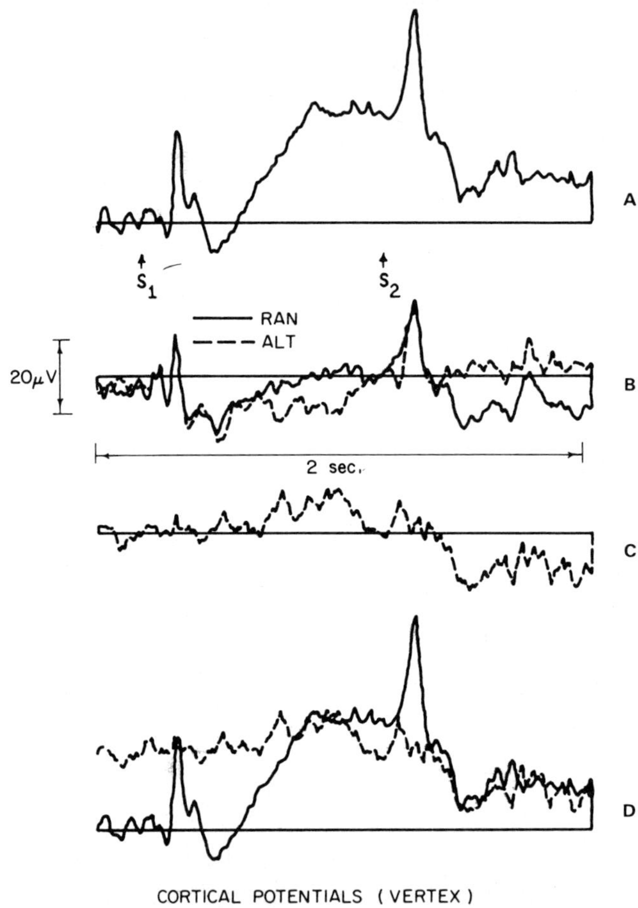

Fig. 3. A. Average Evoked Potential (N=16) recorded at the vertex from subject R, recorded during the Reaction Time (RT) condition under the 'A' paradigm (warning stimulus used) using a random sequence of the two imperative stimuli. The baseline drawn through the data is the mean of the 500 msec just prior to the warning stimulus. B. Data recorded from the same subject from the same electrodes during a Predict (PR) condition with both a RAN and an ALT sequence of choice stimuli. Note the general reduction of the CNV in the PR condition and the enhanced P300 during the RAN condition. C. A curve obtained by point by point subtraction of the random and alternating curves in *Fig.* 1B. D. The difference curve of *Fig.* 1C is superimposed on the RT data of *Fig.* 1A.

It was predicted that the presence of a warning stimulus would enhance a CNV, and the predictability of a stimulus would modify P300. RT sessions preceded Predict runs, and Random presentations preceded Alternate runs. Stimuli were presented by a PDP–12, which also averaged the potentials.

The interpretation of data revealed diversity of opinion about scoring P300. Some felt 'positivity' must be relative to a defined baseline. Others believed 'local direction of polarity' was sufficient since, if a post-CNV 'positive direction' did not cross pre-stimulus baseline, no 'P300' could exist (*Fig.* 1). The working group rejected this method of scoring, and believed that if the *relative* polarity was manipulable, as isolated P300 is, then it can be so identified. Overall relative cortical negativity could 'mask' P300, defined only by baseline. This effect of manipulation is shown in *Fig.* 1B, which superimposes *Predict* Alternate and Random trials; 'Ran' is the larger. Using this as a 'template' (*Fig.* 3C), the occurrence of P300 in other evoked response data could be evaluated. *Fig.* 3D places the template *v*. *Fig.* 3A; P300 is apparent.

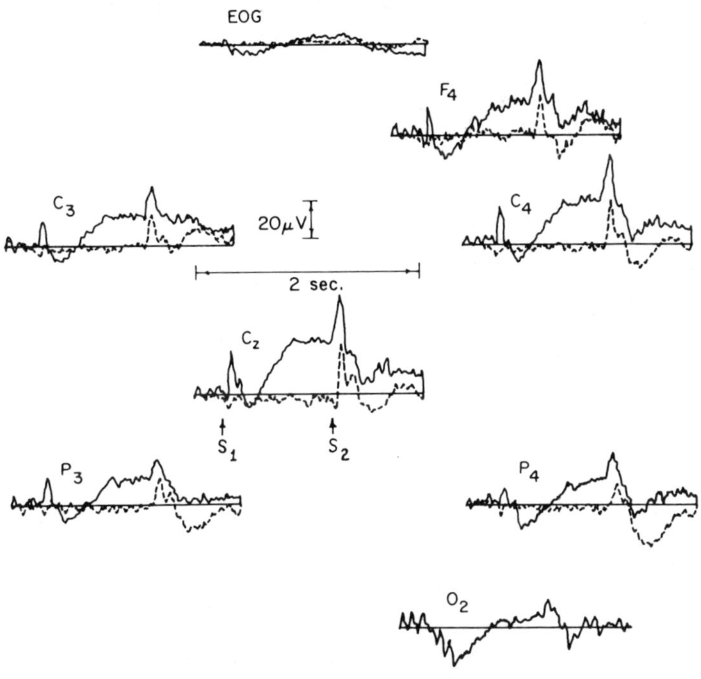

Fig. 4. Cortical potentials recorded during the RT paradigm with random presentation of the choice stimuli with (solid line), and without (dashed line), a warning stimulus. Note the general similarity of the wave form elicited by the imperative stimulus in the two conditions.

The probability of cortical slow activity 'contaminating' EOG records, rather than the opposite, was demonstrated by latency and by field data.

For recording P300, parietal areas yielded higher amplitudes than others (*Fig.* 4). Frontal areas yield greater P300s than occipital areas. Electrodes in the *O* positions maximize visual ERPs and can serve as a type of control.

While CNV is missing in unwarned trials, P300 occurs in both experimental conditions (*Fig*. 5). However, there existed the possibility (in this single case) that in unwarned trials a long-term baseline shift occurred, surface negative in direction.

CNV was greater at vertex and in frontal areas, and least in the occipital area. This confirms previous reports.

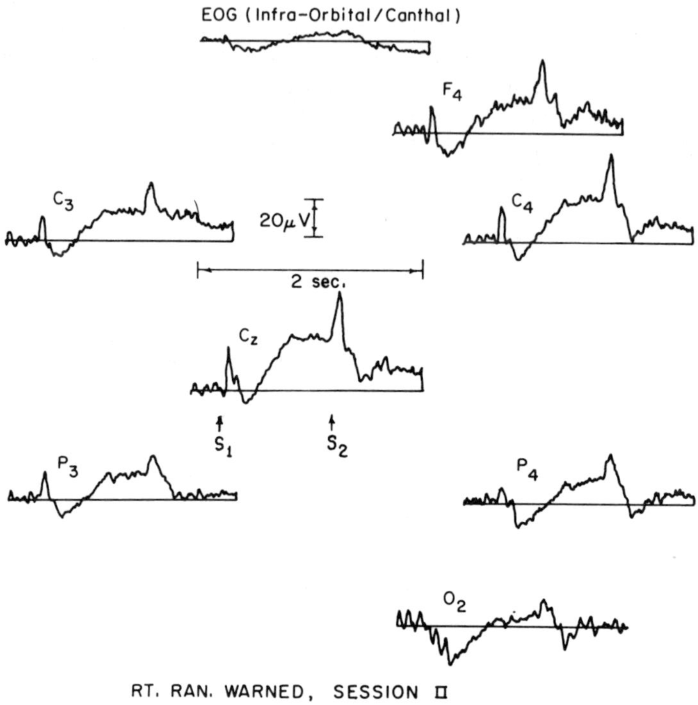

Fig. 5. Data recorded during the RT condition in paradigm A, using random presentation of the stimuli, from 8 different electrode locations. All cortical electrodes referred to a linked mastoid. The EOG is recorded in a bipolar mode.

The working group felt that this demonstration may assist in an eventual resolution of the issue of the relationship between pre- and post-stimulus electrocortical activity. The experimental design was elaborate, so that precautions must be kept in mind.

Two major aspects of this collaborative research emerge: (1) Careful design of experiments is critical, in order precisely to identify and quantitatively manipulate the independent variables. (2) The dependent variables (the electrical events, negative and positive) must be treated in a topographic domain. [Editor's note: e.g., in an electro-anatomical framework, *J.R.K.*]. Multiple recording areas, to include lateral as well as midline placements, are necessary for adequate description of electrocortical processes.

A full experiment based upon 12 subjects was conducted subsequent to the Congress and appears as a paper: 'On the independence of the CNV and P300 components of the human averaged evoked potential.' (Donchin, E., Tueting, P., Ritter, W., Kutas, M. and Heffley, E. (1975). *Electroencephalogr. Clin. Neurophysiol.* **38**, 449–461.) This study confirmed: (1) P300 is not influenced by the presence or absence of a warning stimulus; (2) the distributions of P300 and CNV are topographically different.

Editorial Comment

Space has not permitted us to report on all of these experimental demonstrations in full, but we feel that the international level of co-operation and free interaction between laboratories that was achieved was of a very high order and should be of value in furthering progress in both the scientific and clinical fields.

Similarly, the interaction which emerged between experimenter and Congress participants during the experimental demonstration, resulting in 'on-line' experimental modifications and testing of hypotheses, went a long way towards fulfilling the hope of creative, collective science emerging from an international meeting rather than a somewhat sterile dissemination of established information. Undoubtedly this new concept of interactive experimentation deserves to be developed.

We all tend to select our best cases to illustrate our scientific papers and similarly we tend to expect our public demonstrations to be smooth, clear and unambiguous. However, as scientists we know that real experimentation is rarely like this. We must be prepared for negative results and for technical problems, and we must not allow the prospect of either of these to make us shy of exposing our ideas and methods to the invaluable and constructive comments of our colleagues. [W.C.McC. and J.R.K.]

Appendix: The Relationship between P300 and the CNV

A CORRESPONDENCE CONDUCTED BY
Dr EMANUEL DONCHIN

Participants:

MERLIN DONALD, Department of Psychology, Queen's University, Kingston, Ontario.

*EMANUEL DONCHIN, Department of Psychology, University of Illinois, Champaign-Urbana, Illinois.

STEVEN HILLYARD, Department of Neurosciences, University of California, San Diego, La Jolla, California.

*DEMETRIOS PAPAKOSTOPOULOS, Burden Neurological Institute, Bristol.

†MICHAEL POSNER, Department of Psychology, University of Oregon, Eugene, Oregon.

*WALTER RITTER, Department of Neurology, Queens College, Cuny, New York.

†SAMUEL SUTTON, Biometrics Research, New York State Department of Mental Hygiene, New York.

*PATRICIA TUETING, Biometrics Research, New York State Department of Mental Hygiene, New York.

The account which follows is a compression of this Congress correspondence. Space limitations required that this be done. Direct quotation of individual statements has been used in part, but much has been paraphrased in narrative form. We have attempted to retain individuality of expression on relevant issues. (An extended version may be obtained on request to Dr Donchin.)

[W.C.McC and J.R.K.]

A. A DISCUSSION OF THE SIGNIFICANCE OF P300

Donchin (*4 April, 1973*)‡
The basic assertion guiding this discussion is that the most consistent body of data relating EPs and behaviour are concerned with P300 and the CNV.

The conditions eliciting the CNV are: an *interval* limited by at least two stimuli or events, a *task* associated with the terminal stimulus and a subject *motivated* to perform the task. Under these conditions a surface negative potential is recorded, usually peaking just prior to the 'imperative' or terminal stimulus.

---

* Present at pre-Congress Workshop to finalize experimental plans.
† Did not attend Congress.
‡ The dates are of letters from which the statements were extracted.

It is important to note that the negativity is ordinarily followed by a *rather rapid positive going change* in potential called the P300.

The P300 is a surface positive wave which can be modified (usually enhanced) by a variety of experimental conditions. For example, the presentation of stimuli in a random versus an alternate sequence, a signal detection paradigm, a habituation paradigm, a mixed tasks paradigm and a missing signal paradigm have been used by various investigators to demonstrate enhancement of P300.

There arise questions which relate to the physiology of both the CNV and P300, the identification of and specific locus of their generating mechanisms, the manner in which the waves are generated, etc. These issues are to be discussed in terms of their study in intact human subjects.

A second set of issues involves the psychological circumstances under which P300 can be elicited. The correspondents have at one time or another expressed different interpretations of P300.

a. Sutton and his group have suggested its amplitude to be a function of the resolution of uncertainty by the stimulus, or alternatively the delivery of information by the stimulus.
b. Ritter, with Vaughan, first identified P300 with the orienting reflex, later modifying this view to incorporate cognitive processing in its functional significance.
c. Some, notably Hillyard, have related P300 to the matching and mismatching of perceptual expectancies or templates.
d. Donchin has proposed P300 as a reflection of a decision-making process by the subject, independent of the specific nature of the decision involved.
e. More recently Donchin has associated P300 with the generation of an affirmative decision by the subject.

'It is necessary to distinguish between theories which specify experimental manipulations as a prescription for the enhancement of P300 and theories which postulate intracortical mechanisms descriptive of the generation of P300. The first type of theory has the merit of being operational; the second . . . postulates unobservables. . . . (For) example, how (can) one determine whether or not P300 is a mismatch-detector output without defining the properties of that detector and their effects on observables?'

These questions about P300 (Donchin continues) need specific answers:

1. What variables determine its amplitude?
2. Is there a theory about the functioning of the brain as an information processor which explains the existing data?
3. Are the positions enumerated above mutually exclusive?
4. Can apparently divergent views be integrated?
5. From any given theoretical position, what specific experimental operations will allow a determination of the correctness of that theory?

Sutton and Tueting (*27 April, 1973*) write that while they have used various terms such as 'uncertainty' (Sutton *et al.*, 1965), 'information delivery' (Sutton *et al.*, 1967) and 'salience' (Jenness, 1972a), to characterize the situations in which P300 is obtained, they regard these terms as essentially heuristic. They do not know whether all situations enhancing P300 share some underlying but undefined psychological variable. However, such an underlying variable, if it exists, must have quantitative properties (Tueting *et al.*, 1971; Jenness, 1972a; Paul and Sutton, 1972; Friedman *et al.*, 1973).

Posner (*8 June, 1973*) asserts that most discussion of P300 has been based on 'S–R' relations, P300 being regarded as enhanced by stimulus and instructional conditions, the delivery of information, reduction of information, decision making, etc. He suggests that it may be useful to seek 'R–R' statements. The data show that *S* can 'pay attention' under nearly any set of stimulus and instructional conditions, being more inclined to do so when the signal is novel, has high uncertainty, or involves a decision. He may also 'pay attention' to a routine, predictable signal to which no decision is required. Once he 'pays attention', a number of important consequences occur. For example, he will be less able to deal with new signals.

P300 seems to be closely related to the operations of a brain system which is serial in its operations. Very complex psychological operations, however, may occur entirely outside this system, so that it is reserved for only a small subset of high priority operations. We should be able to describe the consequences of its use.

Hillyard (*19 June, 1973*): 'Donchin has pointed out the difficulty in defining a hypothetical construct such as "mismatch detector" in terms of observables, a problem even more acute for a theoretical construct like "general purpose processor" proposed by Donchin *et al.* (1973b). The notion of "processing" as the reflector of P300 is so general and vague . . . that it appears to remove us one step further from the problem.

'To make a meaningful, testable proposition it is necessary to specify: (1) What is "general purpose processing" and what isn't?; (2) What are the conditions that elicit such processing?; (3) What are its consequences?; and (4) How do we know how much of it has occurred? . . . Concepts like "match" and "mismatch" detectors, "expectancy bias" on decision stages (Squires *et al.*, 1973a), are sufficiently specific to be testable under some circumstances. The literature shows that P300 accompanies some kinds of processing and not others. If it reflects "general purpose processing" why should it be larger for: (1) hits rather than correct rejections of equal confidence in the threshold task with cue lights? (Squires, 1973); (2) No-Go than Go responses in choice RT?; (3) the more improbable signals in a host of detection and feedback paradigms?; (4) signals of intermediate difficulty of detection rather than those of great difficulty of detection?

'It is also clear that P300 does not reflect processing that precedes the identification or recognition of a stimulus, being small or absent for non-signals (Ritter and Vaughan, 1969; Hillyard *et al.*, 1971), highly expected signals (Tueting *et al.*, 1971), etc. Its evocation is *completely dependent* upon the *output* of those processing and memory retrieval stages that result in the stimulus being *identified* as "meaningful", "relevant" or whatever.

'Any proposition that P300 reflects *utilization* of some processor rather than its *output* could be correct but it is unarguable that a necessary condition for evoking P300 is a particular kind of output from a particular kind of processor involving perception and recognition.

'The recent experiment of Donchin *et al.* (1973b), showing late positivity to be independent of S_2 predictability in an RT task, is intriguing. It is of much theoretical importance to know if uncertainty is a necessary condition for enhancing P300 and if increasing uncertainty always increases it. The only other study showing no effect of an uncertainty manipulation on P300 was also in the

RT situation; Karlin *et al.* (1970) found that P300 in simple RT and in choice RT (on the "go" trials) was the same. In the Karlin study, however, there was always temporal uncertainty, despite stimulus certainty. (*See also* Wilkinson and Morlock (1967) and Vaughan and Ritter (1970).)

'Accordingly, before the Donchin *et al.* proposal that uncertainty is not a necessary condition for P300 can be accepted we need to know whether the equality of P300s would hold at shorter warn-S_2 intervals, where temporal uncertainty would be less than with the 1·4 sec i.s.i. Further, may an order effect be present? If the alternate-RT condition was given to all *Ss* as the first condition of Session 2 (the first EP recording session) they would tend to have greater temporal uncertainty regarding the i.s.i. in this initial run than in later runs when the i.s.i. had been better learned. Increased temporal uncertainty might inflate positivity in the alternate condition, so that it matched the P300 on the random condition with less temporal and greater stimulus uncertainty. Secondly, it is necessary to know whether the P300 in the warn-S_2-RT task is the same as in the guess-feedback condition, or whether the post-S_2-RT positivity is composed mainly of the CNV crashing down to base, plus a P_2 of the motor potential. The third question is how does RT change with S_2 uncertainty? Karlin *et al.* (1971) found that an auditory evoked P300 was larger when RTs were faster. Possibly then the faster RTs would in themselves, for whatever reason, enhance positivity otherwise smaller because of the certainty of S_2's identity. These complexities make it difficult to arrive at an unequivocal conclusion that P300 can occur in the absence of uncertainty.

'Various experiments do demonstrate that P300 occurs differentially after a relevant "signal" when signal and non-signal *S*'s are equally probable (*see* Donchin and Cohen, 1967; Harter and Salmon, 1972; Wilkinson and Lee, 1972; Ford *et al.*, 1973). These results, plus our findings that P300 on "hits" is much larger than on equally confident correct rejections, leads towards the view that it is elicited when a stimulus matches a memorized template of a signal that is currently relevant (in that a differential cognitive or motor response must be made to it).'

Hillyard draws support from the findings of Squires *et al.* (1973b) and Squires (1973) that P300 is enlarged on the 'false alarm' trials in which an absent signal of any type elicits an affirmative detection response. Broadbent (1971) calls this variety of selective attention 'response set', as opposed to a 'stimulus set' when stimuli are 'filtered' at a processing stage prior to comparison against memory. Associating P300 with 'response set' selective attention does make explicit the point that it is elicited only after a comparison of sensory data with internal stimulus representations. While P300 may reflect an integral part of this template matching, it may be a sign of a subsequent state, cognitive or non-cognitive.

Hillyard agrees with Sutton and Tueting that in most discrimination-decision type tasks not involving a motor response, the positive component is much more reliably obtained and is usually larger than the CNV. He says his earlier report that larger CNVs accompany correct detections in the threshold task was a very small effect indeed (-4 v. -2 μV), $P<0{\cdot}05$ (*see* Hillyard *et al.*, 1971), although the P300 effect was indubitable. Referring to many unpublished data of his own, he regards the correlation between CNV and perceptual accuracy as at best small and inconsistent, stating that 'It is my current unsubstantiated belief that a subject who concentrates intensely will not show a CNV-perception correlation while one who is drowsy or inattentive enough to produce considerable trial to trial CNV variability will show covarying changes in perceptual accuracy.'

Sutton and Tueting (*25 June, 1973*) refer to some experiments of theirs (*see* Tueting and Sutton, p. 71) suggesting that Donchin's conclusion that in a reaction time task there is little or no increase in P300 as a function of uncertainty is premature.

Donchin (*29 June, 1973*) replies that the new data are 'most interesting' but reaffirms his belief in the statements made in Donchin *et al.* (1973b), denying that in *all* conceivable RT tasks one would obtain a large P300 in a 'certain' condition. The only claim was that uncertainty is not a *necessary* condition for P300, and that 'only *one* negative instance' is required to demonstrate this. Given his *specific* experimental instructions, he finds a P300 in a 'certain' condition. In formulating future accounts of P300 there may be several situations in which there is no uncertainty and yet P300 is large. It may very well be that uncertainty is a *sufficient* but not a *necessary* condition for the elicitation of P300.

Donchin regards Posner's 'R–R' approach favourably, provided all terms are defined operationally (*see also* Donchin, 1973).

Donchin accepts Hillyard's comments about a 'general purpose processor'. He says that 'P300' reflects a processor which performs specific functions, but which can be called, or invoked, by programs performing a variety of tasks, analagous to a 'subroutine', the usage of which 'in a given program is determined not so much by the ultimate task of the program, but by the algorithm which the programmer has selected for achieving his ultimate goal.' Donchin feels that Hillyard's response-set simply describes a candidate subroutine, and comments that 'the fact that False Alarms, where presumably a positive match occurred, and correct rejections, in which an affirmative decision has been taken, fail to elicit a P300, poses serious difficulties for Hillyard's interpretation of his data.'

Donchin regards the subroutine analogy as leading to a possible research strategy and a conceptual framework within which P300 can be considered. He writes: 'In most of the literature concerning these brain events we have been assigning tasks to *Ss* and then trying to correlate the appearance or amplitude of P300 with some features which are defined by the ultimate goal of the assigned task. What I am trying to propose is that there may be some feature of these tasks, some aspect of the algorithms used by the system . . . which leads to the invocation of the P300 "interrupt routine".' Further, 'Speculatively, P300 may be a sign of the occurrence of an interrupt-request, while its amplitude is a measure of the complexity of the interrupt servicing that was required on any given occasion.' In effect, Donchin asks: '*What* processing is accompanied by P300? The answer, to my mind, lies not in the ultimate goal of the *S*'s task but rather in the processing employed on the way to the task.'

Ritter (*20 June, 1973*) raises the relationship to P300 of the negative potential which precedes it. This potential, which is clearest when the P300 is elicited by a missing stimulus, is in some circumstances intimately associated with P300, but has a different scalp distribution and unlike P300 occurs prior to RT. He is also concerned about the absence of P300 to correct rejections, false alarms and misses in signal detection experiments. If P300 reflects an affirmative decision, then in guessing experiments, when *S* is wrong, P300 implies: 'Oh yes, that is *not* the stimulus predicted.' This inversion of the significance of signals and non-signals seems to strain a notion that P300 reflects an affirmative decision.

B. A DISCUSSION OF THE CNV AND P300

Donchin (*7 April, 1973*) recalls that many experiments in which the CNV is recorded have much in common with the situations in which P300 is recorded (Donchin and Smith, 1970), the CNV usually being followed by a substantial positive-going limb, with a peak latency of about 300 msec. In the past he had been concerned with the degree to which the positive-going limb of the CNV is closely related to P300.

Several studies have led to a general consensus that the CNV and P300 may be distinct and that their amplitudes can be dissociated (Donchin *et al.*, 1972a; Donald and Goff, 1971).

If the amplitude of P300 is determined by characteristics of the imperative stimulus (S_2) rather than by those of the preceding CNV, does it follow that P300 is not identical with the positive-going limb of the CNV? Nothing in the CNV literature suggests a correlation between the positive and negative limbs. Any such relationship is unclear and the issue still open.

Donchin points out: 'If we consider two cortical waves "identical" we assume that we are observing the activity of *the same* physiological mechanism generating the same voltages for essentially the same purposes.' Postulation of a relationship between these phenomena imposes limitations on the functions which can be attributed to them.

Other questions are raised by the Donald and Goff study: They investigated the relationship between a CNV and a P300 elicited by a stimulus unrelated to the CNV. This probing technique is indeed useful and elegant; 'Is it possible that the processes reflected by the CNV precede numerous stimulus-triggered events and that we cannot observe these events because we can observe the CNV only when it is time-locked and synchronised with our averaging computers? Possibly, the CNVs we do observe are but a small class of CNV-like processes . . . associated over a short time interval with (one) imperative stimulus. The broader class (would) include all preparatory and anticipatory processes . . .'. Probe stimuli may be preceded by their own CNV, not 'visible' to averaging computers. 'Can we ever determine' asks Donchin, 'if there is CNV activity which always precedes action or is it a moot point which cannot be resolved by experiment?'

Sutton and Tueting (*27 April, 1973*) were intrigued by Donchin's thought that P300 might be the descending limb of the CNV and nevertheless be, to varying degrees, independent. This resolves into the question of the brain generators of the CNV and of P300. They hope data on this will appear at the Congress.

They feel there are some experimental situations in which the CNV influences P300 amplitude, other situations in which the P300 amplitude is independent, and, possibly, other situations in which P300 is the resultant of two processes, one arising from the CNV (Tueting and Sutton, 1973).

'In our hands', say Sutton and Tueting, the CNV has generally been a low amplitude, experimentally somewhat unpredictable, process. Partly this was due to the fact that we were concentrating on guessing designs in which contamination of P300, and even the CNV, by *S*'s motor involvement would be avoided. However, even when we used more classical CNV paradigms we had difficulty in obtaining reliable CNVs within and across *Ss*. Our experimental difficulties with the CNV have been one of the main reasons why we have never been impressed by the hypothesis that P300 is the descending limb of the CNV. We find that most of our P300 experiments are highly reliable across *Ss* and repeatedly within the *S*. Hillyard *et al.* (1971) found larger CNVs associated with accurate than with inaccurate discriminations. Paul and Sutton (1972) found no consistent relationship between accuracy of discrimination and amplitude of the CNV, and Delse *et al.* (1972) reported larger CNVs associated with inaccurate discriminations (in females). These conflicting reports suggest that systematic CNV *v.* P300 relationships need to be worked out in one laboratory rather than inferred from studies in different laboratories.

Posner (*8 June, 1973*) reports being able to obtain P300 without CNV, and also to manipulate P300 amplitude with a constant CNV. The former finding arose when the warning signal was presented on only a portion of the trials and produced a clear P300. P300 usually was not observed to a warning signal given on each trial, with no evidence of CNV preceding the warning and very clear evidence of it following the warning.

Also, comparing warning with no-warning trials, CNV varies greatly but EPs to the imperative signal do not differ in shape. The P300 component was not too clear in either condition, but Posner believes this confirms Donald and Goff in a situation in which the EP occurs to a relevant signal (i.e. one *S* is preparing for).

Further, instructions to 'attend' by counting the stimuli served to increase the size of the P300, while CNV was kept constant by randomization.

Finally, delay of the overt response to a discriminative signal, by not allowing *S* to respond until a later event, caused the CNV not to appear to release after the discriminative signal, although P300 occurred.

Posner's data convince him that CNV and EP are separate, though sometimes superimposed. The use of warning signals which occur only on a portion of the trials is a way to investigate P300 to signals not preceded by a CNV and to compare, at the same time, P300 with and without CNV to a subsequent imperative stimulus.

Ritter (*6 June, 1973*) states: 'The relationship between CNV and P300 are, in Posner's words, very indirect. He reports obtaining P300 without CNV. We have similar data, as apparently do Tueting and Sutton. But is the reverse true' . . . Can CNV terminate without P300? He agrees with earlier comments that examining the relative distributions of CNV and P300 across the scalp could indicate whether the two events come from the same generator. He suggests that 'in any experiment we conduct at Bristol we place several electrodes (for example, F_Z, C_Z and O_Z as a minimum) in order to ascertain whether the simultaneously obtained CNV and P300 have similar topography.' Nevertheless he acknowledges that the CNV may be generated at one locus and its termination at another, the latter being also where the P300 is generated.

Sutton and Tueting (*21 April, 1973*) remark that relations between the CNV and P300 are only one dimension of the problem. The other is whether the P300 findings can be accounted for by a pre-stimulus process other than the CNV, such as a differential degree of preparation, or some excitability cycle, or arousal cycle, which *S* sets up in relation to that stimulus. This aspect of the problem is associated with the selective attention issue and its effects on evoked potential components, P300 among others. When, in their studies conducted at a constant level of attention, *S does not know in advance* which of two stimuli will be presented, P300 is larger for the rarer event.

They continue: 'One formulation on which we would like to solicit comments is Karlin's (1970) formulation that the stimulus "only" acts to release the prior state of readiness—and in the selective attention situation, releases it differentially. It is hard for us to make very much sense of such a formulation since it concedes the battle while at the same time appearing to win it. If the stimulus *differentially* releases the prior state as a function of its identity (which by instructions has been associated with one or another meaning), then necessarily the identity of the stimulus plays a role in determining the characteristics of the evoked potential and of course the amplitude of P300.'

They concur with Ritter that it is critical to know whether, in an S_1–S_2 design, it is ever possible to get a CNV without a P300. Only the negative answer would imply that P300 might be only one of several ways in which the CNV descends to baseline. They comment that they have seen situations, without S_2, in which the CNV slowly goes positive with no clear component resembling a P300. They have also seen a slow negative shift, building up during the first three stimuli of a discrimination trial, and ending in a positive return of long duration after *S*'s response. In this case the very large P300 in response to a feedback click at the end of the train was independent of slow negative and other positive shifts present in the trial (Jenness, 1972a).

Hillyard (*19 June, 1973*) notes 'a general consensus that Karlin's diffuse arousal type of hypothesis regarding the P300 is moribund.' Nevertheless, he says, 'The "reactive change of state" hypothesis of Karlin, while unsatisfactorily vague in many ways (since any brain event can be included under the term 'change of state') still has some life in it . . . evocation of a P300 is *dependent upon* a selective stimulus discrimination and recognition process. . . . This does not necessarily mean, however, that the P300 is a sign or code of the actual signal recognition process or any subsequent specific cognitive act. . . . It could be, as Karlin has suggested, that the P300 reflects non-specific cortical events which *follow* the signal recognition/processing, such as a relaxation or readiness, activation of the reticular system, activation of a response program, etc. As Karlin (1970) puts it, "Some form of cognitive processing is a prior condition of this (non-selective) change of state, but the change does not embody the processing".'

Donchin (*29 June, 1973*) replies that Karlin has 'tried to say two different things simultaneously. His main theme was similar to Näätänen's and assumed P300 to be reflecting primarily the effects of diffuse undifferentiated arousal which may, artifactually, develop prior to the critical relevant stimuli. The other point he made has been formulated as the 'reactive change' hypothesis but, in fact, as long as the stress is on non-cognitive, pre-stimulus processes, this formulation is not strongly differentiated from the first one. Once it is admitted that the resolution of these pre-stimulus processes depends on post-stimulus processing, Karlin's position is indistinguishable from any other position and 'reactive change' becomes a euphemism for cortical potential.

'To summarize where we now stand: On the theoretical question of the nature of P300, we have no consensus though we all seem to agree that P300 is not determined by the pre-stimulus negativity. We all seem to agree that it is affected by a multiplicity of different conditions though we are not agreed on the construct which best explains the common factors in this multiplicity of P300 eliciting conditions.

'It seems that P300-related effects tend to be temporarily coincident with the descending limb of the CNV. What the implications of this temporal coincidence are is yet to be elucidated and apparently we do agree that we need to find methods for separating the two . . . .'

Donald (*24 July, 1973*) sees little point 'in re-emphasizing the search for the ideal psychological determinant of event-related EPs. Each laboratory has evolved its own procedures and theories, and one theory seems about as useful as the next. . . . (We) must attend to the more basic task of classification.'

Lessons can be learned from taxonomists: 'Avoid using the same term at various levels of abstraction; construct a system which accounts for all differences between phenomena, from the most mundane to the most fundamental. Thus, we should stop using P3, P300, CNV, EMR, etc. on different levels of abstraction. The worst offender is P3 or P300. It gives the appearance (as was originally intended) of concrete precision—a surface positive EP at about 300 msec latency. But in the many pages of correspondence in this exchange it has often been used as if (1) it was the only ERP in the sensory evoked response and (2) as if it had been proved that late waves recorded under many experimental conditions of varying latency ranges were identical with P300. We then argue about *the* P300 distribution and *the* CNV–P300 relationship. We do not even know yet whether P300 is distributed the same way in all stimulus modalities, let alone whether highly variable late potentials (LP) of over 400 msec latency, LPs to missing stimuli, post-motor positivities or the original 200 msec Vertex positive wave, can all be classed alongside it.' P300 should be regarded as '. . . a sensory evoked potential of about 300 msec latency and very reliable intra-individual waveform. Until we have a system of classification by function we should use terms only on this restricted, concrete level of description.'

There is much evidence for the independence of P300 from CNV, for example Ritter *et al.* (1968) reported that a large P300 was elicited by the initial stimulus in a repetitive series. 'It is important to determine whether the CNV and P300 interact when they occur simultaneously in the brain. . . . I believe the answer is unequivocal—they do not. Donchin (placed) emphasis on the thesis that our study (Donald and Goff, 1971) did not bear upon the issues he raised because the P300 was elicited by a probe stimulus . . . peripheral to the CNV task. However, as I pointed out in our Vancouver meeting (Donald and Goff, 1973), we analysed in the same way the P300 elicited by the CNV task stimulus and found the same lack of correlation. . . . The amplitude of the P300 could be manipulated (by instructions) independently of the amplitude of the CNV, both being simultaneously present in the brain.

'Virtually everything can be dissociated from everything else . . . and the dissociation can be demonstrated in different ways: by comparing distributions, interactions, behavioural control conditions, and so on. This implies that we can unify neither the P300 literature nor the CNV literature under the aegis of one monolithic "non-specific" system. If we want to establish general laws it will be by demonstrating analogous functional relationships in a variety of paradigms, not by showing that the same neurons are doing the same thing in a variety of paradigms. Perhaps Posner's "serial processor" lies in a *principle* of neural organization, (not) in a specific locus in the CNS. We cannot assume that activity related to the processor should always look the same, only that it should serve the same function for whatever neural networks are on demand in a given experiment.'

Papakostopoulos (*July, 1973*) says that one point on which the correspondents seem to agree is that P300 is a brain event, an article of faith strengthened by recordings of the P300 directly from cortical electrodes in man (Papakostopoulos and Crow, p. 201).

Donchin's (4 April, 1973) distinction of lines of attack on P300 '. . . raises further questions. First, can any one of these be pursued independently? To what extent is P300 subject-dependent, situation-dependent, instrument-dependent, or experimental-theoretical-attitude dependent?'

'The common element in all P300 experiments is that a YES or NO action is retained on every trial . . . what we are energising through our experimental paradigms is a basic mechanism of the nervous system related with the after-effects of binary choice of a "go/no-go" type. The suggestion is that this mechanism is inhibitory in nature. Two further questions have then to be answered: (1) What is the relationship of this basic mechanism with the pre-stimulus electrical state of the brain? (2) What is the common element in "go/no-go" actions which is mediated by this mechanism?

'Different spatial distribution for the pre-action negativity and the following P300 has been established from multichannel cortical recordings (Weinberg and Papakostopoulos, p. 34). P300 does not necessarily depend on pre-stimulus negativity but (it could) depend on the pre-stimulus state of the *S*.

'The answer to the second question would seem to be that in both conditions inhibition occurs. The difference between go and no-go situations could be that in the former, local (parietal lobe) inhibitory mechanisms are involved, whereas in the latter more general and frontally dominant inhibitory systems operate.' Papakostopoulos finds support for this view from animal studies and from human clinical and behavioural evidence, citing experiments by Pribram (1961), Teuber (1966), Luria (1973), conditioning evidence from Clemente *et al.* (1964); Wurtz (1966a); and Marczynski (1972).

Many of the positive shifts described in these experiments seem to be cholinergic in nature and are abolished by atropine and increased by chlorpromazine (Marczynski, 1972), a question which is open for test in humans.

Papakostopoulos concludes that the controversy over the nature of the P300 is a result of diversity in techniques rather than of differences in the experimental paradigms used. In all paradigms a basic mechanism of electrophysiological and behavioural inhibition is activated. This mechanism seems necessary to arrest planned behaviour or to generate the bases for alternation from one behavioural set to another. He maintains that this hypothesis is readily tested experimentally in behavioural, clinical and neurophysiological contexts.

C. NOTES FOR AN EXPERIMENT

Tueting (*21 April, 1973*) says: 'It is obvious that it would be wise to consider the application of every technical method we have to the CNV–P300 problem, in addition to relying on ingenious and simple design. For example, both CNV and P300 could be compared for different experimental conditions on height from baseline, area slope, trial-by-trial variance, between-subject variance, habituation rate, recovery, on differences in averaging from the stimulus versus averaging on the motor response, and on topography using at least six electrode locations. It might even be possible to utilize frequency analysis in order to separate hypothesized overlapping components.'

Sutton and Tueting (*29 June, 1973*) continue 'to be troubled by the assertion by Vaughan and Ritter (1970) (and also Hillyard) that P300 is larger at P_Z than at C_z. In our work using high- *v.* low-pitched clicks, P300 is always clearly larger at C_Z. Vaughan (personal communication) suggests that when P300 is obtained when a stimulus is present . . .' components may overlap. 'It is true that when P300 is obtained to a missing stimulus, it is larger or equally large at P_Z (Ruchkin's recent unpublished data). While this provides some support for Vaughan's suggestion, inspection of our wave forms provides no support for the overlap of components explanation when a stimulus is present.'

Posner (*8 June, 1973*) comments: 'A short S_1–S_2 interval would be very desirable, such as 0·5 sec which is, by the way, nearly optimal for a large class of behavioral tasks.* The use of S_1 on only a proportion of the trials, and the use of an S_2 task which may be varied in difficulty, would help to check on some of the preliminary results cited by Tueting and Sutton (this volume). Return to baseline of the CNV can be manipulated by increasing the difficulty of S_2. . . . the use of the "C reaction method" (e.g. respond to matching pairs, but not to mismatches) allows direct comparison of responses involving both action and inhibition of action without increasing the complexity of the design.'

Hillyard (*17 June, 1973*) considers that 'a useful type of experiment might attempt to partition the late P300-type positivity into separate components on the basis of their differential scalp topography. We all agree that stimuli which elicit P300s tend to be preceded by CNVs, and that the CNV can come crashing down to baseline during the general latency range of "the P300". Thus there is a confounding of these two possible sources of positivity, which has yet to be satisfactorily clarified. Donchin made the key point that CNV cut-off may be governed by different rules than prior CNV level, so that lack of correlation between P300 and prior CNV does not necessarily imply any dissociation, and topographical studies may be the only way to separate the two.

* *Editors' note:* Drs Knott and McCallum feel that this i.s.i. is well below the optimum for CNV work and has the disadvantage that the CNV cannot be effectively differentiated from the late EP components of S_1. However, Dr Posner remains convinced that this interval is preferable. Data he has obtained with Wilkinson suggest that CNVs are largest with this interval and he cites a further paper by Gaillard and Näätänen (1973) as additional support for this view.

'In some published reports the purported "P300" looks like "nothing but" a CNV returning to baseline, while in other situations a large, pointed, "P300" can be seen in the absence of any appreciable phasic D.C. shift, before or after (e.g. Hillyard *et al.*, 1971). Looking over some of my old data recorded with D.C. amplifiers it appears that the P300 which follows an auditory or visual discriminandum (as S_2) has a clearly more posterior scalp distribution than the preceding CNV and its offset. Perhaps we could set up several classic P300 situations while recording (positive and negative) D.C. scalp distributions and observe any spatial dissociation of the two phenomena.'

Donald (*24 July, 1973*) proposes the following 'research strategy which may yield the kind of data we need. The most obvious requirement is that one laboratory (as stated by Sutton and Tueting 27 April, 1973) with a single pool of *Ss* should try to elicit all of the phenomena (controlling stimulus probability, response set, task difficulty, instructions to *Ss*, etc.) recorded in the literature in an unified context, to allow direct comparisons of topography and amplitude between all the different conditions. . . . Although classification schemes can be dreamed up by anyone just reading and speculating, we will not really know whether a scheme works until these data are available.

'Sutton and Tueting, and Hillyard, seem to agree that P300 is more reliable than CNV. My experience is the opposite . . . (while) I have yet to see a *S* who will not develop a CNV . . . I have seen dozens of sessions where a *S* performed a complex vigilance or discrimination task without producing a P300 wave.'

Donchin (*29 June, 1973*) proposes 'a paradigm two of my students and I have been using to resolve the CNV/P300 issue . . . we have (as Posner suggests) to compare the evoked responses to warned and unwarned stimuli. Presumably unwarned stimuli are not preceded by a CNV. . . . Warned stimuli would be preceded by a CNV and (enable us to compare) the evoked response to the stimulus when it is or is not preceded by a CNV.'

Donchin's experimental arrangement was: *S* sat in a darkened, electrically shielded room, facing a 3-field tachistoscope. Two letters, A and B, alternated at randomly controlled intervals, the mean interval being about 3 min, the inter trial intervals ranging from 2 to 6 min. At irregular intervals of around 10 sec, a transilluminated circle was projected, superimposed on the displayed letter, and *S* was required to press a switch as rapidly as he could. The displayed letter indicated to *S* in which type of reaction time régime he was operating. Letter A indicated that the circle would always be preceded by a tone. The interval between the tone and the flash as 1500 msec. The letter B indicated to *S* that the tone would be omitted and the imperative stimuli would be presented without forewarning. If *S* failed to press within a brief interval, about 250 msec, the circle was reflashed and *S* was financially penalized.

The general idea was to compare the response to warned and to unwarned stimuli, while insuring that *S*'s overall state remained relatively the same in both conditions. Intercalating the A and the B régimes insured eliminating major changes in arousal, excitability, motivation, or other factors that might be attributed to session-to-session variability. Also, an analysis of the ongoing EEG activity in the A and the B régime can be made, particularly the background activity during the transition from the A to the B state. To repeat this with DC recording would be invaluable.

Ritter (*17 July, 1973*) responded that Donchin's paradigm had merit and suggested further variations: (1) Record EPs to the changes in the A and B letters, which should yield P300s. There should be no CNV preceding these stimuli. (2) Omit S_2 in a small percentage (10 per cent?) of the trials with a warning stimulus, both for the RT condition and the no-response condition where a warning stimulus is delivered.

Hillyard (*19 July, 1973*) approves Donchin's paradigm. 'The alternation between "A" (warning cue) and "B" (no warning cue) conditions seems an excellent way of getting the CNV in and out of the picture. However, I would urge that we also manufacture P300s by one or more procedures in addition to that of making prompt motor reactions to the light flash. (When motor responses are made) the prior CNV is much larger . . . than before any type of perceptual act, *per se*, and this "motor CNV" may be a "different kind of CNV" from that which precedes a perceptual discrimination or decision (Järvilehto and Fruhstorfer, 1970). This comes crashing down after the flash and confounds itself with the P300 as we have been discussing . . .' perhaps obscuring it entirely so that 'its independent topography cannot be determined. Second, P300s in the simple RT situation are not usually as large as those generated in such paradigms as signal detections, wrong feedbacks, and the "no-go" of discriminative RT. Third, and most important, following a motor response is the "P2" of the motor potential, which Gerbrandt *et al.* (1973) think is a vertex potential evoked by the somatic stimulus entailed in the motor response. This positivity also comes at around 300 msec after the stimulus, so that McAdam and Rubin (1971) have termed the total positive wave accompanying a motor reaction to a stimulus a "P302" composed of a "P300" and a "P2" . . . it would be more straightforward to study the P300 in another situation, perhaps in comparison with the motor-RT task.'

If a motor-RT paradigm is not used, Hillyard believes a discriminative-RT (go/no-go) task, in which one type of randomized light flash means 'respond' and a second type means 'don't respond' would be useful, avoiding the contamination by motor potentials on the no-go trials; a very large P300 would be generated on those trials. Several variant paradigms were suggested but all of these proposals were planned to delineate more clearly the interaction of CNV and P300 and to delineate their respective topographies.

CURRENT TRENDS IN SLOW POTENTIAL RESEARCH

A Critical Appraisal and Concluding Remarks

Chairman: J. R. KNOTT

The intent of the final reflective session of the week was to discover, *via* an Open Forum, in what ways there had been stimulation, cross-fertilization and modification of views, and whether or not what that great educator, Professor George Stoddard (who was one of my own teachers), regarded as the unique peak of intelligent behaviour—'the emergence of originals'—had occurred.

Inevitably, factors of time limited this formal final discussion. However, it has been supplemented by inclusion of remarks made during non-programmed parts of the week's activities. The following comments yield a fairly representative consensus which may stimulate progress until the next collective meeting.

Group Commentary (Narrative prepared by J.R.K.)

Cooper expressed need for continuing emphasis on *time constants*, not only of amplifiers but of electrodes, both surface and intracerebral. He also felt that data processing had been given minimal consideration and that simple amplitude measures were insufficient, 'form' deserving more analytic attention.

Weinberg stated that one of the most important methodological problems facing slow potential investigators was the definition of baseline. He suggestsd need for an automatic (e.g. servo-) re-centering system which could keep a continuous record of voltage level.

Deecke, as well as others, agreed, and said that slope of voltage change was perhaps most relevant, especially as it might be altered during an apparently larger slow potential process. *Tecce* pointed out that a change from -10 to $-15\ \mu V$ might not be neurophysiologically equal to a change from -20 to $-25\ \mu V$. It was said by *Lelord* that it would be important to measure DC shifts between the R of one S_1–S_2–R series and the following S_1 of the next series. *Donchin* commented that his attempts to invcstigate inter-trial intervals by slow potentials and by spectral analysis, showed little profit. *Weinberg*, reverting to the point about new methods of analysis, suggested the program he and Cooper had developed for the Recognition Index could be of use in a situation such as this.

McAdam emphasized that we have, in the present meeting, seen more raw data than in the past, and that these need to be translated into unambiguous statements, which requires a definition of 'zero' however arbitrary this may be. Peters *et al.* (p. 15) have shown that pre-S_1 baseline is meaningful. Different

S_1–S_2–R conditions may nevertheless require different baseline definitions. It may be necessary to have a pre-experimental as well as an intra-experimental electrical context within which to evaluate behaviourally-related SPs.

McCallum was concerned by the lack of truly 'crucial experiments' to test specific hypotheses dealing with cerebral activity.

Donald emphasized that longitudinal changes within an *S* seemed worthy of investigation. [Low has previously—this volume and elsewhere—shown that repeated sessions may enhance some CNV-generation.—*Eds.*]

McCallum stressed need to arrive at common terminology, there being lack of agreement about behavioural terms, in either an operational or hypothetical framework.

Abraham believed that the *Present State Examination* yields precision and a high degree of standardization to behavioural evaluation in the psychiatric field. SPs then had some differential capabilities in the separation of groups of individuals.

Low held that the CNV should be relatable to or give information about how and why the brain functions and what goes on when those functions take place, advocating that CNV recording be restricted to 'meaningful' situations. The finding of lateral CNV differences was most important because it could be used in a clinical situation.

Zappoli endorsed the position of Hillyard and others that multiple site recording would, in clinical studies, be of greater use in obtaining information about specific function than recording vertex-mastoid.

Crow took the view that it is possible to take SP data and put them together with other clinical and behavioural data to form a total gestalt, which would minimize the misuse of a 'diagnostic tag' derived from behavioural data alone. Explicitly, if the behavioural data suggest 'anxiety neurosis' and the SP data in no way fit those of such patients, then a re-appraisal of the first 'tag' would become necessary.

Tecce brought this view into the context of 'normal' experimentation, saying that perhaps information about the *S*'s biographical background could assist in setting up appropriate motivating conditions. He also suggested that we not only account for the 'throw-out' trials due to artefact [a point emphasized in the *Closing Overview* of the Vancouver Congress, and also re-introduced in this volume—*Eds.*], but that these be regarded as a part of the total data of the *S*.

Storm van Leeuwen introduced the matter of experimental paradigms which would utilize more 'life-like' situations. He commented upon the ways in which the use of radiotelemetry could be extended to collect such data.

Hillyard too thought that it was necessary to record from 'a real live human being in a real life-like situation', which would demand serial responses to changing stimuli. While, initially, these experiments might appear complex, later simplification could be introduced as relevant variables became identified. One of the areas of 'complexity' in need of re-emphasis—and which Hillyard thought was one of the emerging points of the Vancouver Congress—was the need for multichannel (6 or 8) recording and field plotting of the various event-related slow potentials.

Donchin pointed out that an averager only treats 'drops of time', just as a microscopist may study drops of water. He suggested further pursuit of the individual trials treatment of data, using pattern recognition to develop a series of templates which will wind up with a discriminant analysis. *Deecke* said that he and *Becker* believed this could provide an approach to fertile fields of research. McAdam's work on speech, as well as Zimmermann's and Knott's, and Low's, was initiating pathways into such fields. More complex and skilled tasks could be utilized for investigation, developing them into hierarchies, so that skilled sequences could be compared in terms of electrocortical events synchronous with or related to them. He pointed out, in this regard, that skills develop a degree of automaticity and that during this development, different slow potentials may then emerge with new areal distributions.

Otto commented that life is not made up of time spent waiting for the next stimulus in a programmed series, but of carrying out continuously shifting complex activity. He pointed out that maturational studies were given some degree of future priority at the Vancouver Congress, but that little had been done, although there was much to be discovered. His view was that surface positive event-related potentials were more dominant in younger children, and negativity more characteristic in older children.

Tueting asked that it be clearly recognized that those who in the past had been primarily concerned with P300 had for some time been aware that this phenomenon could not be looked upon in isolation, that it was now obvious that events preceding (as well as following) P300 required close scrutiny, and that simplistic assumptions about its relationship to preceding negative SPs must be avoided.

McAdam reminded the group that the Vancouver Congress made it clear that not all 'CNVs' were the same, and that the Bristol Congress made it clear that not all P300s are the same.

Concluding Remarks (J.R.K.)

Mark Twain wrote in *Life on the Mississippi:* 'There is something fascinating about science. One can get such wholesale returns of conjecture from such trifling investments of fact.' We can congratulate ourselves that we have for the most part held conjecture in check, while making substantial contributions to fact.

Let us comment upon some of the aspects of the discussion summarized above, as well as certain of the presentations and dialogues occurring throughout the entire week.

First, reflect upon the unique feature of this Congress, the ability to see and hear, through closed circuit television, certain experimental attacks on SP activity. Also, consider the opportunity to delve at depth into the CNV–P300 problem, through sharing in a review of the pre-Congress correspondence which led to an experimental formulation, and an 'on-line' test of the problems identified. The radiotelemetry-vigilance demonstration was perhaps the most dramatic, showing us that obtaining SP data in 'real-life' situations is deserving of wide use. The very thoughtful comments of Dr Harry Crow on the ethical use of implanted electrodes in humans, and the demonstration of SP recordings from such electrodes, gave evidence of the need for positive reappraisal of this area of human research. The CNV–Bereitschaftspotential relationships were, as expected, entirely demonstrable. Psychiatry and SPs are obviously in need of especial emphasis upon individual differences, although the Heisenberg Principle should always be remembered.

The methodological problems of SPs were, of course, not only demonstrated but also discussed at length. We must re-emphasize the necessity for the use of either ultra-long time constant RC amplifiers, or of DC amplifiers, and further emphasize the need for a good hard look at electrode characteristics. Another fact of life which we should come aware of is that, while a single amplifier may have a 20 Mohm input impedance, 16 amplifiers in parallel through a common reference electrode will follow the well-known law of parallel resistances, and the effective impedance of each will be lowered. Asymmetry of electrode impedances can then confound voltage measurement and field plotting.

Certainly the TC problem should be resolved among us, and rapidly. Inadequate TCs not only yield erroneous voltage measures, but distort and change the shapes of the SPs. This was demonstrated, as was the significance of SP shapes, in Vancouver. Some memories seem to have been short.

In a similar vein of constructive reproach, it should be said that to study SPs without having 'hard copy' which one can examine on-line throughout an experiment may invite disaster. Being able to feed unseen 'data' into an exotic computer with an even more exotic statistical program, which 'throws out contaminated trials', can lull one into a feeling of sophisticated security which may be unwarranted. Until the above problems are brought under some type of control the measurement of electrical processes of the brain in relatively long inter-trial intervals, or over the time of an entire experiment, may be neither practical nor meaningful. Baseline problems, much discussed here, and with a report which bears further study, demand hard copy, over long as well as short times.

The very important points raised concerning 'real life situations' deserve further comment. In most of our experiments we may take 50 S_1–S_2–R sequences and average those without artefact, usually over a 4-sec period, of which time one-quarter to one-half is behaviourally critical. Assuming that this is 2 sec, we are sampling under highly artificial circumstances about 0·15 of 1 per cent of a subject's 16-hour waking day. It may be considered remarkable that we have found out as much as we have about SPs of the brain and their relation to even the limited behaviours we have studied.

The questions raised concerning sequential motor activity have challenging consequences. One can develop (program, if you like) hierarchies of skills and observe the SP changes that occur as these combine and compress into a smoothly running, automatic sequence of motor activities, such as playing

Beethoven's 'Hammerklavier' piano sonata. (*See* Donchin's 'Correspondence' comments on 'subroutines.') An approach which studies surface fields and their changes from task to task—perhaps from level to level as tasks are mastered—should have high priority. Pay-offs by such extensions from C_Z, and other sagittal samples, by McAdam and Whitaker, and Ertl and Shaffer, pointed the way. Functional planning can lead to otherwise undetectable results. Low, Wada and Fox, and Zimmermann and Knott used logical planning and came up with significant data of theoretical as well as, possibly, practical value. Hillyard's report on fields of P300 and their relationship to task, even though it suffers from the lack of adequate TCs in the available amplifiers, is proof of the need for many electrodes, sampling widely.

The whole P300 field seems to find itself in a new context. For some years, unfortunately, the 'P300 group' and the 'CNV group' have tried to explain the findings of the other by means of their own. To a degree, the P300 workers isolated themselves by their choices of time constants. It would appear that we now have a 'Slow Potential group', no member of which is overly concerned about polarity or timing, other than to find the basic 'laws' that govern polarity shifts occurring before, throughout the S_1–S_2–R sequences, and after R, not ending the data collection prematurely. There may be an opinion forming that P300 represents (or at least under some experimental conditions represents) an inhibitory process, the functioning of a 'hold' mechanism. By mutual co-operation, adequate technique and well-defined experimental designs (based perhaps upon meaningful hypothetical constructs drawn from physiological and psychological theory), we can resolve some of these issues. This is pretty much what Tueting has been trying to tell us, and she deserves to be heard. If P300s emerge as part of learning, and if negative SPs diminish, what happens when a skilled sequential act is gradually lost or is abruptly blocked by some change in a memory trace? Experimental approaches down this pathway may prove fruitful, even though very careful experimental design will be required. It is precisely this type of approach to brain activity and behaviour which will let us discover, subsequently, the behavioural processes related to SP processes and which will give 'real life' or clinical relevance to SPs. This, in finality, is what our efforts are trying to accomplish: to define the functioning of the responsive brain.

What about the 'emergence of originals'? Have we, in this conference, come up with new leads or concepts? The proposal to develop an on-line voltage scaler and drift recorder which can provide a read-out of cortical DC level and yield a 'zero' for each trial certainly rates high on the list. An already demonstrated fact, which requires much further exploitation, is the recognition that there are gradients of certain event-related SPs, which themselves are task-related. If we can combine over-all DC level gradients with these event related data we may then be able to provide a running 'weather map' of highs and lows—fronts and peaks of excitability and troughs of inhibition—which can be correlated both with abnormal and normal cerebral states. That these maps be developed during activities of a more 'life-like' type we now know can be done by telemetry. The horizons of experimentation are substantially widened.

In the course of such investigations we can seize the opportunity to probe with precision the specific types of polarity changes which occur in relation to specifically defined sequences of behavioural events. To this end we can begin with what we know as the S_1–S_2–R paradigm, but we should carefully study each segment of it, including the afferent consequences, and/or antecedents, and also the efferent antecedents and consequences. Our concept of the sampling of time in relation to this 'simple' paradigm requires, as already suggested, a rather massive extension, both backward (preceding 'S_1', probably so as to include the efferent consequences of the preceding sequence), and also far beyond the immediate motoric activity of a particular sequence.

It then will be practical to apply the Recognition Index to any set of data defined by time and reconstruct the probable functional processes which are involved in cerebral reactivity within that period of time. This should then satisfy the clinician whose principal desire is to utilize SPs in the diagnosis and management of his patients—perhaps preventing his overlooking some crucial factor defining an electro-behavioural syndrome—and at the same time satisfy the theorist whose interest is not whether a brain is sick or well, but only in how it works.

If we can now combine some of these proposals, and continue to emerge with new types of data and new concepts, this Congress will have engaged in a collective process of intelligent activity which can substantially influence, first, our statements of experimental problems, second, our solutions of them, and then lead us to earlier solutions to the mysteries of the reactive processes of the human brain.

References and Bibliography of CNV and other Slow Potentials of the Brain

The following list of references includes all those cited in the papers which make up this volume, together with Part II of the comprehensive bibliography of CNV and other slow potentials prepared by Jon F. Peters. Part I, which appeared as a supplement to the *Proceedings of the Second International CNV Congress* in Vancouver (McCallum and Knott, 1973), covered the period up to 1972. This new, Part II, bibliography covers essentially the period between the Second and Third Congresses and includes some earlier and later supplementary material. The editors, and Mr Peters' professional colleagues throughout the world, continue to be indebted to him for the invaluable service he performs in compiling this excellent source of information. Mr Peters' address is: Department of Psychology, Simon Fraser University, British Columbia, Canada.

Abraham, P. (1971). The investigation of mental disorder by electro-encephalography. *J. R. Army Med. Corps* **117,** 94–97.

Abraham, P. (1973). An approach to standardizing clinical assessments in CNV studies of psychiatric patients. In: A. Fessard et G. Lelord (Ed.). *Activités évoquées et leur conditionnement chez l'homme normale et en pathologie mentale. Colloque INSERM Tours,* pp. 399–404.

Abraham, P. and McCallum, W. C. (1973). The CNV and the spiral after-effect. *Electroencephalogr. Clin. Neurophysiol.* **33,** *Suppl.* 205–208.

Abraham, P., McCallum, W. C., Docherty, T., Newton, P. and Fox, A. (1974). The CNV in schizophrenia. *Electroencephalogr. Clin. Neurophysiol.* **36,** 217–218.

Aladjalova, N. A. (1957). Infra-slow rhythmic oscillations of the steady potential of the cerebral cortex. *Nature* (*Lond.*) **179,** 957–959.

Aladjalova, N. A. (1964). Slow electrical processes in the brain. *Prog. Brain Res.* **7,** 243p.

Alberts, W. W. (1969). Parkinsonian tremor and cerebral potentials. In: F. J. Gillingham and I. M. L. Donaldson (Ed.). *Third Symposium on Parkinson's Disease.* London, Livingstone, pp. 146–149.

Allen, G. I. and Toyama, K. (1973). An automatic DC-level compensation circuit for electrophysiology. *IEEE Trans. Biomed. Eng. BME–20,* 58–60.

Andersen, P., Olsen, L., Skrede, K. and Sveen, O. (1969). Mechanism of thalamocortical rhythmic activity with special reference to the motor system. In: F. J. Gillingham and I. M. L. Donaldson (Ed.). *Third Symposium on Parkinson's Disease.* London, Livingstone, pp. 112–124.

Andreasen, N. C., Peters, J. F. and Knott, J. R. (1973). CNV in normal and hyperactive children with learning disabilities. Paper, Central EEG Society, March.

Annett, M. (1967). The binomial distribution of right, mixed and left handedness. *Q. J. Exp. Psychol.* **19,** 327–333.

Arduini, A. and Pinneo, L. R. (1962). A method for the quantification of tonic activity in the nervous system. *Arch. Ital. Biol.* **100,** 415–424.

Asanuma, H. and Rosen, I. (1972). Topographical organization of cortical efferent zones projecting to distal forelimb muscles in the monkey. *Exp. Brain Res.* **14,** 243–256.

Aubineau, P. F. and Sercombe, R. (1972). Slow oscillatory variations of excitability observed *in vitro* in immature mammalian olfactory cortex. *Brain Res.* **41,** 315–330.

Barry, W. and Jones, G. M. (1965). Influence of eye lid movement upon electro-oculographic recording of vertical eye movements. *Aerospace Med.* **36,** 855–858.

Baudouin, A., Fischgold, H., Caussé, R. and Lerique, J. (1939). An old forgotten notion (1): The corneo-retinal potential difference. Its theoretical and practical interest. *Bull. Acad. Méd.* **121,** 688–693. [Fr.]

Beaumanoir, A., Ballis, Th., Nahory, A. and Genier, M. (1972). VCN and RPG modification with alcohol. *Rev. E.E.G. Neurophysiol.* **2,** 328–332. [Fr.]

Becker, W., Hoehne, O., Iwase, K. and Kornhuber, H. H. (1972). Bereitschaftspotential, prämotorische Positivierung und andere Hirnpotentiale bei sakkadischen Augenbewegungen. *Vision Res.* **12,** 421–436.

Becker, W., Hoehne, O., Iwase, K. and Kornhuber, H. H. (1973). Cerebral and ocular muscle potentials preceding voluntary eye movements in man. *Electroencephalogr. Clin. Neurophysiol.* **33,** *Suppl.* 99–104.

Bendig, A. W. (1962). The Pittsburgh scales of social extroversion-introversion and emotionality. *J. Psychol.* **53,** 199–209.

Bergamasco, B. and Bergamini, L. (1970). 'Contingent negative variation' and attention. Study on patients coming out of post-traumatic coma. *Acta Neurol.* (*Napoli*) **25,** 683–691. [Ital.]

Bertelson, P. and Tisseyre, F. (1969). The time-course of preparation: Confirmatory results with visual and auditory warning signals. In: W. G. Koster (Ed.), *Attention and Performance. Acta Psychol.* **30,** 145–154.

Bickford, R. G. and Gose, E. (1972). Eye-blink elimination in clinical and research EEG. Paper, American EEG Society, October 1972.

Binnie, C. D., Brown, R., Lloyd, D. S. L. and Smith, M. (1973). Some EEG and psychological findings in minimal hepatic encephalopathy. *Electroencephalogr. Clin. Neurophysiol.* **34,** 108.

Blowers, G. (1971). Some amplitude and latency characteristics of the CNV. M.Phil. Thesis, University of Sussex.

Blowers, G., Ongley, C. and Shaw, J. C. (1973). The effect of reducing temporal expectancy on the contingent negative variation. *Electroencephalogr. Clin. Neurophysiol.* **34,** 259–264.

Borda, R. P. (1969). The effect of altered drive states on the contingent negative variation in rhesus monkeys. Doctoral Dissertation, Baylor College of Medicine, Houston.

Borda, R. P. (1970). The effect of altered drive states on the contingent negative variation (CNV) in rhesus monkey. *Electroencephalogr. Clin. Neurophysiol.* **29,** 173–180.

Borda, R. P. (1973). The relationship of CNV to behavior in animals: a review. *Electroencephalogr. Clin. Neurophysiol.* **33,** *Suppl.* 249–256.

Borda, R. P. and Hablitz, J. J. (1972). Cyclic variation in CNV over a twelve-hour period. Paper, American EEG Society, October.

Borda, R. P. and Hablitz, J. J. (1973). Use of a simple visual display to reduce eye movement artifacts in CNV recordings. *Electroencephalogr. Clin. Neurophysiol.* **34,** 433–436.

Boring, E. G. (1950). *A History of Experimental Psychology*. 2nd ed. New York, Appleton-Century-Crofts.

Bostem, F. H. and Champarnaud, J. F. (1971). A system for data gathering and handling in neurophysiology. In gradient spatio-temporal maps. *Proceedings Decus Europ. VIIth Seminar*, 145–153.

Bostem, F. H., Timsit-Berthier, M., Delauncy, J. and Rousseau, J. C. (1973). A topographic study on the 'prolonged CNV'. Communication au VIIIe Congrès International d'Electroencéphalographie et de Neurophysiologie Clinique, Marseille, Septembre.

Bracchitta, H., Lolas, F. and Pinto-Hamuy, T. (1973). Slow cortical potentials (SCP) recorded at different sites of the motor and sensory cortex during avoidance conditioning in the rabbit. *Neuropsychologia* **11,** 67–73.

Brazier, M. A. B. (1961). *A History of the Electrical Activity of the Brain*. London, Pitman.

Broadbent, D. E. (1971). *Decision and Stress*. London, Academic Press.

Brookhart, J. M., Arduini, A., Mancia, M. and Moruzzi, G. (1958). Thalamocortical relations as revealed by induced slow potential changes. *J. Neurophysiol.* **21,** 499–525.

Brooks, V. B. and Stoney, jun., S. D. (1971). Motor mechanisms: the role of the pyramidal system in motor control. *Ann. Rev. Physiol.* **33,** 337–392.

Broughton, R., Regis, H. and Gastaut, H. (1964). Modifications du Potentiel Somesthésique Evoqué pendant une bouffée de rythme mu et par la contraction des poings. *Rev. Neurol.* **111,** 331–332.

Buchwald, N. A. and Hull, C. D. (1967). Some problems associated with interpretation of physiological and behavioral responses to stimulation of caudate and thalamic nuclei. *Brain Res.* **6,** 1–11.

Bures, J. and Buresova, O. (1960). The use of Leao's spreading cortical depression in research on conditioned reflexes. *Electroencephalogr. Clin. Neurophysiol.* **13,** 359–376.

Bures, J., Petran, M. and Zachar, J. (1967) *Electrophysiological methods in Biological Research*. New York, Academic Press, pp. 159–165, 562–567.

Buser, P., Laursen, A. M., Lawrence, D. G. and Weisdanger, M. (Ed.) (1972). Neural control of motor performance. *Brain Res.* **40,** 1–202.

Callaway, E. (1966). Averaged evoked responses in psychiatry. *J. Nerv. Ment. Dis.* **143,** 80–94.

Callaway, E. (1969). Diagnostic uses of the average evoked potential. In: E. Donchin and D. B. Lindsley (Ed.). *Average Evoked Potentials: Methods, Results and Evaluations.* NASA SP–191, U.S. Govt Printing Office, Washington, D.C., 299–332.

Callaway, E. and Naylor, H. (1973). Intelligence and evoked potential asymmetry—a new interpretation. *Psychophysiology,* **10,** 195.

Caspers, H. (1959). Uber die Beziehungen zwischen Dendritenpotential und Gleichspannung an der Hirnrinde. *Pfluegers Arch. Ges. Physiol.* **269,** 157–181.

Caspers, H. (1961). Changes of cortical DC potentials in sleep-wakefulness cycle. In: G. E. W. Wolstenholme and M. O'Connor (Ed.), *The Nature of Sleep,* London: Ciba Foundation, pp. 237–253.

Caspers, H. (1963). Relations of steady potential shifts in the cortex to the wakefulness-sleep spectrum. In: M. A. B. Brazier (Ed.). *Brain Function* Vol. I: Cortical excitability and steady potentials; relations of basic research to space biology. UCLA Forum Med. Sci. No. 1.

Caspers, H., Schütz, E. and Speckmann, E.-J. (1963). Gleichspannungsveränderungen an der Hirnrinde bei Sauerstoffmangel. *Z. Biol.* **114,** 112.

Caspers, H. and Speckmann, E.-J. (1972). Cerebral pO_2, pCO_2 and pH: Changes during convulsive activity and their significance for spontaneous arrest of seizures. *Epilepsia* **13,** 699–725.

Castel-Lucci, V. F. and Goldring, S. (1970). Contribution to steady potential shifts of slow depolarisation in cells presumed to be glia. *Electroencephalogr. Clin. Neurophysiol.* **28,** 109.

Chase, W. G., Graham, F. K. and Graham, D. T. (1968). Components of HR response in anticipation of reaction time and exercise tasks. *J. Exp. Psychol.* **76,** 642–648.

Chiorini, J. R. (1967). Slow potential changes from cat cortex and classical aversive conditioning. *Electroencephalogr. Clin. Neurophysiol.* **23,** 90–91.

Chiorini, J. R. (1969). Slow potential changes from cat cortex and classical aversive conditioning. *Electroencephalogr. Clin. Neurophysiol.* **26,** 399–406.

Clausen, J. (1966). *Ability Structure and Subgroups in Mental Retardation.* Spartan Press, Washington, D.C.

Clemente, D. C., Sterman, M. B., and Wyryicka, W. (1964). Post-reinforcement EEG synchronization during alimentary behavior. *Electroencephalogr. Clin. Neurophysiol.* **16,** 355–365.

Cohen, A. and Longini, R. L. (1971). Self-balancing system for medical and physiological instrumentation. *IEEE Trans. Biomed. Eng. BME-18,* 66–69.

Cohen, H. J., Taft, L. T., Mahadeviah, M. S. and Birch, H. G. (1967). Developmental changes in overflow in normal and aberrantly functioning children. *J. Pediatr.* **71,** 39–47.

Cohen, J. (1969). Very slow brain potentials relating to expectancy: the CNV. In: E. Donchin and D. B. Lindsley (Ed.). *Average Evoked Potentials: Methods, Results and Evaluations.* NASA SP-191, U.S. Govt Printing Office, Washington, D.C. 143–198.

Cohen, J. (1973a). Developmental aspects of the CNV: a review. *Electroencephalogr. Clin. Neurophysiol.* **33,** *Suppl.* 133–137.

Cohen, J. (1973b). The CNV in children with special reference to learning disabilities. *Electroencephalogr. Clin. Neurophysiol.* **33,** *Suppl.* 151–154.

Cohen, J. (1973c). The CNV and visual recognition. *Electroencephalogr. Clin. Neurophysiol.* **33,** *Suppl.* 201–207.

Cohn, R. (1963). DC potential variations in paroxysmal discharges. *Electroencephalogr. Clin. Neurophysiol.* **15,** 151.

Cohn, R. (1966). DC recordings of visual evoked summated cortical potentials. *Proc. Symp. Biomed. Eng.* **1,** 68–69.

Cohn, R. (1973). The influence of task-irrelevant stimulus—variations on the reliability of auditory evoked responses in schizophrenia. In: A. Fessard et G. Lelord (Ed.). *Activités évoquées et leur conditionnement chez l'homme normale et en pathologie mentale. Colloque INSERM Tours,* pp. 373–388.

Connor, W. H. and Lang, P. J. (1969). Cortical slow-wave and cardiac rate responses in stimulus orientation and reaction time conditions. *J. Exp. Psychol.* **82,** 310–320.

Cooper, R. (1971). Recording changes in electrical properties in the brain: the EEG. In: R. D. Myers (Ed.). *Methods in Psychobiology, Vol. 1.,* Ch. 6. London & New York, Academic Press.

Cooper, R., Osselton, J. W. and Shaw, J. C. (1974). *EEG Technology,* 2nd Ed. London, Butterworths.

Cooper, R., Papakostopoulos, D., Gombi, R. and Crow, H. J. (1973). Measurement of the steady and slowly changing potentials in the human brain. *Electroencephalogr. Clin. Neurophysiol.* **34,** 721–723.

Cooper, R. and Weinberg, H. (1972). The recognition index: a pattern recognition technique suitable for noisy signals. *Electroencephalogr. Clin. Neurophysiol.* **33,** 447.

Cooper, R., Winter, A. L., Crow, H. J. and Walter, W. G. (1965). Comparison of subcortical, cortical and scalp activity using chronically indwelling electrodes in man. *Electroencephalogr. Clin. Neurophysiol.* **18,** 217–228.

Coquery, J-M. et Coulmance, M. (1971). Variations d'amplitude de réflexes monosynaptiques avant un mouvement volontaire. *Physiology and Behavior*, **6,** 65–69.

Coquery, J-M., Coulmance, M. et Leron, M-C. (1972). Modification of somaesthetic cortical evoked potentials during active and passive movements in man. *Electroencephalogr Clin. Neurophysiol.* **33,** 269–276. [Fr.]

Corby, J. C. and Kopell, B. S. (1972). Differential contributions of blinks and vertical eye movements as artifacts in EEG recording. *Psychophysiology*, **9,** 640–644.

Costell, R. M., Lunde, D. T., Kopell, B. S. and Wittner, W. K. (1972). Contingent negative variation as an indicator of sexual object preference. *Science*, **177,** 718–720.

Cowen, M. A. (1973). Scalp: direct current amplification phenomena. *Psychophysiology* **10,** 194–195.

Cowen, M. A., Nishi, H. and Hammack, D. (1972). An electrophysiological analysis of hallucinogens. *Biol. Psychiat.* **5,** 239–256.

Creutzfeldt, O. D. (1969). Neuronal mechanisms underlying the EEG. In: H. H. Jasper, A. A. Ward jun. and A. Pope (Ed.). *Basic Mechanisms of the Epilepsies.* Boston, Little, Brown & Co., 397–420.

Creutzfeldt, O. D., Watanabe, S. and Lux, H. D. (1966). Relation between EEG phenomena, and potentials of single cells. I. Evoked responses after thalamic and epicortical stimulation. *Electroencephalogr. Clin. Neurophysiol.* **20,** 1–18.

Crow, H. J. (1973). Interacerebral polarisation and multifocal leucocoagulation in some psychiatric illnesses. *Psychiat. Neurol. Neurochir. (Amst.)* **76,** 365–381.

Crow, H. J., Cooper, R. and Phillips, D. G. (1961). Controlled multifocal leucotomy for psychiatric illness. *J. Neurol. Neurosurg. Psychiatry* **24,** 353–360.

Dargent, J. et Dongier, M. (Ed.) (1969). Variations contingentes négatives. Congrès et Colloquesde l'Université de Liège. Vol. 52, 249 p.

Darrow, C. W. and Gullickson, G. R. (1967). Phase reversal of EEG during attentive, associative and reactive conditions. *Electroencephalogr. Clin. Neurophysiol.* **23,** 92.

David, E., Finkenzeller, P., Kallert, S. und Keidel, W. D. (1971). Correlation analysis of averaged and unaveraged evoked potentials. *Kybernetik* **9,** 22–26. [Ger.]

Davis, H. (1964). Attention, vigilance, and cortical evoked potentials in humans. *Science* **145,** 180–183.

Davis, H. and Heninger, G. R. (1972). Comparison of response plasticity between the eyeblink and vertex potential in humans. *Electroencephalogr. Clin. Neurophysiol.* **33,** 283–293.

Davis, M., Osterhammel, P. A., Wier, C. C. and Gjerdingen, D. B. (1972). Slow vertex potentials: interactions among auditory, tactile, electric and visual stimuli. *Electroencephalogr. Clin. Neurophysiol.* **33,** 537–545.

Deecke, L., Scheid, P., Kornhuber, H. H. (1969). Distribution of readiness potential, premotion positivity, and motor potential of the human cerebral cortex preceding voluntary finger movements. *Exp. Brain Res.* **7,** 158–168.

Deecke, L., Becker, W., Grozinger, B., Scheid, P. and Kornhuber, H. (1973). Human brain potentials preceding voluntary limb movements. *Electroencephalogr. Clin. Neurophysiol.* **33,** *Suppl.* 87–94.

Delaunoy, J. (1971). Etude longitudinale des potentiels lents cerebraux chez les malades mentaux. Communication au Congrès Mondial de Psychiatrie, Novembre.

DeLong, M. R. (1973). Putamen: Activity of single units during slow and rapid arm movements. *Science* **179,** 1240–1242.

DeLong, M. R. (1974). Motor functions of the basal ganglia, single unit activity during movement. In: F. O. Schmitt and F. G. Warden (Ed.): *The Neurosciences III Study Program*, MIT-press, Cambridge and London, pp. 319–325.

Delse, F. C. (1972) CNV correlates of task difficulty and accuracy of pitch discrimination. *Psychophysiology* **9,** 53–62.

Delse, F. C., Marsh, G. R. and Thompson, L. W. (1972). CNV correlates of task difficulty and accuracy of pitch discrimination. *Psychophysiology*, **9,** 53–62.

Dichter, M., Herman, C. and Selzer, M. (1973). Extracellular unit analysis of the hippocampal penicillin focus. *Electroencephalogr. Clin. Neurophysiol.* **34,** 619–629.

Dinges, D. F. and Klingaman, R. L. (1972). Effects of induced muscle tension upon the visual evoked potential and motor potential: a replication. *Psychon. Sci.* **28,** 303–305.

Dobronravo, I. S. and Gundarov, V. P. (1965). Correlational analysis of the human electroencephalogram during the elaboration of a rhythmic motor stereotype. (in Russian.) *Zh. Vyssh. Nerv. Deyat. Pavlova* **15,** 624–630.

Dolce, G. and Sannita, W. (1973). A CNV-like negative shift in deep coma. *Electroencephalogr. Clin. Neurophysiol.* **34,** 647–650.

Donald, M. W. (1972). Parallel specific and non-specific late components in the somatic evoked response. *Electroencephalogr. Clin. Neurophysiol.* **33,** 450.

Donald jun., M. W. and Goff, W. R. (1971). Attention-related increases in cortical responsivity dissociated from the contingent negative variation. *Science* **172,** 1163–1166.

Donald, M. W. and Goff, W. R. (1973). Contingent negative variation and sensory evoked responses: their interaction and relationship to auditory discrimination. *Electroencephalogr. Clin. Neurophysiol.* **33,** *Suppl.* 109–117.

Donchin, E. (1973). Methodological issues in CNV research: a review. *Electroencephalogr. Clin. Neurophysiol.* **33,** *Suppl.* 3–17.

Donchin, E. and Cohen, L. (1967). Average evoked potentials and intramodality selective attention. *Electroencephalogr. Clin. Neurophysiol.* **22,** 537–546.

Donchin, E. and Cohen, L. (1969). Further reply to Näätänen. *Percept. Mot. Skills* **29,** 270.

Donchin, E., Gerbrandt, L. K., Leifer, L. and Tucker, L. (1972a). Is the contingent negative variation contingent on a motor response? *Psychophysiology* **9** 178–188.

Donchin, E., Gerbrandt, L. K., Leifer, L. and Tucker, L. R. (1973a) Contingent negative variations and motor responses. *Electroencephalogr. Clin. Neurophysiol.* **33,** *Suppl.* 187–190.

Donchin, E., Kubovy, M., Kutas, M., Johnson, R. and Herning, R. (1973b). Graded changes in evoked response (P300) amplitude as a function of cognitive activity. *Perception and Psychophysics* **14,** 319–324.

Donchin, E., Kutas, M. and Johnson jun., R. (1972b). The CNV does not behave like a 'motor' potential. Paper, American EEG Society, October.

Donchin, E., Otto, D., Gerbrandt, L. K. and Pribram, K. (1973c). Studies in the physiology of Electrocortical events recorded during the foreperiod of a reaction time study. *Electroencephalogr. Clin. Neurophysiol.* **31,** 115–127.

Donchin, E., Otto, D., Gerbrandt, L. K. and Pribram, K. (1973). Studies in the physiology of CNV. *Electroencephalogr. Clin. Neurophysiol.* **33,** *Suppl.* 257–261.

Donchin, E. and Smith, D. D. B. (1970). The contingent negative variation and the late positive wave of the average evoked potential. *Electroencephalogr. Clin. Neurophysiol.* **29,** 201–203.

Dongier, M. (1969). Separation of the various phenomena among the slow potential changes (contingent negative variations). *Electroencephalogr. Clin. Neurophysiol.* **27,** 108–109.

Dongier, M. (1973). Clinical applications of the CNV: a review. *Electroencephalogr. Clin. Neurophysiol.* **33,** *Suppl.* 309–315.

Dongier, M. and Koninckx, N. (1970). Present-day neurophysiological models of mind-body interaction. *Psychother. Psychosom.* **18,** 123–129.

Dongier, M., Timsit-Berthier, M., Koninckx, N. and Delaunoy, J. (1973). Compared clinical significance of CNV and other slow potential changes in psychiatry. *Electroencephalogr. Clin. Neurophysiol.* **33,** *Suppl.* 321–326.

Dubrovsky, B., Garcia-Rill, E., Tepper, J. and Vanagas, J. (1973a). Some aspects of the methodology of recording slow potential changes. In: S. Bogoch (Ed.). *Biological Diagnosis of Brain Disorders. The Future of the Brain Sciences.* Spectrum Publications, Inc. 300–309.

Dubrovsky, B., Garcia-Rill, E. and Dongier, M. (1973b). Effect of hyperventilation on abnormally prolonged CNV. Paper, 8th International Congress of EEG and Clinical Neurophysiology, September.

Durkovic, R. G. and Cohen, D. H. (1968). Spontaneous, evoked and defensively conditioned steady potential changes in the pigeon telencephalon. *Electroencephalogr. Clin. Neurophysiol.* **24,** 474–481.

Dustman, R. E. and Beck, E. C. (1969). The effects of maturation and aging on the wave form of visually evoked potentials. *Electroencephalogr. Clin. Neurophysiol.* **26,** 2–11.

Eason, R. G., Aiken, L. R., jun., White, C. T. and Lichtenstein, M. (1964a). Activation and behavior: II. Visually evoked cortical potentials in man as indicants of activation level. *Percept. Mot. Skills* **19,** 875–895.

Eason, R. G., Harter, M. R. and Storm, W. F. (1964b). Activation and behavior: I. Relationship between physiological 'indicants' of activation and performance during memorization of nonsense syllables using differing induced tension conditions. *Percept. Mot. Skills* **19,** 95–110.

Eccles, J. C. (1967a). The inhibitory control of spinal reflex action. *Electroencephalogr. Clin. Neurophysiol.* **25,** *Suppl.* 20–34.

Eccles, J. C. (1967b). The way in which the cerebellum processes sensory information from muscle. In: Yahr, M. D. and Purpura, D. P. (Ed.). *Neurophysiological Basis of Normal and Abnormal Motor Activities.* New York, Raven Pres, Hewlett.

Eccles, J. C. (1973). The cerebellum as a computer: patterns in space and time. *J. Physiol.* **229,** 1–32.

Eisengart, M. A. and Symmes, D. (1971). Effect of eye blink on the visual evoked response in children. *Electroencephalogr. Clin. Neurophysiol.* **31,** 71–75.

Ellis, L. (1971). Slow electroencephalographic changes preceding speech production. Dissertation, University of Illinois (DAI V.32(8–B)4920).

Emde, J. W. (1964). A time locked low level calibrator. *Electroencephalogr. Clin. Neurophysiol.* **16,** 616–618.

Eroshkin, V. F. and Bugaev, S. A. (1971). A device for automatic calculation of the drift value in direct current amplifier during prolonged recording (in Russian). *Biull. Eksp. Biol. Med.* **71,** 123–124.

Ertl, J. and Schafer, E. W. P. (1967). Cortical activity preceding speech. *Life Sciences* **6,** 473–479.

Ertl, J. and Schafer, E. W. P. (1969). Erratum. *Life Sciences* **8,** 559.

Evarts, C. R. and Mulholland, T. B. (1969). *Attention in Neurophysiology.* London, Butterworths, 447p.

Evarts, E. V. (1970). Activity of ventralis lateralis neurons prior to movement in the monkey. *Physiologist* **13,** 191.

Evarts, E. V. (Ed.) (1971). Central Control of Movement. *Neurosciences Res. Program Bull.* **9,** No. 1.

Evarts, E. V. (1972a). Contrasts between activity of precentral and postcentral neurons of cerebral cortex during movement in the monkey. *Brain Res.* **40,** 25–31.

Evarts, E. V. (1972b). Activity of motor cortex neurons in association with learned movement. *Int. J. Neuroscience* **3,** 113–124.

Evarts, E. V. (1972c). Studies on cerebral mechanisms of motor control. *Int. J. Neuroscience* **4,** 89–92.

Evarts, E. V. (1973). Brain Mechanisms in Movement. *Sci. Am.* **229,** 96–103.

Farley, B. G. (1965). A neural network model and the 'slow potentials' of electrophysiology, In: R. W. Stacy and D. B. Waxman (Ed.). *Computers in Biomedical Research.* Vol. 1. New York, Academic Press, pp. 265–294.

Ferguson, J. H. and Jasper, H. H. (1971). Laminar DC studies of acetylcholine-activated epileptiform discharge in cerebral cortex. *Electroencephalogr. Clin. Neurophysiol.* **30,** 377–390.

Fessard, A. et Lelord, G. (Ed.) (1973). Average evoked responses and their conditioning in normal subjects and psychiatric patients. *Colloque INSERM Tours.*

Ford, A. (1959). Significance of terminal transients in electro-oculographic recordings. *Arch. Ophthalmol.* **61,** 87/899–94/906.

Ford, J. M., MacPherson, L. and Kopell, B. S. (1972). Differences in readiness potential associated with push-button construction. *Psychophysiology* **9,** 564–567.

Ford, J. M., Roth, W. T., Dirks, S. J. and Kopell, B. S. (1973). Evoked potential correlates of signal recognition between and within modalities. *Science* **181,** 465–466.

Fox, S. S. and Rosenfeld, J. P. (1972). Recording evoked potentials. In: R. D. Myers (Ed.). *Methods in Psychobiology, Specialized Laboratory Techniques in Neuropsychology and Neurobiology.* Vol. 2. London, Academic Press.

Franzen, O. and Offenloch, K. (1969). Evoked response correlates of psychophysical magnitude estimates for tactile stimulation in man. *Exp. Brain Res.* **8,** 1–18.

Friedman, D., Hakererm, G., Sutton, S. and Fleiss, J. L. (1973). Effect of stimulus uncertainty on the pupillary dilation response and the vertex evoked potential. *Electroencephalogr. Clin. Neurophysiol.* **34,** 475–484.

Fruhstorfer, H. (1966). Der Einfluss von Reizfolgefrequenz, Reizanzahl und Reizort auf die Intensitatsabhangigkeit vibratorisch ausgeloster corticaler Antworten beim Menschen. Dissertation, Erlangen.

Fruhstorfer, H., Guth, H. and Pfaff, U. (1972). Thermal reaction time as a function of stimulation site. *Pfluegers Arch.* **335,** *Suppl.* R 49.

Fuster, J. M. (1973). Unit activity in prefrontal cortex during delayed-response performance: neuronal correlates of transient memory. *J. Neurophysiol.* **36,** 61–78.

Gaillard, A. W. and Näätänen, R. (1973). Slow potential changes and choice RT as a function of interstimulus interval. *Acta Psychol.* (*Amst.*) **37,** 173–186.

Gale, A., Haslum, M. and Lucas, B. (1972). Arousal value of the stimulus and EEG abundance in an auditory vigilance task. *Br. J. Psychol.* **63,** 515–522.

Gale, A., Haslum, M. and Penfold, V. (1971). EEG correlates of cumulative expectancy and subjective estimates of alertness in a vigilance type task. *Q. J. Exp. Psychol.* **23,** 245–254.

Ganglberger, J. A., Groll-Knapp, E. and Haider, M. (1973). Computer analysis of electrophysiological phenomena during stereotaxic operations in man. *Electroencephalogr. Clin. Neurophysiol.* **35,** 99–100.

Ganglberger, J. A. and Haider, M. (1969). Computer analysis of cortical responses to thalamic stimulation and of thalamo-cortical relationship of contingent negative variation in man. In: F. J. Gillingham and I. M. L. Donaldson (Ed.), *Third Symposium on Parkinson's Disease*. Edinburgh, Livingstone, 138–141.

Gazzaniga, M. S. and Hillyard, S. A. (1972). Attention mechanisms following brain bisection. In: S. Kornblum (Ed.). *Attention and Performance*, Vol. IV, New York, Academic Press, 221–238.

Gerbrandt, L. K., Goff, W. R. and Smith, D. B. (1973). Distribution of the human average movement potential. *Electroencephalogr. Clin. Neurophysiol.* **34,** 461–474.

Gerbrandt, L. K., Spinelli, D. N. and Pribram, K. H. (1970). The interaction of visual attention and temporal cortex stimulation on electrical activity evoked in the striate cortex. *Electroencephalogr. Clin. Neurophysiol.* **29,** 146–155.

Girton, D. G. and Kamiya, J. (1973). A simple on-line technique for removing eye movement artifacts from the EEG. *Electroencephalogr. Clin. Neurophysiol.* **34,** 212–216.

Goldring, S. and O'Leary, J. L. (1951). Summation of certain enduring sequelae of cortical activation in the rabbit. *Electroencephalogr. Clin. Neurophysiol.* **3,** 465–471.

Goldring, S. and O'Leary, J. L. (1954). Correlation between steady transcortical potential and evoked response. I. Alterations in somatic receiving area induced by veratrine, strychnine KCl and novacaine. *Electroencephalogr. Clin. Neurophysiol.* **6,** 189–200.

Goldring, S. and O'Leary, J. L. (1954). Correlation between steady transcortical potential and evoked response. II. Effect of veratrine and strychnine upon responsiveness of visual cortex. *Electroencephalogr. Clin. Neurophysiol.* **6,** 201–212.

Gombi, R., Cooper, R., Papakostopoulos, D. and Crow, H. J. (1973). Measurement of slowly changing potentials in the human brain. *Electroencephalogr. Clin. Neurophysiol.* **34,** 109.

Gottlieb, G. L., Agarwal, G. C. and Stark, L. (1970). Interactions between voluntary and postural mechanisms of the human motor system. *J. Neurophysiol.* **33,** 365–381.

Granit, R. (ed.). (1970). *The Basis of Motor Control.* London and New York, Academic Press.

Gross, C. G., Chorover, S. L. and Cohen, S. M. (1965). Caudate, cortical hippocampal and dorsal thalamic lesions in rats: Alternation and Hebb-Williams maze performance. *Neuropsychologia* **5,** 53–68.

Grözinger, B., Kornhuber, H. H. and Kriebel, J. (1975). Methodological problems in the investigation of cerebral potentials preceding speech: determining the onset and suppressing artefacts caused by speech. *Neuropsychologia* **13,** 263–270.

Grözinger, B., Kornhuber, H. H., Kriebel, J. and Murata, K. (1972). Menschliche Hirnpotentiale vor dem Sprechen. *Pfluegers Arch.* **332,** *Suppl.* 332: R100.

Guilford, J. P. (1950). *Fundamental Statistics in Psychology and Education.* New York, McGraw-Hill.

Gullickson, G. R. (Ed.) (1973a). *The Psychophysiology of Darrow.* New York, Academic Press.

Gullickson, G. R. (1973b). CNV and behavioural attention to a glide-tone warning of interesting non-moving or kaleidoscopic visual or auditory patterns in 2- and 3-year-old children. *Electroencephalogr. Clin. Neurophysiol.* **33,** *Suppl.* 145–150.

Gullickson, G. R. and Darrow, C. W. (1973). Contingent negative variation modified by respiratory phase. *Electroencephalogr. Clin. Neurophysiol.* **33,** *Suppl.* 295–297.

Gumnit, R. J. (1961). The distribution of direct current responses evoked by sounds in the auditory cortex of the cat. *Electroencephalogr. Clin. Neurophysiol.* **13,** 889–895.

Hablitz, J. J. (1973). Operant conditioning and slow potential changes from monkey cortex. *Electroencephalogr. Clin. Neurophysiol.* **34,** 399–408.

Hablitz, J. J. (1972). Operant conditioning and slow potential changes from monkey cortex. Doctoral Dissertation, University of Houston.

Hablitz, J. J. and Borda, R. P. (1973). The effects of Dalmane® (Flurazepam Hydrochloride) on the contingent negative variation. *Electroencephalogr. Clin. Neurophysiol.* **33,** *Suppl.* 317–320.

Hagbarth, K-E. and Vallbo, A. B. (1968). Discharge characteristics of human muscle afferents during muscle stretch and contraction. *Exp. Neurol.* **22,** 674–693.

Haider, M. (1970). Neuropsychology of attention, expectation and vigilance. In: D. I. Mostofsky (Ed.). *Attention: Contemporary Theory and Analysis.* New York, Appleton-Century-Crofts, 419–432.

Haider, M., Ganglberger, J. A. and Groll-Knapp, E. (1968). Computer analysed thalamic potentials and their relation to expectancy waves in man. *Acta Neurol. Lat.* **14,** 132–137.

Haider, M., Ganglberger, J. A. and Groll-Knapp, E. (1969). Thalamo-cortical components of reaction time. *Acta Psychol. (Amst.)* **30,** *Suppl.* 378–381.

Haider, M., Ganglberger, J. A. and Groll-Knapp, E. (1972). Computer analysis of subcortical and cortical evoked potentials and of slow potential phenomena in humans. *Confin. Neurol.* **34,** 224–229.

Haider, M., Groll, E. and Studynks, G. (1968). Orientierungs- und Bereitschafts-potentials bei unerwarteten Reizen. *Exp. Brain Res.* **5,** 45–54.

Halas, E. S., Beardsley, J. V. and Sandlie, M. E. (1970). Conditioned neuronal responses at various levels in conditioning paradigm. *Electroencephalogr. Clin. Neurophysiol.* **28,** 468–477.

Hall, K. M., Hicks, R. A. and Hopkins, H. K. (1970). The effects of low level DC scalp positive and negative current on the performance of various tasks. *Br. J. Psychiat.* **117,** 689–691.

Hamilton, C. E., Peters, J. F. and Knott, J. R. (1972). Task initiation and CNV amplitude. Paper, American EEG Society, October.

Hamilton, C. E., Peters, J. F. and Knott, J. R. (1973). Task initiation and amplitude of the contingent negative variation (CNV). *Electroencephalogr. Clin. Neurophysiol.* **34,** 587–592.

Hanley, J., Nirenberg, L. M. and Adey, W. R. (1973). Phase-locked loop detection of motor phenomena in the spontaneous EEG. Paper, American EEG Society, June.

Hart, S. and Huff, E. (1972). Influence of dependency structure in four-alternative learning. Paper, Western Psychological Association.

Harter, T. R. and Salmon, L. E. (1972). Intra-modality selective attention and evoked cortical potentials to randomly presented patterns. *Electroencephalogr. Clin. Neurophysiol.* **32,** 605–613.

Hays, W. L. (1963). *Statistics for Psychologist.* New York, Holt, Rinehart & Winston.

Heinemann, U. (1973). Time variability of slow negative potentials and changes in alpha rhythm. *Electroencephalogr. Clin. Neurophysiol.* **33,** *Suppl.* 125–126.

Hensel, H. (1973). Cutaneous thermoreceptors. In: *Handbook of Sensory Physiology,* Vol. II, p. 75–106. Berlin-Heidelberg-New York, Springer.

Hensel, H. and Iggo, A. (1971). Analysis of cutaneous warm and cold fibres in primates. *Pfluegers Arch.* **329,** 1–8.

Hillyard, S. A. (1973). The CNV and human behavior: a review. *Electroencephalogr. Clin. Neurophysiol.* **33,** *Suppl.* 161–171.

Hillyard, S. A. and Galambos, R. (1967). Effects of stimulus and response contingencies on a surface negative slow potential shift in man. *Electroencephalogr. Clin. Neurophysiol.* **22,** 297–304.

Hillyard, S. A. and Galambos, R. (1970). Eye movement artifact in the CNV. *Electroencephalogr. Clin. Neurophysiol.* **28,** 173–182.

Hillyard, S., Krausz, H. and Picton, T. (1973a). Scalp distribution of late positive wave (P3) of auditory evoked potential. Paper, Western EEG Society, February.

Hillyard, S. A., Hink, R. F., Schwent, V. L. and Picton, T. W. (1973b). Electrical signs of selective attention in the human brain. *Science* **182,** 177–180.

Hillyard, S. A., Squires, K. C., Bauer, J. W. and Lindsay, P. H. (1971). Evoked potential correlates of auditory signal detection. *Science* **172,** 1357–1360.

Hubbard, J. H., Corrie, W. S., Thompson, H. K. and Marshall, W. H. (1971). Beta-adrenergic mechanisms influencing brain steady potential in cats and rhesus monkeys. *Int. J. Neuroscience* **2,** 57–68.

Huff, E. M. (1970). Probability learning: First-order Markov structures of quarternary events. *NASA TN D-5684.*

Hull, C. D., Buchwald, N. A. and Ling, G. (1967). Effects of direct cholinergic stimulation of forebrain structure. *Brain Res.* **6,** 22–35.

Hurwitz, S. L. and Bremner, F. J. (1972). Hippocampal correlates of a conditioned sniffing response. *Neurophysiologia,* **10,** 163–170.

Irwin, D. A., Knott, J. R., McAdam, D. W. and Rebert, C. S. (1966). Motivational determinants of the 'contingent negative variation'. *Electroencephalogr. Clin. Neurophysiol.* **21,** 538–543.

Irwin, D. A. and Rebert, C. S. (1970). Slow potential changes in cat brain during classical appetitive conditioning of jaw movements using two levels of reward. *Electroencephalogr. Clin. Neurophysiol.* **28,** 119–126.

Ivanova, M. P. and Gundarov, V. P. (1969). Correlational analysis of EEG during elementary voluntary movements (in Russian). *Vopr. Psikhol.* **6,** 60–69.

Jami, L. (1972). Patterns of cortical population discharges during metrazol-induced seizure in cats. *Electroencephalogr. Clin. Neurophysiol.* **32,** 641–654.

Järvilehto, T. and Fruhstorfer, H. (1970). Differentiation between slow cortical potentials associated with motor and mental acts in man. *Exp. Brain Res.* **11,** 309–317.

Järvilehto, T. and Fruhstorfer, H. (1973). Is the sound-evoked DC potential a contingent negative variation? *Electroencephalogr. Clin. Neurophysiol.* **33,** *Suppl.* 105–108.

Jeffreys, D. A. (1971). Cortical source locations of pattern-related visual evoked potentials recorded from the human scalp. *Nature* (*Lond.*) **229,** 502–504.

Jenness, D. (1972a). Auditory evoked-response differentiation with discrimination learning in humans. *J. Comp. Physiol. Psychol.* **80,** 75–90.

Jenness, D. (1972b). Stimulus role and gross differences in the cortical evoked response. *Physiol. Behav.* **9,** 141–146.

Jennings, J. R., Averill, J. R., Opton, E. M. and Lazarus, R. S. (1971). Some parameters of heart rate change: perceptual versus motor task requirements, noxiousness, and uncertainty. *Psychophysiology,* **7,** 194–212.

Johnson, L. C. and Macleod, W. L. (1973). Sleep and awake behavior during gradual sleep reduction. *Percept. Mot. Skills* **36,** 87–97.

Joyner, R. and Somjen, G. (1972). A model simulating the contribution of glia to sustained potential shifts of central nervous tissue. Paper, American EEG Society, October.

Kamp, A. (1963a). An eight channel electro-encephalograph telemetering system. *Proc. Int. Telemetering Conf., London* **1,** 204–209.

Kamp, A. (1963b). Eight channel EEG telemetering. *Electroencephalogr. Clin. Neurophysiol.* **15,** 164P.

Kamp, A., Schrijer, C. F. M. and Storm van Leeuwen, W. (1972). Occurrence of 'beta bursts' in human frontal cortex related to psychological parameters. *Electroencephalogr. Clin. Neurophysiol.* **33,** 257–267.

Karlin, L. (1970). Cognition, preparation, and sensory-evoked potentials. *Psychol. Bull.* **73,** 122–136.

Karlin, L., Martz, M. J. and Mordkoff, A. M. (1970). Motor performance and sensory evoked potentials. *Electroencephalogr. Clin. Neurophysiol.* **28,** 307–313.

Karlin, L., Martz, M., Brauth, S., and Mordkoff, A. (1971). Auditory evoked potentials, motor potentials and reaction time. *Electroencephalogr. Clin. Neurophysiol.* **31,** 129–136.

Karrer, R., Kohn, H. and Ivins, J. (1973a). Effects of varying the stimulus and response contingencies on the CNV. *Electroencephalogr. Clin. Neurophysiol.* **33,** *Suppl.* 39–43.

Karrer, R., Kohn, H. and Ivins, J. (1973b). Large steady potential shifts accompanying phasic arousal during CNV recording in man. *Electroencephalogr. Clin. Neurophysiol.* **33,** *Suppl.* 119–124.

Kato, M. and Tanji, J. (1972). Cortical motor potentials accompanying volitionally controlled single motor unit discharge in human finger muscles. *Brain Res.* **47,** 103–111.

Kawamura, H. and Pompeiano, O. (1969). Tonic and phasic DC potential shifts in the cat spinal cord and brain during desynchronized sleep. *Brain Behav. Evol.* **2,** 263–287.

Kawamura, H. and Sawyer, C. H. (1964). D-C potential changes in the rabbit brain during slow-wave and paradoxical sleep. *Am. J. Physiol.* **207,** 1379–1386.

Kawamura, H., Whitmoyer, D. I. and Sawyer, C. H. (1967). DC potential changes recorded between brain and skull in the rabbit after eating and drinking. *Electroencephalogr. Clin. Neurophysiol.* **22,** 337–347.

Keidel, W. D. (1972). The DC potential in the human cortical evoked response. A new contribution to the objective audiometry. *J. Fr. Otorhinolaryngol.* **21,** 153–158.

Khachaturian, Z. S., Chisholm, R. and Kerr, J. L. (1973). The effects of arousal on evoked potentials to relevant and irrelevant stimuli. *Psychophysiology* **10,** 194.

Klass, D. and Bickford, R. G. (1960). Glossokinetic potentials appearing in the electroencephalogram. *Electroencephalogr. Clin. Neurophysiol.* **12,** 239.

Klemm, W. R. (1971). EEG and multiple-unit activity in limbic and motor systems during movement and immobility. *Physiol. Behav.* **7,** 337–343.

Knott, J. R. (1973). Closing overview. *Electroencephalogr. Clin. Neurophysiol.* **33,** *Suppl.* 349–355.

Knott, J. R. and Irwin, D. A. (1967) Anxiety, stress and the contingent negative variation. *Electroencephalogr. Clin. Neurophysiol.* **22,** 188.

Knott, J. R. and Irwin, D. A. (1968). Anxiety, stress and the contingent negative variation (CNV). *Electroencephalogr. Clin. Neurophysiol.* **24,** 286–287.

Knott, J. R. and Irwin, D. A. (1973). Anxiety, stress and the contingent negative variation. *Arch. Gen. Psychiat.* **29,** 538–541.

Knott, J. R. and Peters, J. F. (1973a). Sex, stress and interstimulus interval. *Electroencephalogr. Clin. Neurophysiol.* **33,** *Suppl.* 191–193.

Knott, J. R. and Peters, J. F. (1973b). Effects of stress variation on CNV in males and females. Paper, American EEG Society, June.

Knott, J. R., Peters, J. F., Cohen, S. I., Van Veen, W. and Miller, L. H. (1971). Sex, stress and the contingent negative variation. *Electroencephalogr. Clin. Neurophysiol.* **31,** 104.

Knott, J. R., Van Veen, W. J., Miller, L. H., Peters, J. F. and Cohen, S. I. (1973). Perceptual mode, anxiety, sex and the contingent negative variation. *Biol. Psychiatry* **7,** 43–52.

Knott, J. R. and Zimmerman, G. (1973). CNV and nominal aphasia: a case report. Paper, Central EEG Society, March.

Koninckx, N. (1972). CNV and psychosomatics. Communication, Colloque sur les 'Acquisitions recentes en Medecine Psychosomatique', Liège, October.

Konishi, M. (1963). The role of auditory feedback in the vocal behaviour of the domestic fowl. *Z. Tierpsychol.* **20,** 349–367.

Kopell, B. S., Tinklenberg, J. R. and Hollister, L. E. (1972). Contingent negative variation amplitudes. *Arch. Gen. Psychiat.* **27,** 809–811.

Kornhuber, H. H. (1971). Motor functions of cerebellum and basal ganglia: The cerebello-cortical saccadic (ballistic) clock, the cerebello-nuclear hold regulator and the basal ganglia ramp (voluntary speed smooth movement) generator. *Kybernetik*, **8,** 157–162.

Kornhuber, H. H. (1974). Cerebral cortex, cerebellum and basal ganglia: An introduction to their motor functions. In: F. O. Schmitt and F. G. Warden (Ed.). *The Neurosciences III. Study Program.* MIT-Press, Cambridge and London, pp. 267–280.

Kornhuber, H. H. and Deecke, L. (1964). Hirnpotentialanderungen beim Menschen vor und nach Willkürbewegungen, dargestellt mit Magnetbandspeicherung und Rückwärtsanalyse. *Pfluegers Arch. Ges. Physiol.* **281,** 52.

Kornhuber, H. H. and Deecke, L. (1965). Hirnpotentialänderungen bei Willkürbewegungen und passiven Bewegungen des Menschen: Bereitschaftspotential und reafferente Potentiale. *Pfluegers Arch. Ges. Physiol.* **284,** 1–17.

Kostandov, E. A. and Arzumanov, Y. L. (1971). Averaged evoked potentials of the human cortex to emotional visual stimuli (in Russian). *J. Higher Nervous Activity* (Moscow) **21,** 811–819.

Kostandov, E. A. and Arzumanov, Y. L. (1971). Changes of cortical evoked potentials in response to emotional visual stimuli in man under the action of amyzil (in Russian). *J. Higher Nervous Activity* (Moscow) **21,** 1247–1255.

Kostandov, E. A. and Arzumanov, I. U. L. (1972). Dependence of conditioned slow negative potential of the human cerebral cortex on motivation and emotion (in Russian). *Zh. Vyssh. Nerv. Deiat.* **22,** 679–687.

Lacey, J. I. (1972). Some cardiovascular correlates of sensorimotor behaviors: Examples of visceral afferent feedback? In: C. H. Hockman (Ed.)., *Limbic System Mechanisms and Autonomic Function.* Springfield, Ill., Thomas, pp. 175–201.

Lacey, J. I. and Lacey, B. C. (1970). Some autonomic-central nervous system interrelationships. In P. Black (Ed.), *Physiological Correlates of Emotion.* New York, Academic Press, pp. 205–227.

Lacey, J. I. and Lacey, B. C. (1973). Experimental association and dissociation of phasic bradycardia and vertex-negative waves: a psychophysiological study of attention and response-intention. *Electroencephalogr. Clin. Neurophysiol.* **33,** *Suppl.* 281–285.

Lader, M. H. and Montagu, J. D. (1962). The psycho-galvanic reflex: a pharmacological study of the peripheral mechanism. *J. Neurol. Neurosurg. Psychiat.* **25,** 126–133.

Laffont, F. (1973). Role of movement evoking slow cortical activities. In: A. Fessard et G. Lelord (Ed.). *Activités évoquées et leur conditionnement chez l'homme normale et en pathologie mentale. Colloque INSERM Tours,* pp. 405–420.

Landoll, J. R. (1969). Use of automatic baseline correcting amplifiers. *IEEE Trans. Biomed. Eng. BME-16,* 167–168.

Lassen, N. A. and Ingvar, D. H. (1972). Radioisotopic assessment of regional cerebral blood flow. *Prog. Nuc. Med.* **1,** 376–409.

Leifer, L. J. and Otto, D. (1973). Eye movement artifact compensation for evoked response electroencephalography. Paper, American EEG Society, June.

Lenhardt, M. L. (1973). Effects of verbal association on auditory evoked response. *Audiol.* **12,** 174–179.

Libby, jun., W. L., Lacey, B. C. and Lacey, J. I. (1973). Pupillary and cardiac activity during visual attention. *Psychophysiology* **10,** 270–294.

Lickey, M. E. and Fox, S. S. (1966). Location and habituation of sensory evoked DC responses in cat cortex. *Exp. Neurol.* **15,** 437–454.

Lille, F. (1973). Potentiels Evoqués et Taches psychologiques complexes. In: A. Fessard et Lelord (Ed.). *Activités évoquées et leur conditionnement chez l'homme normal et en pathologie mentale.* INSERM, Publ., Paris, pp. 23–34.

Lindsley, D. B. (1957). Psychophysiology and motivation. In: M. R. Jones (Ed.). *Nebraska Symposium on Motivation.* Univ. Neb. Press, Lincoln, Nebraska.

Lindsley, D. B. (1969). Neurophysiological mechanisms of attention and perception: their role in information processing in the visual system. In: D. P. Kimble (Ed.). *Readiness to Remember, Vol. 2.* New York, Gordon & Breach, pp. 345–507.

Lindsley, D. B., Wilson, G. F. and Seales, D. M. (1973). Evoked potentials and information processing. Paper, Western EEG Society, February.

Loveless, N. E. and Sanford, A. J. (1973). The CNV baseline: considerations of internal consistency of data. *Electroencephalogr. Clin. Neurophysiol.,* **33,** *Suppl.* 19–23.

Low, M. D. (1969). In: E. Donchin and D. B. Lindsley (Ed.). *Average Evoked Potentials: Methods, Results and Evaluations.* NASA SP-191, U.S. Govt Printing Office, Washington, D.C., 163–170.

Low, M. D., Borda, R. P., Frost, jun., J. D. and Kellaway, P. (1966). Surface-negative slow potential shift associated with conditioning in man. *Neurology,* **16,** 771–782.

Low, M. D., Coats, A. C., Rettig, G. M. and McSherry, J. W. (1967). Anxiety, attentiveness—alertness: a phenomenological study of the CNV. *Neuropsychologia* **5,** 379–384.

Low, M. D. and McSherry, J. W. (1968). Further observations of psychological factors involved in CNV genesis. *Electroencephalogr. Clin. Neurophysiol.* **25,** 203–207.

Low, M. D. and Stoilen, L. (1973). CNV and EEG in children: maturational characteristics and findings in the MCD syndrome. *Electroencephalogr. Clin. Neurophysiol.* **33,** *Suppl.* 139–143.

Low, M. D., and Swift, S. J. (1971). The contingent negative variation and the 'resting' d.c. potential of the human brain: Effects of situational anxiety. *Neuropsychologia* **9,** 203–208.

Low, M. D., Wada, J. A. and Fox, M. (1972). Electroencephalographic localization of conative aspects of language production in the human brain. Paper, American EEG Society, October.

Lunde, D. T. and Hamburg, D. A. (1972). Techniques for assessing the effects of sex hormones on affect, arousal and aggression in humans. *Recent Progr. Horm. Res.* **28,** 627–663.

Luria, A. R. (1963). Psychological studies of mental deficiency in the Soviet Union. In: N. R. Ellis (Ed.). *Handbook of Mental Deficiency,* New York, McGraw-Hill.

Luria, A. R. (1973). *The Working Brain.* Harmondsworth, Penguin.

Lyman, R. S. (1942). Eye movements in the electroencephalogram. *Bull. Johns Hopkins Hosp.* **68,** 1–31.

McAdam, D. W. (1973). Physiological mechanisms: a review. *Electroencephalogr. Clin. Neurophysiol.* **33,** *Suppl.* 79–86.

McAdam, D. W. and Rubin, E. H. (1971). Readiness potential, vertex positive wave, contingent negative variation and accuracy of perception. *Electroencephalogr. Clin. Neurophysiol.* **30,** 511–517.

McAdam, D. W. and Whittaker, H. A. (1971). Language production: Electroencephalographic localization in the normal human brain. *Science* **172,** 499–502.
McCallum, W. C. (1967). New waves in the brain. *New Sci.* **36,** 592–594.
McCallum, W. C. (1969). The contingent negative variation as a cortical sign of attention in man. In: C. R. Evans and T. B. Mulholland (Ed.), *Attention in Neurophysiology*, London, Butterworths, pp. 40–63.
McCallum, W. C. (1973). Some psychological, psychiatric and neurological aspects of the CNV. In: A. Fessard et G. Lelord (Ed.). *Activités évoquées et leur conditionnement chez l'homme normale et en pathologie mentale. Colloque INSERM Tours*, pp. 295–326.
McCallum, W. C. (1973). CNV and conditionability in psychopaths. *Electroencephalogr. Clin. Neurophysiol.* **33,** *Suppl.* 337–343.
McCallum, W. C. and Abraham, P. (1973). The contingent negative variation in psychosis. *Electroencephalogr. Clin. Neurophysiol.* **33,** *Suppl.* 329–335.
McCallum, W. C. and Cummins, B. (1973). The effects of brain lesions on the contingent negative variation in neurosurgical patients. *Electroencephalogr. Clin. Neurophysiol.* **35,** 449–456.
McCallum, W. C. and Knott, J. R. (Ed.) (1973). Event-related slow potentials of the brain: their relations to behavior. *Electroencephalogr. Clin. Neurophysiol.* **33,** *Suppl.* 390p.
McCallum, W. C. and Papakostopoulos, D. (1972). The effects of sustained motor activity on the contingent negative variation. *Electroencephalogr. Clin. Neurophysiol.* **33,** 446.
McCallum, W. C. and Papakostopoulos, D. (1973). The CNV and reaction time in situations of increasing complexity. *Electroencephalogr. Clin. Neurophysiol.* **33,** *Suppl.* 179–185.
McCallum, W. C., Papakostopoulos, D., Gombi, R., Winter, A. L., Cooper, R. and Griffith, H. B. (1973). Event related slow potential changes in human brain stem. *Nature* (*Lond.*) **242,** 466–467.
McCallum, W. C. and Walter, W. G. (1968). The effects of attention and distraction on the contingent negative variation in normal and neurotic subjects. *Electroencephalogr. Clin. Neurophysiol.* **25,** 319–329.
McCallum, C., Walter, W. G., Winter, A., Scotton, L. and Cummins, B. (1970). The contingent negative variation in cases of known brain lessions. *Electroencephalogr. Clin. Neurophsyiol.* **28,** 210.
McKee, G., Humphrey, B. and McAdam, D. W. (1973). Scaled lateralization of alpha activity during linguistic and musical tasks. *Psychophysiology* **10,** 441–443.
McSherry, J. W. (1971). Intracortical origin of the contingent negative variation in the rhesus monkey. Doctoral Dissertation, Baylor College of Medicine, Houston.
McSherry, J. W. (1973a). Multiple-contact probes for intracortical CNV research. *Electroencephalogr. Clin. Neurophysiol.* **33,** *Suppl.* 25–27.
McSherry, J. W. (1973b). Physiological origins: a review. *Electroencephalogr. Clin. Neurophysiol.* **33,** *Suppl.* 53–61.
McSherry, J. W. and Borda, R. P. (1973). The intracortical distribution of the CNV in rhesus monkey. *Electroencephalogr. Clin. Neurophysiol.* **33,** *Suppl.* 69–74.
Marczynski, T. J. (1972). Electrophysiological correlates of positive reinforcement: Post-reinforcement synchronization, modulation of sensory input, and steady potentials. In: A. G. Karczmar and J. C. Eccles (Ed.), *Brain and Human Behavior*, New York, Springer-Verlag, pp. 303–321.
Marsden, C. D., Merton, P. A. and Morton, H. B. (1972). Servo action in human voluntary movement. *Nature* (*Lond.*) **238,** 140–143.
Marsh, G. R. and Thompson, L. W. (1973). Effect of verbal and non-verbal psychological set on hemispheric asymmetries in the CNV. *Electroencephalogr. Clin. Neurophysiol.* **33,** *Suppl.* 195–200.
Marsh, G. R., Wheeler, G. and Thompson, L. W. (1973). Regional differences in averaged evoked potentials (P300) during manipulations of attentional set. *Psychophysiology* **10,** 210.
Matsuo, F., Peters, J. F. and Reilly, E. L. (1973). Frontal electrical potentials associated with blinking and eye movement. Paper, Central EEG Society, March.
Matthews, P. B. C. (1972). *Mammalian Muscle Receptors and their Central Actions*. London, Edward Arnold.
Miller, L. H., Van Veen, W., Sandman, C. A. and Knott, J. R. (1973). Intermodal and intramodal distraction and CNV. *Electroencephalogr. Clin. Neurophysiol.* **33.** *Suppl.* 209–211.
Milstein, V., Small, J. G., Gans, G. E. and Moore, J. E. (1972). Risk taking and the CNV. Paper, American EEG Society, October.
Morrell, L. K. (1973). LOKLIN: On-line averaging of potentials prior to and following events

not subject to experimental control. *Psychophysiology*, **10,** 184–185.
Morrell, L. K. (1973). On-line averaging of potentials antecedent to voluntary behavior. *Psychophysiology* **10,** 307–310.
Morrell, L. K. and Huntington, D. A. (1972). Cortical potentials time-locked to speech production: Evidence for probable cerebral origin. *Life Sci.* **11,** 921–929.

Näätänen, R. (1967). Selective attention and evoked potentials. *Ann. Acad. Sci. Fenn.* **151,** 1–226.
Näätänen, R. (1969). Anticipation of relevant stimuli and evoked potentials: a comment on Donchin's and Cohen's 'Average evoked potentials and intramodality selective attention'. *Percept. Mot. Skills* **28,** 639–646.
Näätänen, R. (1972). Time uncertainty and occurrence uncertainty of the stimulus in a simple reaction time task. *Acta Psychol.* **3,** 492–503.
Näätänen, R. (1973). On what is the contingent negative variation (CNV) contingent in reaction-time experiments? In: A. Fessard et G. Lelord (Ed.). *Activités évoquées et leur conditionnement chez l'homme normale et en pathologie mentale. Colloque INSERM Tours,* pp. 121–152.
Naitoh, P., Johnson, L. C. and Lubin, A. (1973). The effect of selective and total sleep loss on the CNV and its psychological and physiological correlates. *Electroencephalogr. Clin. Neurophysiol.* **33,** *Suppl.* 213–215.
Nelligan, D. P. (1964). Eye movement artefacts and electrical recording of eye position. *Proc. Electrophysiol. Technol. Assoc.* **11,** 25–43.
Netchine, S. (1969). L'activité électrique cérébrale chez l'enfant normal de 6 a 10 ans. In: *Croissance de l'Enfant. Genese de l'Homme*, Vol. 2, Paris, P.U.F.
Norman, D. A. (1969). *Memory and Attention.* New York, Wiley.

Obrist, P. A., Webb, R. A., Sutterer, J. R. and Howard, J. L. (1970). Cardiac deceleration and reaction time: an evaluation of two hypotheses. *Psychophysiology* **6,** 695–706.
Otto, D. A. and Leifer, L. J. (1972a). The effect of modifying response and performance feedback parameters on the CNV in humans. *NASA Technical Memorandum*, X-62, 097.
Otto, D. A. and Leifer, L. J. (1972b). Positive and negative slow potential changes in the human brain related to sustained motor response. Paper, American EEG Society, October.
Otto, D. A. and Leifer, L. J. (1973). The effect of modifying response and performance feedback parameters on the CNV in humans. *Electroencephalogr. Clin. Neurophysiol.* **33,** *Suppl.* 29–37.

Pandya, D. and Kuypers, H. C. J. M. (1969). Cortico-cortical connections in the rhesus monkey. *Brain Res.* **13,** 13–36.
Papakostopoulos, D. (1973). CNV and autonomic function: a review. *Electroencephalogr. Clin. Neurophysiol.* **33,** *Suppl.* 269–280.
Papakostopoulos, D. and Cooper, R. (1973a). The contingent negative variation (CNV) and spinal motor neurone excitability. *Electroencephalogr. Clin. Neurophysiol.* **34,** 109.
Papakostopoulos, D. and Cooper, R. (1973b). The contingent negative variation and the excitability of the spinal monosynaptic reflex. *J. Neurol. Neurosurg. Psychiatry* **36,** 1003–1010.
Papakostopoulos, D. and McCallum, W. C. (1973). The CNV and autonomic change in situations of increasing complexity. *Electroencephalogr. Clin. Neurophysiol.* **33,** *Suppl.* 287–293.
Papakostopoulos, D., Walter, W. G. and Crow, H. J. (1970). The effect of stimulus context on evoked responses in human frontal cortex. *Electroencephalogr. Clin. Neurophysiol.* **29,** 327–328.
Papakostopoulos, D., Winter, A. and Newton, P. (1973). New techniques for the control of eye potential artefacts in multichannel CNV recordings. *Electroencephalogr. Clin. Neurophysiol.* **34,** 651–653.
Papini, M. and Zappoli, R. (1972). The contingent negative variation (CNV) in children: 'experimental paradigms'. *Electroencephalogr. Clin. Neurophysiol.* **33,** 356.
Papini, M. and Zappoli, R. (1972). Onda d'aspettativa (contingent negative variation): aspetti neurofisiologici, psicologici e clinici di un nuovo metodo d'indagine elettroencefalografica nell'uomo. (Contingent negative variation: Psychological, neurophysiological and clinical aspects. A new method of analyzing electroencephalographic change in man.) *Riv. Neurol.* **42,** 380–419.

Pasik, P., Pasik, T. and Bender, M. B. (1965). Recovery of the electro-oculogram after total ablation of the retina in monkeys. *Electroencephalogr. Clin. Neurophysiol.* **19,** 221–297.

Patterson, M. M., Cegavska, C. F. and Thompson, R. F. (1973). Effects of a classical conditioning paradigm on hind-limb flexor nerve response in immobilized spinal cats. *J. Comp. Physiol. Psychol.* **84,** 88–97.

Paul, D. D. (1971). Late components of the auditory evoked potential and binary decisions in signal detection tasks. Doctoral Dissertation, Columbia University.

Paul, D. D. and Sutton, S. (1972). Evoked potential correlates of response criterion in auditory signal detection. *Science* **177,** 362–364.

Peters, J. F., Hamilton, C. E. and Knott, J. R. (1972). Stimulus certainty, subject set and event related potentials. Paper, American EEG Society, October.

Peters, J. F. (1973). Bibliography of CNV and other slow potentials of the brain. *Electroencephalogr. Clin. Neurophysiol.* **33,** *Suppl.* 357–383.

Peters, J. F., Knott, J. R., Miller, L. H., VanVeen, W. J. and Cohen, S. I. (1970). Response variables and magnitude of the contingent negative variation. *Electroencephalogr. Clin. Neurophysiol.* **29,** 608–611.

Pfurtscheller, G. and Cooper, R. (1975). Selective averaging of the intracerebral click evoked responses in man: an improved method of measuring latency and amplitudes. *Electroencephalogr. Clin. Neurophysiol.* **38,** 187–190.

Phillips, C. G. (1969). Motor apparatus of the baboon's hand. *Proc. R. Soc. Lond.* (*Biol.*) **173,** 141–174.

Picton, T. W. and Hillyard, S. A. (1972). Cephalic skin potentials in electroencephalography. *Electroencephalogr. Clin. Neurophysiol.* **33,** 419–424.

Picton, T., Hillyard, S. A. and Galambos, R. (1973). Auditory evoked potentials during vigilance. Paper, Western EEG Society, February.

Picton, T. W., Hillyard, S. A. and Galambos, R. (1974). Evoked responses to omitted stimuli. In: M. N. Livanov (Ed.). *Major Problems of Brain Electrophysiology*. USSR Academy of Science.

Picton, T. W. and Low, M. D. (1971). The CNV and semantic content of stimuli in the experimental paradigm: effects of feedback. *Electroencephalogr. Clin. Neurophysiol.* **31,** 451–456.

Pierrot-Deseilligny, E., Lacert, P. and Cathala, H. P. (1971). Amplitude et variabilité des reflexes monosynaptiques avant un mouvement volontaire. *Physiol. Behav.* **7,** 495–508.

Pinto-Hamuy, T., Bracchitta, H. and Lagarrigue, I. (1969). Slow cortical potentials (SCP) during habituation conditioning and extinction in rabbit cortex. *Neuropsychologia*, **7,** 335–347.

Pirch, J. H. and Barnes, P. R. (1972). Steady potential responses from the rat cortex during conditioning. *Experientia* **28,** 164–165.

Porter, R. (1972). Relationship of the discharges of cortical neurones to movement in free-to-move-monkey. *Brain Res.* **40,** 39–43.

Posner, M. I. and Boies, S. J. (1971). Components of attention. *Psychol. Rev.* **78,** 391–408.

Posner, M. I., Klein, R., Summers, J. and Buggie, S. (1973). On the selection of signals. *Memory and Cognition* **1,** 2–12.

Prevec, T. S., Lokar, J. and Cernelc-Durlajeva, S. (1973). Use of CNV in audiometry. *Electroencephalogr. Clin. Neurophysiol.* **35,** 100–101.

Pribram, K. H. (1961). A further experimental analysis of the behavioural deficit that follows injury to the primate frontal cortex. *Exp. Neurol.*, **3,** 432–466.

Pribram, K. H. (1969). The amnestic syndromes: Disturbances in coding? In: G. A. Talland and N. C. Waugh (Ed.). *The Pathology of Memory*, New York: Academic Press, pp. 127–157.

Pribram, K. H. (1970). Looking to see: Some experiments on the brain mechanisms of attention in perception. *Res. Publ. Ass. Nerv. Ment. Dis.* **48,** 150–162.

Pribram, K. H. (1971). *Languages of the Brain.* New Jersey, Prentice-Hall.

Puchinskaya, L. M. (1972). EEG of the sensorimotor area in the state of motor readiness and during the movement of the hand. (in Russian.) *J. Higher Nervous Activity* (Moscow) **22,** 368–375.

Ransom, B. R. and Goldring, S. (1973). Slow depolarisation in cells presumed to be glia in cerebral cortex of cat. *J. Neurophysiol.* **36,** 869–878.

Rebert, C. S. (1972a). Further analysis of CNV and reaction time. Paper, American EEG Society, October.

Rebert, C. S. (1972b). The effect of reaction time feedback on reaction time and contingent negative variation. *Psychophysiology* **9,** 334–339.

Rebert, C. S. (1972c). Cortical and subcortical slow potentials in the monkey's brain during a preparatory interval. *Electroencephalogr. Clin. Neurophysiol.* **33,** 389–402.

Rebert, C. S. (1973a). Elements of a general cerebral system related to CNV genesis. *Electroencephalogr. Clin. Neurophysiol.* **33,** *Suppl.* 63–67.

Rebert, C. S. (1973b). A technique for simultaneous measurement of DC and multiple unit responses. *Electroencephalogr. Clin. Neurophysiol.* **34,** 326–328.

Rebert, C. S. (1973c). Slow potential correlates of neuronal population responses in the cat's lateral geniculate nucleus. *Electroencephalogr. Clin. Neurophysiol.* **35,** 511–515.

Rebert, C. S. and Irwin, D. A. (1969). Slow potential changes in cat brain during classical appetitive and aversive conditioning of jaw movement. *Electroencephalogr. Clin. Neurophysiol.* **27,** 152–161.

Rebert, C. S. and Irwin, D. A. (1973). Simple electrode configurations for chronic or acute recording of D.C. potentials from subcortical nuclei of the brain. *Electroencephalogr. Clin. Neurophysiol.* **34,** 440–442.

Rebert, C. S. and Knott, J. R. (1970). The vertex non-specific evoked potential and latency of contingent negative variation. *Electroencephalogr. Clin. Neurophysiol.* **28,** 561–565.

Rebert, C. S. and Sperry, K. G. (1973). Subjective and response-related determinants of CNV amplitude. *Psychophysiology* **10,** 139–144.

Rebert, C. S. and Tecce, J. J. (1973). A summary of CNV and reaction time. *Electroencephalogr. Clin. Neurophysiol.* **33,** *Suppl.* 173–178.

Rebert, C. S., McAdam, D. W., Knott, J. R. and Irwin, D. A. (1967). Slow potential change in human brain related to level of motivation. *J. Comp. Physiol. Psychol.* **63,** 20–23.

Regan, D. (1972). *Evoked Potentials in Psychology, Sensory Physiology, and Clinical Medicine.* London, Chapman & Hall.

Requin, J. (1969). Some data on neurophysiological processes involved in the preparatory motor activity to reaction time performance. *Acta Psychol. (Amst.)* **30,** 358–367.

Ritter, W., Simson, R. and Vaughan, H. G. jun. (1972). Association cortex potentials and reaction time in auditory discrimination. *Electroencephalogr. Clin. Neurophysiol.* **33,** 547–555.

Ritter, W. and Vaughan, H. G. jun. (1969). Average evoked responses in vigilance and discrimination: a reassessment. *Science* **164,** 326–328.

Ritter, W., Vaughan, H. G. jun. and Costa, L. D. (1968). Orienting and habituation to auditory stimuli: a study of short term changes in average evoked responses. *Electroencephalogr. Clin. Neurophysiol.* **25,** 550–556.

Rodin, E., Wasson, S., Porzak, J., Grisell, J. and Gudobba, R. (1970). The cerebral somatosensory response induced by passive joint movement. *Electroencephalogr. Clin. Neurophysiol.* **28,** 91.

Rohrbaugh, J., Donchin, E. and Eriksen, C. W. (1972). On the differential effect of cognitive variables on early or late components of the visual evoked response. Paper, American EEG Society, October.

Rosen, S. C. and Stamm, J. S. (1972). Transcortical polarization: facilitation of delayed response performance by monkeys. *Exp. Neurol.* **35,** 282–289.

Rosenfeld, J. P. and Fox, S. S. (1972). Sequential representation of voluntary movement in cortical macropotentials: direct control of behavior by operant conditioning of wave amplitude. *J. Neurophysiol.* **35,** 879–891.

Roth, W. T. (1973). Auditory evoked responses to unpredictable stimuli. *Psychophysiology* **10,** 125–138.

Roth, W. T. and Kopell, B. S. (1973). P300—an orienting reaction in the human auditory evoked response. *Percept. Mot. Skills* **36,** 219–225.

Rousseau, J. C. (1969). Le procédé de mesure de l'amplitude de la variation contingente négative. In: J. Dargent and M. Dongier (Ed.). *Variations Contingentes Négatives,* Congrès et colloques de l'Université de Liège, Vol. 52, pp. 45–47.

Rowland, V. (1968). Cortical steady potential (direct current potential) in reinforcement and learning. In: E. Stellar and J. M. Sprague (Ed.), *Progress in Physiological Psychology,* Vol. 2, New York, Academic Press, pp. 2–77.

Rowland, V. and Dines, G. (1973). Cortical steady potential shift in relation to the rhythmic electrocorticogram and multiple unit activity. In: R. F. Thompson and M. M. Patterson (Ed.). *Bioelectric Recording Techniques. Vol. 1-A.* New York, Academic Press, pp. 369–385.

Rusinov, V. S. (1972). *The electrical activity of the brain during the development of simple forms of temporal connection.* (in Russian.) *Nauka, Moscow.*

Rusinov, V. S. (1973). *The Dominant Focus. Electrophysiological Investigations.* Trans. by B. Haigh, New York, Plenum Press.

Rusinov, V. S. and Grindl, O. M. (1967). Correlation analysis of central EEG rhythms of the healthy human cortex. (in Russian.) In: V. S. Rusinov (Ed.). *Electrophysiology of the Central Nervous System.* Nauka, Moscow. Trans. by B. Haigh and R. W. Doty, 1970, New York, Plenum Press, pp. 375–389.

Sanders, A. F. and Wertheim, A. H. (1973). The relation between physical stimulus properties and the effect of foreperiod duration on reaction time. *Q. J. Exp. Psychol.* **25,** 201–206.

Sandler, L. S. and Schwartz, M. (1973). Differential effects of selective attention and stimulus probability upon the evoked potential and the contingent negative variation. *Psychophysiology* **10,** 195.

Schafer, E. W. P. (1967). Cortical activity preceding speech: semantic specificity. *Nature (Lond.)*, **216,** 1338–1339.

Schaltenbrand, G. and Bailey, P. (1959). *Introduction to Stereotaxis with an Atlas of the Human Brain.* New York, Grune & Stratton.

Schulman, C. A. (1973). Eye movements in infants using DC recording. *Neuropadiatrie* **4,** 76–87.

Semmes, J., Weinstein, S., Ghent, L. and Teuber, H. L. (1963). Correlates of impaired orientation in personal and extra-personal space. *Brain,* **86,** 747–772.

Shagass, C. (1972). Electrical activity of the brain. In: N. S. Greenfield and R. A. Sternbach (Ed.). *Handbook of Psychophysiology.* New York, Holt, Rinehart & Winston, Inc. pp. 263–328.

Shagass, C., Straumanis, J. J. jun. and Overton, D. A. (1972). Electrophysiological recordings in the reaction time experiment: exploratory studies for possible psychiatric research application. *Biol. Psychiatry* **5,** 271–287.

Sharbrough, F. W. and Karnes, W. E. (1973). An EEG study of a fast midline rhythm evoked by movement and peripheral nerve stimulation: a case report. *Electroencephalogr. Clin. Neurophysiol.* **34,** 103.

Sharrard, G. A. W. (1973). Further conclusions regarding the influence of word meaning on the cortical averaged evoked response in audiology. *Audiology* **12,** 103–115.

Sheafor, P. and Rowland, V. (1973). Cortical steady potential shift and integrated mass unit activity in forms of temporal conditioning. Paper, Society for Neurosciences.

Shucard, D. W. (1973). Effects of instructional set on visual evoked potentials. *Psychophysiology* **10,** 194.

Skinner, J. E. (1971). Abolition of a conditioned, surface-negative, cortical potential during cryogenic blockade of the non-specific thalamocortical system. *Electroencephalogr. Clin. Neurophysiol.* **31,** 197–209.

Skinner, J. E., Yingling, C. D. and Naman, J. (1973). Correlated changes in the waveform of frontal cortex steady potentials and heart rate responses recorded during acquisition of tone-shock conditioning in the cat. Paper, Western EEG Society, February.

Small, J. G. (1968). Sensory evoked responses of autistic children. *Colloquium on Infantile Autism.* Indianapolis.

Small, J. G. (1971). Sensory evoked responses of autistic children. In: D. W. Churchill, G. D. Alpern and M. K. DeMyer (Ed.). *Infantile Autism.* Springfield, Ill., C. C. Thomas.

Small, J. G., DeMyer, M. K. and Milstein, V. (1971). CNV responses of autistic and normal children. *J. Autism Child Schizo.* **1,** 215–231.

Small, J. G. and Small, I. F. (1969). Practical experience with CNV research in psychiatry. 2nd Annual Winter Conf. Brain Res. pp. 73–74.

Small, J. G. and Small, I. F. (1971). CNV correlations with psychiatric diagnosis. *Arch. Gen. Psychiatry* **25,** 550–554.

Small, J. G. and Small, I. F. Expectancy waves in affective psychoses. In: T. A. Williams, (Ed.). *Recent Advances in the Psychobiology of Depression.* U.S. Govt Printing Office.

Smirnov, V. M. and Speranski, M. M. (1972). Slow bioelectric processes of the cortex and deep structures of the human brain and emotional behavior. (in Russian.) *Vopr. Psikhol.* **18,** 21–38.

Smith, D. B. D., Donchin, E., Cohen, L. and Starr, A. (1970). Auditory evoked potentials in man during selective binaural listening. *Electroencephalogr. Clin. Neurophysiol.* **28,** 146–152.

Sokolov, E. N. (1963). *Perception and the Conditioned Reflex.* Oxford, Pergamon.

Sologub, E. B. (1964). Autocorrelation and crosscorrelation analysis of 'labelled ryhthms' in the human EEG during muscular work. (in Russian.) *Fiziol. Zh. SSSR.* **6,** 681–689.

Sologub, E. B. (1965). EEG cross correlation analysis during the formation of motor dynamic stereotype in human. (in Russian.) *Zh. Vyssh. Nerv. Deyat. Pavlova* **12,** 32–41.

Somjen, G. G. (1973). Electrogenesis of sustained potentials. *Prog. Neurobiol.* **1,** 199–237.

Speckmann, E. J., Caspers, H. and Janzen, R. W. (1972). Relations between cortical DC shifts and membrane potential changes of cortical neurons. In: H. Petsche and M. A. B. Brazier (Ed.). *Synchronization of EEG Activities in Epilepsies.* New York, Springer-Verlag, pp. 93–111.

Sperry, R. W. (1952). Neurology and the mind-brain problem. *Am. Sci.* **40,** 291–312.

Spitz, H. H. (1966). The role of input organization in the learning and memory of mental retardates. In: N. R. Ellis (Ed.). *International Review of Research in Mental Retardation.* New York, Academic Press.

Squires, K. C. (1973). Evoked potentials during auditory signal detection. Doctoral Dissertation. University of California, San Diego.

Squires, K. C., Hillyard, S. A. and Lindsay, P. L. (1973a). Cortical potentials evoked by confirming and disconfirming feedback following an auditory discrimination. *Perception and Psychophysics* **13,** 25–31.

Squires, K. C., Hillyard, S. A. and Lindsay, P. H. (1973b). Vertex potentials evoked during auditory signal detection: relation to decision criteria. *Perception and Psychophysics* **14,** 265–272.

Stamm, J. S. and Rosen, S. C. (1972). Cortical steady potential shifts and anodal polarization during delayed response performance. *Acta Neurobiol. Exp.* (*Warsz*) **32,** 193–209.

Starobinetz, M.Kh. and Pshedetskaya, A. D. (1971). Steady potentials of the human brain after thermal action on the trigeminal zone. (in Russian.) *Fiziol. Zh. SSSR.* **57,** 956–961. Abst. in *Int. J. Neuroscience* (1972) **1,** 251.

Steinmetz, G. T. and Rebert, C. S. (1972). Postreinforcement changes of steady potentials in premotor cortex of monkeys. *Physiol. Behav.* **9,** 769–772.

Stern, J. A. and Plapp, J. M. (1969). Psychophysiology and clinical psychology. In: C. D. Spielberger (Ed.). *Current Topics in Clinical and Community Psychology.* New York, Academic Press.

Storm van Leeuwen, W. and Kamp, A. (1973). Occurrence of beta bursts in frontal cortex during CNV paradigm. *Electroencephalogr. Clin. Neurophysiol.* **33,** *Suppl.* 95–97.

Stowell, H. (1972). Signal averaging the human somatosensory evoked response: a focal evoked response? *Psychophysiology* **9,** 634–639.

Sutton, S., Braren, M., John, E. R. and Zubin, J. (1964). Evoked potential correlates of stimulus uncertainty. Paper presented at the meetings of the American Psychological Association, Los Angeles, September.

Sutton, S., Braren, M., Zubin, J. and John, E. R. (1965). Evoked-potential correlates of stimulus uncertainty. *Science* **150,** 1187–1188.

Sutton, S. and Paul, D. D. (1973) Evoked potential correlates of the detectability of low intensity signals. In: A. Fessard et G. Lelord (Ed.). *Activités évoquées et leur conditionnement chez l'homme normale et en pathologie mentale. Colloque INSERM Tours*, pp. 79–92.

Sutton, S., Tueting, P., Zubin, J. and John, E. R. (1967). Information delivery and the sensory evoked potential. *Science* **155,** 1436–1439.

Taub, E. and Berman, A. J. (1968). Movement and learning in the absence of sensory feedback. In: S. J. Freedman (Ed.). *The Neuropsychology of Spatial Oriented Behavior.* Homewood, Ill., Dorsey Press, pp. 173–192.

Tecce, J. J. (1972). Contingent negative variation (CNV) and psychological processes in man. *Psychol. Bull.* **77,** 73–108.

Tecce, J. J. and Cole, J. O. (1972). Effects of amphetamine on CNV and behavior in normal adults. Paper, Society for Psychophysiological Research, November.

Tecce, J. J. and Hamilton, B. T. (1973). CNV reduction by sustained cognitive activity (distraction). *Electroencephalogr. Clin. Neurophysiol.* **33,** *Suppl.* 229–237.

Tecce, J. J. and Scheff, N. M. (1969). Attention reduction and suppressed direct-current potentials in the human brain. *Science* **164,** 331–333.

Teuber, H. L. (1966). The frontal lobes and their function: Further observations on rodents, carnivores, subhuman primates, and man. *Int. J. Neurol.* **5,** 282–300.

Thach, W. T. (1970a). Discharge of cerebellar neurons related to two maintained postures and two prompt movements. I. Nuclear cell output. *J. Neurophysiol.* **33,** 527. (1970b) II. Purkinje cell output and input, *Ibid.*, **33,** 537.

Thor, D. H. and Thor, C. J. (1970). The dark interval threshold and intelligence. *J. Exp. Psychol.* **85,** 270–274.

Timsit, M., Koninckx, N., Dargent, J., Fontaine, O. and Dongier, M. (1968). Etude de la variation contingente négative chez les psychotiques et les pré-psychotiques. *Ann. Med. Psychol.* **126,** 424–435.

Timsit, M., Koninckx, N., Dargent, J., Fontaine, O. and Dongier, M. (1970). Variations contingentes négatives en psychiatrie. *Electroencephalogr. Clin. Neurophysiol.* **28,** 41–47.

Timsit-Berthier, M. (1973). CNV, slow potentials and motor potential studies in normal subjects and psychiatric patients. In: A. Fessard et G. Lelord (Ed.). *Activités évoquées et leur conditionnement chez l'homme normale et en pathologie mentale. Colloque INSERM Tours*, pp. 327–366. [Fr.]

Timsit-Berthier, M., Delaunoy, J., Koninckx, N. and Rousseau, J. C. (1973a). Slow potential changes in psychiatry. I. Contingent negative variation. *Electroencephalogr. Clin. Neurophysiol.* **35,** 355–361.

Timsit-Berthier, M., Delaunoy, J. and Rousseau, J. C. (1973b). Slow potential changes in psychiatry. II. Motor potentials. *Electroencephalogr. Clin. Neurophysiol.* **35,** 363–367.

Timsit-Berthier, M. and Hausman, J. (1973). Study of the contingent negative variation (CNV) and the motor preparation phenomenon in children aged 5–15 years. *Electroencephalogr. Clin. Neurophysiol.* **34,** 217.

Timsit-Berthier, M., Koninckx, N., Delaunoy, J. and Rousseau, J. C. (1972). New statistical studies of the negative variation in psychiatry. *Rev. EEG Neurophysiol.* **2,** 333–339. [Fr.]

Timsit-Berthier, M., Rousseau, J. C. and Delaunoy, J. (1971). Reactivité de l'onde d'attente et des ondes négatives post impératives. *Rev. EEG Neurophysiol.* **1,** 245–248.

Timsit-Berthier, M. and Timsit, M. (1972). Functional electroencephalography in psychiatry: evaluation and perspectives. *Evol. Psychiatr.* (*Paris*) **37,** 567–584.

Treisman, A. (1964). Selective attention in man. *Br. Med. Bull.* **20,** 12–16.

Trumbo, D. A. and Gaillard, A. W. K. (1974). Drugs, time uncertainty, signal modality and reaction time. In: P. M. A. Rabbitt (Ed.). *Attention and Performance. V.* New York, Academic Press.

Tueting, P. and Sutton, S. (1973). The relationship between pre-stimulus negative shifts and post-stimulus components of the averaged evoked potential. In: S. Kornblum (Ed.). *Attention and Performance. IV.* New York, Academic Press, pp. 185–207.

Tueting, P., Sutton, S. and Zubin, S. (1971). Quantitative evoked potential correlates of the probability of events. *Psychophysiology*, **7,** 385–394.

Uttal, W. (1965). Do compound evoked potentials reflect signs or codes? *Psychol. Bull.* **64,** 377–392.

Van Doorne, H. and Sanders, A. F. (1968). PSARP, a programmable signal and response processor. *Behav. Res. Method. Instr.* **1,** 29–32.

Van Veen, W. J., Knott, J. R., Peters, J. F., Miller, L. H. and Cohen, S. I. (1973). CNV shape differences and perceptual mode. *Electroencephalogr. Clin. Neurophysiol.* **33,** *Suppl.* 327–328.

Van Veen, W. J., Peters, J. F., Knott, J. R., Miller, K. H. and Cohen, S. I. (1973). Temporal characteristics of the contingent negative variation: relationship with anxiety, perceptual mode, sex and stress. *Biol. Psychiatry* **7,** 101–111.

van Weerden, T. W. (1973). Artefacts in electro-oculography. *Electroencephalogr. Clin. Neurophysiol.* **35,** 105.

Vastola, E. F. (1955). Steady potential responses in the lateral geniculate body. *Electroencephalogr. Clin. Neurophysiol.* **7,** 557–567.

Vaughan, H. G. jun. (1974). The analysis of scalp recorded brain potentials. In: R. F. Thompson and M. M. Patterson (Ed.). *Bioelectric Recording Techniques, Vol. 1-B.* New York, Academic Press, pp. 158–207.

Vaughan, H. G. jun. and Corey, S. R. (1973). Some remarks on the psychological and physiologic interpretation of human event related potentials. In: A. Fessard et G. Lelord (Ed.). *Activités évoquées et leur conditionnement chez l'homme normale et en pathologie mentale. Colloque INSERM Tours*, pp. 93–110.

Vaughan, H. G. and Ritter, W. (1970). The sources of auditory evoked responses recorded from the human scalp. *Electroencephalogr. Clin. Neurophysiol.* **28,** 360–367.

Vaughan, H. G. and Ritter, W. (1973). Physiologic approaches to the analysis of attention and performance: Tutorial review. In: S. Kornblum (Ed.). *Attention and Performance. IV.* New York, Academic Press, pp. 129–154.

Venables, P. M. (1960). The effect of auditory and visual stimulation on the skin potential response of schizophrenics. *Brain* **83,** 77.

Vladimirova, G. (1972). Asymmetry of front duration of the EEG-waves during performance of motor tasks in normal 3- and 6-year old children. *2nd Int. Symp. Motor Control*, Varna, pp. 73–79.

Wallach, M. B. and Wallach, S. S. (1964). Involuntary eye movements in certain schizophrenics. *Arch. Gen. Psychiatry* **11,** 71–73.

Walter, W. G. (1956). Slow potential changes in epilepsy and other conditions. *Electroencephalogr. Clin. Neurophysiol.* **8,** 162.

Walter, W. G. (1964). Slow potential waves in the human brain associated with expectancy, attention and decisions. *Arch. Psychiatr. Nervenkr.* **206,** 309–322.

Walter, W. G. (1965a). Effects on anterior brain responses of an expected association between stimuli. *J. Psychosomat. Res.* **9,** 45–49.

Walter, W. G. (1965b). Brain responses to semantic stimuli. *J. Psychosomat. Res.* **9,** 51–61.

Walter, W. G. (1966). Electrophysiologic contributions to psychiatric therapy. In: J. H. Masserman (Ed.). *Current Psychiatric Therapies*, New York, Grune & Stratton, Vol. VI, pp. 13–35.

Walter, W. G. (1967). Slow potential changes in the human brain associated with expectancy, decision and intention. *Electroencephalogr. Clin. Neurophysiol.* **26,** *Suppl.* 123–130.

Walter, W. G. (1972). Viewpoints of mental illness: neurophysiologic aspects. *Semin. Psychiatry* **4,** 211–231.

Walter, W. G., Aldridge, V. J., Cooper, R., O'Gorman, G., McCallum, C. and Winter, A. L. (1971). Neurophysiological correlates of apparent defects of sensori-motor integration in autistic children. In: D. W. Churchill, G. D. Alpern and M. K. DeMyer (Ed.). *Infantile Autism*. Springfield, Ill., C. C. Thomas.

Walter, W. G., Cooper, R., Aldridge, V. J., McCallum, W. C. and Winter, A. L. (1964). Contingent negative variation: An electric sign of sensorimotor association and expectancy of the human brain. *Nature (Lond.)* **203,** 380–384.

Walter, W. G., Cooper, R., Crow, H. J., McCallum, W. C., Warren, W. J., Aldridge, V. J., Storm van Leeuwen, W. and Kamp, A. (1967). Contingent negative variation and evoked responses recorded by radiotelemetry in free-ranging subjects. *Electroencephalogr. Clin. Neurophysiol.* **23,** 197–206.

Walter, W. G., Weinberg, H. and Aldridge, V. J. (1972). The influence of response information about performance on brain potentials related to real and imaginary stimuli. *Electroencephalogr. Clin. Neurophysiol.* **33,** 446.

Warren, J. M. (1964). The behavior of carnivores and primates with lesions in the prefrontal cortex. In: J. M. Warren and K. Akert (Ed.). *The Frontal Granular Cortex and Behavior*. New York, McGraw-Hill, pp. 168–191.

Waszak, M. and Obrist, W. D. (1969). Relationship of slow potential changes to response speed and motivation in man. *Electroencephalogr. Clin. Neurophysiol.* **27,** 113–120.

Webb, W. B. (1969). Partial and differential sleep deprivation. In: A. Kales (Ed.). *Sleep: Physiology and Pathology*. Philadelphia, Lippincott, pp. 221–231.

Weerts, T. C. and Lang, P. J. (1973). The effect of stimulus and response location on eye movements and the CNV. *Psychophysiology* **10,** 214.

Weerts, T. C. and Lang, P. J. (1973). The effects of eye fixation and stimulus and response location on the contingent negative variation (CNV). *Biol. Psychol.* **1,** 1–19.

Weinberg, H. (1972). The contingent negative variation: its relation to feedback and expectant attention. *Neuropsychologia* **10,** 299–306.

Weinberg, H. (1973). The contingent negative variation: its relation to feedback and expectant attention. *Electroencephalogr. Clin. Neurophysiol.* **33,** *Suppl.* 219–228.

Weinberg, H. and Cooper, R. (1972a). The use of correlational analysis for pattern recognition. *Nature (Lond.)* **238,** 292.

Weinberg, H. and Cooper, R. (1972b). The recognition index: a pattern recognition technique for noisy signals. *Electroencephalogr. Clin. Neurophysiol.* **33,** 608–613.

Weinberg, H., Walter, W. G., Cooper, R. and Aldridge, V. J. (1974). Emitted cerebral events. *Electroencephalogr. Clin. Neurophysiol.* **36,** 449–456.

Weinberg, H., Walter, W. G. and Crow, H. J. (1970). Intracerebral events in humans related to real and imaginary stimuli. *Electroencephalogr. Clin. Neurophysiol.* **29,** 1–9.

Welford, A. T. (Ed.) (1968). *Fundamentals of Skill*. London, Methuen.

Wilke, J. T. (1973). Variations in the 'motor potential' with the force exerted during voluntary arm movements in man. Paper, Western EEG Society, February.

Wilkinson, R. T. (1970). Methods for research on sleep deprivation and sleep function. In: E. Hartmann (Ed.). *Sleep and Dreaming*. Boston, Little, Brown & Co., pp. 369–381.

Wilkinson, R. T. and Ashby, S. M. (1974). Selective attention, contingent negative variation and the evoked potential. *Biol. Psychol.* **1,** 167–179.

Wilkinson, R. T. and Lee, M. V. (1972). Auditory evoked potentials and selective attention. *Electroencephalogr. Clin. Neurophysiol.* **33,** 411–418.

Wilkinson, R. T. and Morlock, H. C. (1967). Auditory evoked response and reaction time. *Electroencephalogr. Clin. Neurophysiol.* **23,** 50–56.

Wilkinson, R. T. and Spence, M. V. (1973). Determinants of the post-stimulus resolution of contingent negative variation (CNV). *Electroencephalogr. Clin. Neurophysiol.* **35,** 503–509.

Williams, E. J. (1949). Experimental design balanced for the residual effects of treatments. *Aust. J. Sci. Res.* **A, 2,** 149–168.

Wilson, L. E., Harter, M. R. and Wells, H. H. (1973). Evoked cortical potentials and discrimination problem-solving in humans. *Electroencephalogr. Clin. Neurophysiol.* **34,** 15–22.

Wing, J. K., Birley, J. L. T., Cooper, J. E., Graham, P. and Isaacs, A. D. (1967). Reliability of a procedure for measuring and classifying 'Present Psychiatric State'. *Br. J. Psychiat.* **113,** 499–515.

Winter, A. L. (1974). A simple eye movement monitoring system for clinical electroencephalography. *Electroencephalogr. Clin. Neurophysiol.* **37,** 182–184.

Winter, A. L. and Kellenyi, L. (1971). Eye movement monitoring. *Proc. Electrophysiol. Technol. Ass.* **18,** 121.

Wood, C. (1973). On the nature of unilateral neural activity during speech reception. Paper, American EEG Society, 1973.

Woodworth, R. S. and Schlosberg, H. (1954). *Experimental Psychology*. New York, Holt.

Wurtz, R. H. (1965a). Steady potential shifts during wakefulness and sleep. *Electroencephalogr. Clin. Neurophysiol.* **18,** 528–529.

Wurtz, R. H. (1965b). Steady potential shifts during arousal and deep sleep in the cat. *Electroencephalogr. Clin. Neurophysiol.* **18,** 649–662.

Wurtz, R. H. (1966a). Steady potential correlates of intracranial reinforcement. *Electroencephalogr. Clin. Neurophysiol.* **20,** 59–67.

Wurtz, R. H. (1966b). Steady potential fields during sleep and wakefulness in the cat. *Exp. Neurol.* **15,** 274–292.

Zappoli, R. (1970). On relation between spontaneous epileptic discharge, conditioned reflex responses of the first and second order and 'expectancy wave' (CNV). *Estratto da gli stati di coscienza Conference di aggiornamento della Societa Italiana di Elettroencefalografia e Neurofisiologia Rimini*, pp. 63–82. [Ital.]

Zappoli, R. and Papini, M. (1968). Avoidance conditioning and contingent negative variation in epileptic subjects. *Excerpta Med. Int. Cong.* Series 171, pp. 65–66.

Zappoli, R., Papini, M., Cabras, P. L., Benvenuti, P. and Binazzi, A. (1971). 'Expectancy wave' (CNV), operant responses and autonomic reactions in patients with Parkinson's disease. *Riv. Patol. Nerv. Ment.* **92,** 240–263. [Ital.]

Zappoli, R., Papini, M., Cabras, P. L., Benvenuti, P. and Binazzi, A. (1972). 'Expectancy wave' (CNV), operant responses and autonomic reactions in patients with Parkinson's syndrome. *Riv. Neurol.* **42,** 310–312. [Ital.]

Zappoli, R., Papini, M., Cabras, M., Benvenuti, P. and Binazzi, A. (1973). Relationships between CNV, operant responses and sphygmic wave patterns. *Electroencephalogr. Clin. Neurophysiol.* **33,** *Suppl.* 299–302.

Zeaman, D. and House, B. J. (1963). The role of attention in retardate discrimination learning. In: N. R. Ellis (Ed.). *Handbook of Mental Deficiency*, New York, McGraw-Hill.

Zenz, M., Fruhstorfer, H., Nolte, H. and Hensel, H. (1973). Dissociated loss of cold and warm sensibility during regional anesthesia. *Pfluegers Arch.* **339,** *Suppl.* R 86.

Zigler, E. (1967). Familial mental retardation: a continuing dilemma. *Science* **155,** 292–298.

Zimmermann, G. and Knott, J. R. (1973). CNVs related to spoken words in stutterers and non-stutterers. Paper, Western EEG Society, February.

Zimmermann, G. and Knott, J. R. (1973). Classical CNV, lateralized cerebral negativity and speech: a study of normal speakers and stutterers. Paper, Central EEG Society, March.